# Megacities and the Coast

Based on a major international study, this volume provides a synthesis of scientific knowledge on the urbanisation processes, environmental impacts, and policy response options, and disaster risk management challenges that are associated with coastal megacity development. It is the primary output of a major international scientific project sponsored by the International Geosphere Biosphere Programme, the Land-Ocean Interactions at the Coastal Zone programme of IHDP/IGBP and others. It brings together the work of over 60 contributing authors and an international review board.

This volume presents the international policy and academic community with an unbiased and high-quality assessment of social–ecological systems interaction in coastal megacities. One of its main messages is that while we know a great deal about megacities of more than ten million people and about urban processes, and about coasts and their physical and ecological processes (aquatic, physical and atmospheric), there is relatively little work that focuses primarily at points of intersection between these. The book responds to this gap by providing the first global synthesis of megacity and large urban region urbanisation on the coast. Its focus is on environmental and development challenges, climate change and disaster. It is interdisciplinary and brings together world-recognised scientists (including many IPCC lead authors) on urban climate and atmosphere, disaster risk management, demography and coastal environments.

**Mark Pelling** is Professor of Geography in the Department of Geography at King's College London, UK.

**Sophie Blackburn** is a Research Associate in the Department of Geography, King's College London, UK.

# Megacities and the Coast

## Risk, Resilience and Transformation

Edited by
Mark Pelling and
Sophie Blackburn

Routledge
Taylor & Francis Group
LONDON AND NEW YORK

First published 2013
by Routledge
2 Park Square, Milton Park, Abingdon, Oxon, OX14 4RN

Simultaneously published in the USA and Canada
by Routledge
711 Third Avenue, New York, NY 10017

*Routledge is an imprint of the Taylor & Francis Group, an informa business*



*British Library Cataloguing in Publication Data*
A catalogue record for this book is available from the British Library

*Library of Congress Cataloging-in-Publication Data*
A catalog record has been requested for this book

ISBN13: 978-0-415-81504-8 (hbk)
ISBN13: 978-0-415-81512-3 (pbk)
ISBN13: 978-0-203-06642-3 (ebk)

Typeset in Sabon and Gill Sans by
Florence Production Ltd, Stoodleigh, Devon, UK

# Contents

# Plates

(Top row) mean temperature for June 7–10, 2011 with the NYC UHI visible; and (bottom) 1500 h on June 8, when a sea breeze cooled southern Long Island, on the right side of the map. Data courtesy of NOAA (NOS-PORTS, NWS-ASOS, NWS-HADS, Urbanet), Rutgers NJ Weather and Climate Network, APRSWXNET, AWS Convergence Technologies, Inc. (WeatherBug), and Weatherflow, via Mark Arend (NYCMetNet).

8 *COAMPS 6-h forecast of the Tokyo Bay region*

Notes: (06 UTC 30 June 2005) (1.67-km deterministic ensemble member) of condition at 10-m (a) wind speed and air temperature (contours every 0.4 °C), and (b) dew point depression (conoturs every 0.4 °C) plus vertical velocity greater than 5 cm $s^{-1}$ at 210-m (black dashed contours) and station observations of air and dew point temperature (°C) and wind (full barb=5 m $s^{-1}$). Solid line shows urban extent. In (a) the heavy solid line is deterministic model location of sea breeze front, box – shows the ensemble spread of sea breeze front location (Holt *et al.* 2009).

9 *Tokyo Metropolitan Area 103 dosage contour*

Notes: $10^3$ dosage contour from 12-hour COAMPS Nest 4 (1.67 km) ensemble members initiated June 30, 2005.

10 *The HadRM3 climate model for coastal megacities/regions in Europe*

Notes: Difference between the HadRM3 climate model (coastal megacities of London, in addition Paris, Ruhr Valley region, Europe) with and without urban land surface for temperature (°C), convective precipitation (%), 10 m wind speed (%), and net primary productivity (NPP, %). The NPP data have been weighted to represent the productivity per unit area of vegetation, rather than total area.

11 *The NCAR urban canyon model*

Notes: Modelled annual temperature differences using NCAR urban canyon model (Oleson *et al.*, 2008a, 2011; Jackson *et al.*, 2010) (a) the difference between urban and rural diurnal temperature range (DTR) in the Vancouver parameter simulation; (b) the difference between urban and rural DTR in the global parameter simulation; and (c) the difference between urban and rural DTR between the Global and Vancouver parameter simulations.

12 *Change in the UHI strength between the periods 1981–2000 and 2081–2100 simulated by HadGEM2-ES*

Notes: The 1%, 5%, and 10% urban land cover contours are overlain for (black) 2005 and (red) 2100. For this experiment the urban land cover remained fixed at 2005 levels.

13 *Ensemble mean results for the North East Coast of North America (coastal megacity of New York City)*

Notes: Ensemble mean results for the North East Coast of North America (coastal megacity of New York City) from 16 GCMs for the A1B emission scenario for the 2080s relative to the 1970–99 base period for (a) temperature and (b) precipitation (Horton *et al.* 2011). A1B emission scenario (http://www.ipcc.ch/ipccreports/tar/wg1/029.htm#storya1) assumes rapid economic growth with the peak in global population mid-century and technological advances are efficient but balanced across energy sources. (© American Meteorological Society. Used with permission.)

# Tables

# Acronyms

| | |
|---|---|
| ADB | Asian Development Bank |
| CBDP | community-based disaster preparedness |
| CCA | climate change adaptation |
| CCM | climate change mitigation |
| CCN | concentration of condensation nuclei |
| CCSM | Community Climate System Model |
| COAMPS | Coupled Ocean-Atmosphere Mesoscale Prediction System |
| DKI | Special Capital Territory of Jakarta |
| DRM | disaster risk management |
| DRR | disaster risk reduction |
| DTR | diurnal temperature range |
| EAC | Eko Atlantic City |
| ENSO | El Nino Southern Oscillation |
| ESS | ecosystem services |
| FAO | Food and Agriculture Organisation |
| GCM | global climate model |
| GDP | gross domestic product |
| GHG | greenhouse gas |
| GEC | global environmental change |
| GECAFS | Global Environmental Change and Food Systems |
| ICZM | integrated coastal zone management |
| IPC | integrated food security phase classification |
| IPCC | Intergovernmental Panel on Climate Change |
| LASG | Lagos State Government |
| LECZ | low elevation coastal zone |
| LLGHG | long-lived greenhouse gases |
| LOICZ | Land-Ocean Interactions on the Coastal Zone |
| MCGM | Municipal Corporation of Greater Mumbai |
| MM | Metro Manila |
| MSP | maritime spatial planning |
| MEA | Millennium Ecosystem Assessment |
| MENA | Middle East and North Africa |
| NCAR | National Center for Atmospheric Research |
| NPCC | New York City Panel on Climate Change |
| NYC | New York City |

| | |
|---|---|
| PM | particulate matter |
| RCP | representative concentration pathways |
| SLR | sea-level rise |
| SREX | IPCC Special Report for Managing the Risks of Events and Disasters to Advance Climate Change Adaptation |
| TOA | top of the atmosphere |
| UBL | urban boundary layer |
| UET | urban environmental transition |
| UHI | urban heat island |
| UN-DESA | United Nations Department for Economic and Social Affairs |
| UNCLOS | United Nations Law of the Sea |
| UNEP | United Nations Environment Programme |
| UNESCO | United Nations Educational, Scientific and Cultural Organisation |
| UNFPA | United Nations Population Fund |
| UN-HABITAT | United Nations Human Settlements Programme |
| UNISDR | International Secretariat for Disaster Reduction |
| WHO | World Health Organisation |

# List of contributors

**Editors**
Mark Pelling
Sophie Blackburn

**Review editors**
Gordon McGranahan
Robert Nicholls
Michail Fragkias
Antonia Yulo Loyzaga

**Assistant editor**
Nina Qaiem Maqami

**Lead authors**

| | |
|---|---|
| Chapter 1 | Sophie Blackburn (King's College London, UK) and César Marques (UNICAMP, Brazil) |
| Chapter 2 | Antje Bruns (Humboldt-University Berlin, Humboldt Universität zu Berlin, Germany) |
| Chapter 3 | Andrea Ferraz Young (UNICAMP, Brazil) |
| Chapter 4 | C.S.B. Grimmond (King's College London, UK) |
| Chapter 5 | Jessica Lamond (University of the West of England, UK) |
| Chapter 6 | Joern Birkmann (United Nations University, Germany), Matthias Garschagen (United Nations University, Germany), Adriana Lopez (University del Pacifico, Colombia), Mark Pelling (King's College London, UK), Nina Qaiem Maqami (King's College London, UK) and Qingnian Yu (Hohai University, China) |

**Contributing authors**
Albert Adu-Gyamfi
Alexander Baklanov
Fabiana Barbi
Martin Best
Richard Betts
Sophie Blackburn
Colin Booth

Antje Bruns
Harriet Bulkeley
Katrin Burkart
Samuel Carpenter
Carlos Castellanos Perez Bolde
Wen-Yan Chiau
Chao-Yi Chung
Rafael D'Almeida Martins
Leila da Costa Ferreira
Alex de Sherbinin
Pete Falloon
Johannes Feddema
Gerd Folberth
Rob Francis
Bruce Glavovic
Kapil Gupta
Teddy Holt
Claudia Hosch
Jessica Jache
Trisha Jackson
Andreas Kannen
Justus Kithiia
Dietmar Kraft
Elizabeth Larson
Yangfan Li
Adriana Lopez
Qingshui Lu
Mohsen Makki
Philippa Mason
Mark McCarthy
Gordon McGranahan
Tom Measham
Talmor Meir
Naho Mirumachi
Francine Modesto
Magnus Moglia
Veronique Morin
Ricardo Ojima
Keith Oleson
Akeem Olaniyi
Sebastien Oliveau
Philip Orton
Julie Pullen
Ivonne Radjawane
Ramesh Ramachandran
Marcello Sano
Guoqing Shi

Evi Sofiyah
William Solecki
Chia-Yang Tsai
Nassos Vafeidis
Ralf Weisse
Poh Poh Wong
Maida Zahid
Xin Zhou

# Acknowledgements

Huge thanks to each of the contributing authors for their invaluable input to this synthesis report. The report is the principal outcome of an international, collective assessment of scientific knowledge on the interaction of megacities and the coastal environment. This is completed as a core task of the Land-Sea Interactions at the Coastal Zone (LOICZ) project and the International Geosphere Biosphere Programme (IGBP). Work was coordinated from the Urbanization in Coastal Zones Theme of LOICZ (see www.loicz.org/, last accessed 20 March 2013). The review process has been coordinated with advice from a senior steering and review committee including members of LOICZ and IGBP. Lead authors and a number of contributing authors were selected through this process and via the LOICZ Open Science Congress, Yantai, China, 2011. Additional opportunities for input from the international scientific community were encouraged through sessions at the IHDP Open Science Meeting, London, 2012, and the second US Coastal City Summit, St Petersburg, Florida, 2012.

The editors particularly wish to thank the lead authors for their dedication to this synthesis report, without whose expertise and leadership the work could not have been completed. We are very grateful for financial support from the School of Social Science and Public Policy, King's College London, the Taiwan Environmental Protection Administration, IHDP-IGBP core project Land-Sea Interactions at the Coastal Zone (LOICZ), and the International Geosphere Biosphere Programme (IGBP). We also wish to thank Professor Chen-Tung Arthur Chen, National Sun Yat-sen University for his assistance.

The report has undergone internal review, review by the expert review editors and external peer review. The views expressed in the report remain those of the contributing authors and editors.

# Executive summary

*Mark Pelling and Sophie Blackburn*

The coast represents a highly dynamic interface between land, sea and atmosphere, subjecting urban development in this zone to a unique set of pressures and opportunities. However, whilst offering many benefits, a coastal location is also exposed to varied sources of risk – many of which are being exacerbated in the context of the new and uncertain pressures associated with global environmental change. Such concerns are most salient in megacities – defined as having a population exceeding 10 million – where the concentration of human life and assets is greatest, and where consequences for failure as well as opportunities for innovative solutions remain high. Global trends in urban geography have acquired heightened significance as the majority of the world's population is now recognised as urban-dwelling. Simultaneously, megacity development itself exerts pressures on coastal ecosystems and geomorphology, with both short and long-term implications for ecological and human wellbeing and sustainability. The degree to which megacity residents, property and ecologies are exposed and vulnerable to environmental hazards is an outcome not only of technological and economic capacity, but – more importantly – of governance systems, dominant development priorities and values. How far current trajectories for environmental and social change are shifting the balance between opportunity and risk, and for whom, are thus important questions of our time.

Responding to this globally strategic concern and opportunity for sustainable development, this volume is the principle outcome of an international, interdisciplinary assessment of global scientific knowledge on the interaction of megacities and the coastal environment. It is the product of collaboration between the Land-Sea Interactions at the Coastal Zone (LOICZ) hotspot theme on Urbanization in Coastal Zones, and the International Geosphere Biosphere Programme (IGBP). It aims to begin the process of building an international community of researchers that can lead trans-disciplinary expertise, and frame future research on megacities and urban regions on the coast. The completion of this volume in itself has brought together an international group of more than 60 scientists from a multitude of subject specialisms, from across North and South America, Asia, Africa, Europe and Australasia. Lead authors and a number of contributing authors were selected via the LOICZ Open Science Congress, Yantai, China, 2011, with additional opportunities for inputs from the international scientific community encouraged through sessions at the IHDP Open Science Meeting, London, 2012, and the 2nd US Coastal City Summit, St Petersburg, Florida, 2012. The review process has been coordinated with advice from a senior steering and review committee including members of LOICZ and IGBP.

## Context

Megacities are large and highly dynamic systems, and represent concentrated sites of human life and assets as well as of pollution and ecological stress. Due to this combination, as neighbourhoods and cities grow in size and resource capacity, transformations in social-environmental relations are also observed. Over time, local environmental hazards associated with inadequate sanitation and services tend to be replaced by risks that accrue at the city scale and globally, as cities progress through industrial and post-industrial stages. Furthermore the increasingly interconnected nature of urban places at a global scale alters the nature and rate of urban change. Accountability for this can be difficult to trace, since the sites of risk production and impact are often far removed. For example, there are clear justice consequences for globalised megacities that are important nodes in the production of pollution and greenhouse emissions which have negative impacts on distant ecologies and populations at regional and global scales. Historically, coastal megacities have recorded some of the highest human and economic losses to disaster events, however this is not a trend which need necessarily continue into the future: simultaneous with being sites of extreme risk, megacities are also centres of capacity, ingenuity and resource.

The complexity and reach of urbanisation processes is matched by those of coastal environmental systems, which are amongst the most diverse ecological systems worldwide. Beyond their intrinsic value, these are systems that offer significant ecosystem services ranging widely from coastal protection to fisheries and recreation. Management of hazards, vulnerability and environmental management in megacities is inherently complex, and governance responses require attention to multiple scales of impact, and negotiation between many competing interests.

## The Aim: an integrated agenda for research and policy on large coastal, urban systems

The scientific community understands a good deal of the drivers and constraints acting in megacity systems and coastal systems, and work from social and natural sciences and integrated and trans-disciplinary programmes of research have made substantial recent advances. However, there is one over-riding and critical gap: the lack of integration between urban and coastal research. As yet, we know much less about the interaction of coastal and large urban systems than we do of the constituent parts. Knowledge of the dynamic two-way interactions between megacities and the coast is, we argue, a large and dangerous gap in our collective knowledge.

This global review seeks to address this gap, adopting an interdisciplinary approach that is unique in drawing on both the physical and social sciences to explore the causes, impacts and management of environmental degradation, human vulnerability, and feedbacks between the two. In an attempt to delineate the contours of the relationship between mega-urbanisation and coasts that is explored in detail in the rest of the volume, this Executive Summary highlights seven key messages arising from this synthesis review project.

*Key Message 1*

## Fragmentation, agglomeration and disproportionately rapid expansion in less wealthy countries are major global trends in coastal megacity development

Megacities present important lessons on planning and risk management at scale. There are currently 23 worldwide, 16 of which are in the coastal zone – defined here as the area within 100km and 50m elevation from the coast. Of these, ten are in Asia (Tokyo, Mumbai, Shanghai, Guangzhou, Shenzhen, Kolkata, Karachi, Manila, Osaka-Kobe, Jakarta), two each in Latin America (Buenos Aires, Rio de Janeiro) and North America (New York City, Los Angeles), and one each in Africa (Lagos) and Europe (Istanbul). However there are many problems inherent in defining the above list. Most significant among these are: 1) the use of multiple conflicting parameters in defining the boundary between 'urban' and 'rural', a dichotomy which is increasingly rejected in favour of a 'continuum' concept; 2) ambiguity over uses of the term 'urbanisation', which range from demographic to structural interpretations; and 3) the intuitive difficulty of excluding rapidly growing cities lying below the 10 million threshold, and those lying slightly outside the 'coastal' limits but whose growth nonetheless has relied heavily on the features of a coastal or estuarine location (for example Cairo, London and Sao Paulo).

Further complicating megacity identification are global trends of urban sprawl, fragmentation, and agglomeration, which serve to blur city boundaries. Recent reports highlight the growing importance of peri-urban areas in regional economies, which are expanding largely as a result of urban sprawl. In order to incorporate such 'grey areas' – which nonetheless are a key feature of contemporary megacities – a joint focus on megacities and 'urban regions' is preferred in this volume. This also seeks to take account of 'mega-regions', where several megacities have agglomerated through urban expansion, or where large urban areas have developed with multiple foci rather than a single epicentre. Currently the largest of such agglomerations is the Hong Kong-Shenhzen-Guangzhou region in China, which has a total population of approximately 120 million.

Set against these trends is the relatively static growth of megacities in Europe, the Americas and East Asia, relative to the very rapid urban growth observed in Central Asia and Africa. Rapid urbanisation is especially consequential in Africa where demography, economies and environments are set to be transformed in the next decade. The implications for international trade networks and distributions of global capital is still a young field, and is an important area for future inquiry that should focus on the politics of internal trade-offs between economic development, human well-being and environmental sustainability. The importance of this research agenda is magnified by the risks to trade associated with a coastal location and additional pressures from global environmental change anticipated to have a disproportionately significant impact on low- and middle-income countries.

### *Key Message 2*

## Large-scale urbanisation in the coastal zone causes transformations in the geosphere, biosphere, atmosphere and hydrosphere, which can be damaging to the environment over both short and long timescales

The environmental impacts of coastal megacity growth are complex and intersecting, with many feedbacks between spheres which can exacerbate or ameliorate outcomes. Environmental impacts are largely an outcome of urban processes of structural development, waste and pollution, and resource consumption, and on such a large scale, these are greatly magnified. Furthermore, due to the complex and delicate nature of coastal ecosystems, the environmental impacts of coastal megacities can also be disproportionate compared to non-coastal cities.

The most significant impacts of megacities and urban regions on the geosphere include: subsidence due to excessive building load and groundwater extraction, pollution of soils and groundwater reserves, and resource extraction. Pollution and habitat disturbance are the most damaging bi-products of megacity development on the coastal biosphere, causing changes in species composition, population and resilience. Coral reefs and wetlands including temperate salt-marsh and tropical mangrove are particularly vulnerable ecosystems, due to their ecological and structural sensitivity. Degradation of wetlands also has important feedback implications for human vulnerability, as both provide a buffer against storm surge and tsunami. Given the range of ecosystem services on which cities depend, seeking to reduce these negative impacts by improving sustainability is therefore a vital consideration for megacity development.

Aside from aquatic pollution reducing marine and estuarine water quality, the hydrosphere is affected through structural changes to estuaries and the coastal strip. This can alter regimes of erosion and sedimentation in such a way that disturbs ecosystems and damages existing structures. Ecosystems can also be damaged as a result of extractive industries, for example along the Jakarta coastline, coral has been mined for use in cement. Finally, megacities impact on the atmosphere as a result of polluting industries and their intense concentration of motor vehicles, and the creation of an urban microclimate is common (for example, the 'heat island effect'). However interaction between land and coastal air can ameliorate local pollution and temperature extremes. A more sophisticated analysis is required to track the impacts of megacity consumption on local, regional and global environments.

The movement not only of goods but of the energy (ie carbon footprint), resource and water embedded within them brings new responsibilities for consumers. As yet science has only a limited view on the pathways and dynamics through which benefits for some bring risk to others and the ways in which such distributions are moderated – some would say distorted – by the action of the global economy, and financial speculation (for example) on commodity prices. New ways of conceptualising these relationships and of tracking flows as a precursor to a more nuanced allocation of responsibilities and rights in the global social contract is well overdue. These are questions that go beyond coastal megacities and represent where some focussed research effort can generate wider benefit.

*Key Message 3*

## Climate change is affecting both the sources of environmental hazard and their impacts on coastal megacities, although the scale and nature of these changes is highly uncertain

Evidence from the IPCC Special Report on Managing the Risks of Extreme Events and Disasters to Advance Climate Change Adaptation (SREX 2011) sheds new light on existing predictions of how tropical cyclones will be affected by global environmental change. Whilst previous studies suggested that both the frequency and intensity of storms were increasing and would continue to do so in the future, SREX indicates that cyclone frequency is in fact stable. The proportion of these which are at the high end of the intensity spectrum (level 4 or 5) is, however, increasing. This has key implications for how megacities and urban regions can best prepare for cyclone events, and the preparations made in New York City in anticipation of Hurricanes Irene in 2011 and Sandy in 2012, thus serve as useful simulations.

Whilst the rate and magnitude of mean sea level rise is also debated, there is consensus about the negative implications of rising sea levels for marine hazards such as increased wave height, storm surge and coastal flooding. Additional hazards of concern to coastal megacities are heat wave, tsunami, monsoonal rains and landslide. These hazards threaten human life and livelihoods both directly – i.e. causing deaths, structural damage and interruption to baseline economic activity – or indirectly, through hazards reducing the productivity or availability of ecosystem services (e.g. damage to reefs) which support particular livelihoods (e.g. fisheries). The anticipated threat to Tokyo from the Fukushima nuclear plant following the 2011 Japanese tsunami is another apt illustration of indirect hazard. In order to pinpoint more specifically the social-ecological and socio-economic impacts likely to affect coastal megacities in the future, further research is needed to narrow the margins of error surrounding estimates of projected shifts in climate-forced hazards.

*Key Message 4*

## Coastal megacities have direct and indirect impacts on the atmospheric composition, climate, hydrology and ecology, however current models are inadequate in simulating these dynamics

Observations and numerical modelling are just starting to yield the first insights into the dynamic processes by which coastal megacities impact the environment at a variety of scales. These impacts are largely the product of megacities' high concentration of industrial activity and emissions, coinciding with proximity to the sensitive of hydrological and ecological systems associated with a coastal location. Data also indicate that megacities can play a very significant role in reducing global emissions of greenhouse gases and pollutants.

This report also finds that to date, atmospheric (aerosol/chemical) and meteorological models have tended to treat different environmental impacts in isolation from one another, failing to acknowledge and therefore fully understand the complex interactions and feedbacks between them. The report therefore calls for the development of coupled models addressing the non-linear interactions between urban environmental phenomena. This will improve understanding of urban-environmental interactions at a variety of scales, assisting in the simulation of outcomes – particularly in understanding the scaled-up 'macro' impacts of coastal megacities. A step change in understanding may be possible with the further integration across natural and social science modelling. In particular analysis that can demonstrate feedback between natural and social systems is important for identifying the processes of social amplification that can take on considerable scale effects in megacities, but where, as yet, there is a limited analytical and methodological base.

### *Key Message 5*

### The politics of environmental risk management in megacities is complex due to the high number of stakeholders and interest groups. There is increasing pressure and good opportunity to mainstream risk management into other core functions of urban governance

Joined up adaptation approaches that bring together engineering and institutional projects of reform or transformation are the underpinning resources for sustainable cities, however integration requires concerted effort and new skills. The multilevel characteristics of megacities where local, city region, city and wider scale initiatives need coordination make for a particular challenge. This is especially so where different ways of understanding vulnerability, hazard and its drivers are influenced by specific cultural viewpoints fed through political agendas.

Contemporary approaches to risk management acknowledge the need for soft risk reduction measures such as education, good governance, and risk communication, being applied in conjunction with hard (i.e. technocratic) engineering solutions such as sea walls, groynes and levees seeking to protect the coastline. However applying these principles in megacities is challenging, requiring collaboration between a multitude of stakeholders. These include actors operating at all scales from the international (in the case of large land-owning corporations and supranational donor organisations funding climate change mitigation measures) to the individual (public consultation is now a standard component of the majority of urban development projects). Good governance is key to risk management, but there is much debate over precisely what this entails and its limitations.

The International Council for Local Environmental Initiatives (ICLEI) 2011 White Paper argues convincingly that the above challenges would be more easily overcome if resilience was treated as a mainstream concern in urban planning. This is based on the observation that resources are often wasted where urban problems are addressed with an isolated agenda in mind. For example, road surfaces ought to be maintained routinely with an in-built consideration of flood risk. ICLEI argue that therefore, climate change adaptation finance should be demand rather than supply-driven because this is less prescriptive and more responsive to community needs. Further research is needed to examine the logistical, financial and political implications of such an approach.

### *Key Message 6*

### There are many aspects of disaster risk reduction (DRR) and climate change adaptation (CCA) which overlap, whilst also having distinct and conflicting implications for urban economic development and human well-being

Whilst CCA shares DRR concerns about reducing human vulnerability to extreme events, it also goes beyond this to include managing the incremental stresses to everyday livelihoods and shifting 'baseline' conditions that are associated with climate change. It also places additional emphasis on forward-looking assessments of risk and the ability of risk actors to reorganise entitlements and resources.

Both CCA and DRR have direct implications for economic development, because both extreme events and incremental livelihood pressures – particularly for the urban poor – are known to slow or even 'reverse' development. However, in practice the introduction of proactive adaptive strategies at the level of individuals, organisations and governments is difficult because initial outlay costs tend to exceed tangible benefits in the short-term. Decision-makers face a very significant challenge in negotiating these various pressures. Further work is needed on the progress of existing practical attempts to mainstream DRR and CCA into the sustainable development agenda for coastal megacities.

### *Key Message 7*

### Global trends in urban development and environmental change are permanently altering the ways in which coastal megacities interact with the environment, necessitating policy and research agendas that are innovative and forward-thinking

The material in this synthesis report paints a picture of contemporary mega-urbanisation on the coast. It highlights both new and existing sources of risk as they relate to populations, urban infrastructure, and the environment, outlining in each case their construction, impact and changing nature. Amongst the most significant trends contributing to new sources of risk on the coast are:

- the increasing inter-connectedness between urban centres, which simultaneously exposes cities to new 'domino-effect' global cascades of risk whilst also offering fresh opportunities for collaborative and complementary adaptation responses;
- the disproportionately rapid growth of megacities in low- and middle-income country contexts, accompanied by rising urban inequality in richer and poorer countries;
- poor understanding of complex feedback mechanisms between various environmental processes and urban development – including coastal realignment, subsidence, environmental health – which are mutually constitutive and culminate in the construction of 'feedback hazards' which tend to be inadequately addressed in planning policy

Whilst representing a 'knowledge-capture' of existing published and peer review evidence on these subjects, this report also seeks to identify gaps and areas of weakness or contradiction in received wisdoms. Centrally the process of developing this book has been successful in building a community of practice surrounding the most pressing questions for better understanding the interface between megacities and the coastal zone, by stimulating fresh enquiry and debate in this vitally important area of research.

## Conclusion

The reach of coastal megacities is global. As hubs of global trade, centres for technological innovation and political leadership, and sites of consumption and social reproduction, large cities occupy a pivotal position in struggles towards global sustainability. Recent trends in the urbanisation of poverty and inequality, local and global environmental degradation and disaster risk and loss suggest that mega-urbanisation not only makes the ambition of sustainable development increasingly remote, but that these characteristics of dominant 'development' pathways are compounding risk and loss in large urban systems, and through them to the global.

As a first international effort in synthesising scientific knowledge on megacities and the coast, this report highlights the great capacity for resilience inherent in large cities, but also the barriers that have tended to deny this possibility – certainly for the majority of urban residents in low and middle income countries and for ecological systems. Some have observed an urban environmental transition in large cities, one where economic growth allows for the exporting of environmental externalities. This can be seen in the transition in the nature of urban environmental risks from sanitation to local air quality and finally global environmental impacts. Is this dominant approach to environment-development relationships sustainable? Reflexivity in social-ecological systems experienced in coastal megacities suggests not. Feedback mechanisms are now impacting on the urban through increased hazard, and exacerbated by high levels of exposure and susceptibility. Vulnerability is expressed at the household level through poverty and the occupation of low-lying, hazard prone locations, but also systemically as critical infrastructure systems (energy, water, transport, communication and security) in large cities become increasingly interdependent, opening scope for compound crisis. Storm Sandy in New York is the most recent indicator of this threat with damage to transport and energy systems arguably generating greater loss than direct storm impacts.

Hard decisions lie ahead for megacities – and other areas of the urbanised coast – as aspirations for resilient and sustainable development are tested by growing economic inequality and environmental risk. Continuing failure in climate change mitigation, and the spectre of 2° Celsius warming, requires planning decisions to be made now and in the next decade in anticipation of a planet living with the consequences of dangerous climate change by mid-century. Cities have already transformed their local environments. Adapting to live with climate change will likely usher in a new era of environmental and social transformation, as the desires for growth and security are rebalanced. Understanding better the interactions between ecological, physical, socio-economic and political processes is a starting point for making transformations that can be deliberate – rather than forced – and through this process for enhancing informed and transparent decision-making.

Chapter 1

# Mega-urbanisation on the coast

## Global context and key trends in the twenty-first century

Lead authors: *Sophie Blackburn (King's College London, UK) and César Marques (UNICAMP, Brazil)*

Contributing authors: *Alex de Sherbinin (CIESIN, USA), Francine Modesto (UNICAMP, Brazil), Ricardo Ojima (UFRN, Brazil), Sebastien Oliveau (CICRED, France), Carlos Castellanos Perez Bolde (LOICZ, Germany)*

### 1 Introduction

The aim of this book is to explore the dynamic interactions that exist between urban and environmental systems in coastal megacities. This interaction is highly significant because of the coinciding concentrations of biophysical risk, ecological sensitivity, and human life and assets that exist in these areas. In the context of the global increase in the number and size of coastal megacities worldwide, the negative impact of intense urban expansion and population pressures on sensitive coastal environments, and the expectation of new and uncertain hazards associated with the onset of climate change, improving our understanding of the dynamic pressures at work on the coast has never been more important. This chapter sets the scene for the in-depth exploration of these interactions throughout this volume, outlining some of the key trends, shifts and concerns facing coastal megacities in the twenty-first century.

The coastal zone is characterized by high biological diversity and richness, and highly dynamic geomorphology. The coast offers a range of opportunities for industry, including access to valuable geological resources (particularly oil and gas reserves), national and international trade routes, and opportunities for tourism. These qualities provide coastal settlements a plethora of advantages, based on the coast's economic, ecological, aesthetic and scientific values. As a result, it is estimated that in more than half of the world's coastal countries, at least 80 per cent of the national population lives within 100 kilometres of the coastline (Martínez *et al.* 2007). The largest coastal settlements are coastal megacities – defined as cities with ten million or more inhabitants – which are the focus of this report.

This intensity of urban development along the coastal strip – which has accelerated markedly in low- and middle-income countries over the last half-century (UN-HABITAT 2008) – is understood to be strongly influenced by increased openness to trading opportunities (Henderson and Wang 2007). Urbanisation is a critical driver of environmental transformation on the coast, and the manner in which urban centres evolve is a core determinant of coastal environmental sustainability in the immediate and long-term (Klein *et al.* 2003). Urbanisation on the coast also leads to increased exposure of human life and assets to a wide range of hazards associated with a coastal location – including coastal flooding, cyclones and tsunami – and environmental health concerns. In the context of global environmental change, this interaction between human and geo-ecological spheres is heightened due to the added pressures faced by both the environment and society. These include resource pressures, sea level rise and changes in hazard behaviour and impact.

In seeking to explore these various dynamics, this volume fills a vital gap in existing literature surrounding dynamic two-way interactions on the coast between human and natural systems. It draws on a very wide range of literatures including urban geography, global environmental change, natural hazards, environmental science, ecology and development studies. The breadth of knowledge captured in this volume thus provides a vital resource for academics and decision makers associated with urban and environmental planning on the coast.

This chapter provides an introduction to the rest of the book, by outlining a number of key themes to contextualize the ensuing discussion of specific aspects of coastal megacities. This chapter will focus on the following questions: what is a coastal megacity, and where are they found? How do we define a coastal megacity and what are the challenges inherent in this? What are the key global trends in coastal megacity development worldwide? What are the risks and opportunities for resilience to be found in coastal megacities? And finally, what significance does the growth of coastal megacities have on the global economy and human well-being of megacity dwellers?

## 2 Locating coastal megacities

It is estimated that in 2010 the global population became predominantly urban (UN-HABITAT 2010), with an estimated 360 million residing in megacities in 2011 (UN-DESA 2012a). Unless otherwise stated, all other population statistics in the remainder of this chapter are drawn from UN-DESA (2012b) *World Urbanization Prospects 2011*, including Table 1.1.

This volume adopts the UN-DESA (2012a) definition of megacities as those with more than ten million inhabitants. However, our definition is expanded to include 'urban regions', defined as city-regions of comparable magnitude (i.e. aggregate population of at least ten million) but where some parts lie outside of administratively designated urban zones. This includes areas with multiple overlapping jurisdictions (i.e. municipal, city or local governments) and areas dominated by a high number of small, intervening settlements linked by semi-rural or peri-urban areas that, individually, would escape the megacity characterization. Our inclusion of urban regions seeks to take account of the economic and infrastructural significance of these inter-connected areas. We define the coast as the area within a 50-metre elevation and a 100-kilometre distance of mean high water, after UNEP (2005).

Whilst half the world's population is urban, only 2.8 per cent of global land cover is urbanized (McGranahan *et al.* 2006). This intense concentration of population is most acute on the coast: two-thirds of cities with a population exceeding five million are located at least partially within 0–10 metres above sea level (McGranahan *et al.* 2007).

*Table 1.1* Population and geographical characteristics of coastal megacities in 2011

| *City* | *Distance from coast (km)* | *Significant portions below 20 m above mean sea level?* | *Population of urban agglomerations (millions)* | *Coastal type* |
|---|---|---|---|---|
| Tokyo | 0 | Yes | 37.2 | Coastal plain, some hills |
| New York-Newark | 0 | Yes | 20.4 | Island and coastal plain |
| Shanghai | 0 | Yes | 20.2 | Delta |
| Mumbai | 0 | Yes | 19.7 | Islands, delta, coastal plain |
| Kolkata | 78 | Yes | 14.4 | Delta |
| Karachi | 0 | Yes | 13.9 | Coastal plain, delta |
| Buenos Aires | 0 | Yes | 13.5 | Coastal plain, part delta |
| Los Angeles-Long Beach-Santa Ana | 0 | Yes | 13.4 | Coastal plain |
| Rio de Janeiro coastal plain | 0 | Yes | 12.0 | Mountainous, narrow |
| Manila | 0 | Yes | 11.9 | Coastal plain |
| Osaka-Kobe | 0 | Yes | 11.5 | Coastal plain, mountainous |
| Lagos | 0 | Yes | 11.2 | Low-lying coastal plain |
| Istanbul | 0 | Yes | 11.3 | Mountainous |
| Guangzhou | 59 | Yes | 10.9 | Delta |
| Shenzhen | 22 | Yes | 10.6 | Delta, coastal plain, islands |
| Jakarta* | 0 | Yes | 9.8 | Coastal plain |

Population statistics sourced from UN-DESA (2012b) World Urbanisation Prospects 2011 table on the 30 largest urban agglomerations; typology of major coastal cities sourced from UN-DESA (2009) World Urbanization Prospects 2009; distance from coast calculated using *SEDAC TerraViva! Viewer*, by drawing a straight line from the centre of each urban agglomeration (as defined by CIESIN et *al.* 2011) to the coast; portions below 20 metres sea level were estimated using *SEDAC TerraViva! Viewer* Sea Level Rise data set based on Shuttle Radar Topography Mission (McGranahan et *al.* 2007).

*Whilst these official population statistics for Jakarta place it at just under ten million Jakarta is included here to reflect the cumulative population of the Greater Jakarta Metropolitan Area and the uncertainty of the city boundaries. The city is also understood to have a greatly inflated daytime population of up to 12 million (*The Jakarta Post* 2011).

Table 1.1 identifies 16 coastal megacities in 2011, out of a total 23 worldwide. Geographically, ten coastal megacities are in Asia; two each in Latin America and North America; and one each in Africa and Europe. These 16 cities have a combined population of 242 million out of a total global population of seven billion – thus almost 3.5 percent of the world's population lives in coastal megacities. It must be noted that the list presented in this table is subject to many contradictions and uncertainties, and is therefore subject to significant debate as well as rapid change. For example, this list based on 2011 data already has two additional Chinese megacities than that presented in von Glasow *et al.* (2012), which was based on 2010 data (see also Section 4.1 of the present chapter). Section 2 explores the difficulties of attempting to compile definitive megacity data in some detail.

Table 1.1 identifies a range of environmental contexts where coastal megacities have emerged, highlighting significant diversity in physical form, geology (geomorphology) and climate. These differences determine, amongst other things, the range of biophysical hazards the city is affected by – outlined in greater detail in Section 4. Large urban areas have tended to be constructed near the mouths of major rivers, where locational and transportation advantages emerge through the linking of interior hinterlands with global trade through shipping (de Sherbinin *et al.* 2007). Of the cities outlined in Table 1.1, five (Guangzhou, Shenzhen, Shanghai, Kolkata and Lagos) are found in delta areas, which have significant land areas under 20 metres above mean sea level. Being low-lying, these are particularly prone to subsidence owing to sediment compaction (Vörösmarty *et al.* 2009). Other coastal urban types identified in Table 1.1 include those situated on steeply rising inland topography or even mountainous zones, such as characterizes many cities along the Mediterranean (Istanbul) and along the Brazilian coast (Rio de Janeiro). For some of these cities, there is a narrow coastal shelf that is densely populated, while in others the elevation rises abruptly from the coastline.

## 3 Defining coastal megacities

It is vital to acknowledge that any list of coastal megacities (including that presented in Table 1.1) is not definitive but open for discussion. This is for several reasons associated with the reliability, comparability and accuracy of the data sets, because the timing, methodology and quality of data collection vary widely between cities. For example, many data sources (including the UN-DESA data) distinguish between 'urban proper' (i.e. city centre) and 'urban agglomeration' (i.e. the wider or 'greater' city area), but others do not, which can cause inappropriate comparisons to be drawn. Satterthwaite (2010) outlines in detail the weaknesses of population data sets, including those associated with delineating city boundaries. A significant methodological challenge is identifying the limits within which the population data is included. Official city- or national-level urban data tend to be based on administrative boundaries and may exclude the significant number of people who migrate into the city for work, those who live in the suburbs and populations living in informal or 'slum' settlements who may be excluded from city population data but make substantial contributions to the city economy. Thus depending on methodology, there may be significant over- or underestimates of megacity populations (Cohen 2006, Satterthwaite 2010).

An additional barrier to definitively identifying coastal megacities is lack of definitional clarity. The challenges inherent in defining our two most significant parameters – what constitutes a city, and what constitutes the coast – are outlined below.

### *3.1 Definitional Challenge 1: urban, rural and beyond*

The seemingly most obvious way in which to define a city is relative to what it is not: intuitively, cities are not rural. However, the flaws of a simplified rural–urban dichotomy, and the challenges associated with defining both terms, are now well-established (see for example Uzzell 1979, Pumain and Robic 1996, Beguin 1996). This stems from the rejection of assumptions about essential cultural differences between urban and rural societies (Uzzell 1979), recognition of the complexity of rural capital accumulation systems, and acknowledging that the tendency to define rural simply as 'not urban' denies its agency and uniqueness (Halfacree 1993). As a result the treatment of 'rural' and 'urban' as two extremes of a continuum rather than discrete categories is now widely preferred (Champion and Hugo 2004). Thus settlement classifications now tend to be based on relative, measurable parameters including population size and density, diversity of economic activity, level of 'built-up-ness' (related to either density of construction or average building height), and level of capital generation – which are all generally (but not always) higher in urban areas.

Application of the continuum concept in land-use classifications in practice varies in complexity, and can incorporate multiple dimensions of urban measurement. For example in Indonesia, each sub-community is ranked according to three parameters (population density, percentage of population engaged in agriculture and number of 'urban' facilities present) and the sum total determines its categorization as rural or urban (Champion and Hugo 2004, p. 14). However, the persistent essentialism of such systems, incorporating multiple parameters but continuing to culminate in either/or categorizations (i.e. urban *or* rural), has attracted criticism. As a result, over time more elaborate conceptualizations of interconnections of the urban/ rural interconnections have emerged. Partly due to the rapidity of suburban expansion and urban sprawl, what was previously considered 'rural' is now interspersed with partially built-up areas that link together adjacent cities. This makes the demarcation of individual cities more difficult and strengthens arguments about the inadequacy of the rural–urban dichotomy (Cohen 2006). Research into the nature of and transitions within this 'peri-urban interface' has burgeoned as a result (see for example Allen 2003). Simultaneously there has been growing appreciation of the validity of ecological research in urban centres (a field formerly reserved for rural areas, within which 'nature' was assumed to be bounded), stimulating the growing field of urban ecology (Pickett *et al.* 2001).

Challenges also arise when attempting to define 'urbanisation' empirically, because it too relies on being able to clearly capture what it means to be urban. Put simply, urbanisation refers to the process of becoming increasingly urban. Thus a population can be said to be urbanising if the share that is urban is increasing, while an area is urbanising if the share of its land that is urban is increasing. In practice, however, urbanisation is often used to refer to different processes, which do not necessarily coincide, and indeed are becoming increasingly distinct. Thus, whereas once the urbanisation of population, space and culture could legitimately be treated as different aspects of the same transformation, this is now no longer the case. To ensure continuity and clarity in this volume, three alternative definitions of urbanisation are summarized below:

### *Demographic urbanisation*

This is the 'standard' definition of urbanisation that will serve as the default in this synthesis report. Demographic urbanisation refers to an increasing proportion of the population living in urban areas, accompanied by a relative decline in rural populations and enterprise. Thus urban population growth is the sum of natural increase plus the rate of urbanisation.

*Urbanisation as urban expansion*

Here urbanisation is equated to the spatial expansion of built-up areas, or of land administratively designated as such, and the commensurate loss of rural land. This can be useful in evaluating environmental impacts. However, current trends of declining urban density and fragmentation mean that urban expansion often takes place without demographic urbanisation, and involves a process of declining settlement density. It is misleading to refer to this declining settlement density as urbanisation, since demographic urbanisation refers to a process of increasing settlement density, and this has historically been seen as one of the key features of urbanisation.

*Urbanisation as process of socio-economic transition*

This definition refers to a transition from the social, economic, cultural and political systems assumed to be characteristic of rural areas, to those associated with urban centres. This definition is now contested due to the aforementioned debate over definitions of rural and urban 'characteristics' that are now viewed as overly simplistic.

It is due to these complexities that this report includes 'urban regions' in addition to megacities, to avoid excluding 'borderline' cities that escape classification as megacities as a result of administrative (rather than qualitative) boundaries. It is for this reason that Jakarta, for example, has been included in the report as an 'urban region', to reflect its uncertain urban boundaries, very large peripheral population and rapid growth trajectory. This approach reflects the intention to engage primarily in analysis of urban–marine interactions, rather than rural–urban. As Champion and Hugo (2004) argue, the key in any study of urban phenomena is to be clear, and to avoid definitive statistics in the absence of clear qualification as to their meaning.

## 3.2 Definitional Challenge 2: urbanisation on the coast

As with urban areas, there is no single definition of the coastal zone and use of this term varies significantly across disciplines and between locations. Generally, the coastal zone is classified as an interface between land and sea, and contains unique ecological, geologic and biologic domains, vital for the maintenance of a multitude of life forms (Beatley, Brower and Schwab 2002). As stated, this volume adopts a simple definition of the area located within a 50-metre elevation and a 100-kilometre distance from mean high water. However, once again, the diversity of definitions in circulation leads to a degree of subjectivity in the designation of coastal zones (Klein *et al.* 2003).

Kay and Alder (2005) identify four typologies of coastal definitions in the policy realm: fixed distance definitions, variable distance definitions, definition according to use and hybrids of these (p. 4). Others still have defined it in terms of elevation in proximity to the coast (McGranahan *et al.* 2007), or combined distance and elevation criteria (Klein *et al.* 2003, UNEP 2005). Such measures are useful, but fail to capture the changing spatiality of land–ocean interactions. They also fail to delineate the inclusion of three-dimensional dynamics such as atmospheric flows and exchanges, which have fundamental implications at ground level (e.g. controlling the hydrological cycle). Furthermore, strict use of the elevation-based definitions excludes some cities from the 'coastal megacity' bracket that would intuitively be included – for example São Paulo (Brazil), which lies only 50 kilometres from the coast but 800 metres above sea level (Nicholls 1995) and is included by von Glasow *et al.* (2012).

Rather than attempting to come up with a more rigorous definition of the coastal zone (i.e. one based on a complex set of quantitative parameters), it is also possible to identify coastal megacities according to certain typical or exclusively coastal geomorphic and economic qualities – more in line with Kay and Alder's third typology: (definition according to use). For example, typical coastal landscapes include tidal, deltaic and estuarine features, which often coincide with economic reliance on large-scale fishing or freight industries (Klein *et al.* 2003). Use of a more qualitative definition based on the presence of such features would include Cairo (11.17 million, 119 kilometres from coast) and Dhaka (15.39 million, 140 kilometres from coast), which interact strongly with the coast despite being located more than 100 kilometres from the shoreline (see Plate 3). Von Glasow *et al.* (2012) acknowledge and address the difficulty of delineating between 'coastal' and 'not coastal', by identifying both 'coastal agglomerations' and 'those with coastal influence'. Such a 'third way' could arguably also include Greater London (although von Glasow *et al.* do not). Furthermore the Thames – which historically and currently has been a hugely significant source of both economic opportunity and environmental hazard in London – is highly tidal, necessitating flood defence structures such as the Thames Barrier. These arguments also support the inclusion of those parts of urban regions that lie beyond the coastal plain (in places such as Los Angeles, Seoul and Istanbul) in accounts of coastal urbanisation.

## 4 Global trends in megacity development

The purpose of this section is to highlight some of the most significant trends in coastal megacity development, which have bearing on the nature of risk and opportunity in these areas. Note that due to the large degree of contradiction and debate which surrounds megacity growth statistics, what is presented here is an overview of key narratives. It should also be noted that many of the issues identified are not limited to the spatial fixidity of coastal megacities – that is, they are more to do with cities than with the coast and also apply in non-coastal contexts – however, they are of particular significance in coastal megacities due to the concentration of risk, assets and environmental sensitivity in these areas as previously outlined. Given the difficulties outlined above, a broad definition of megacities and urban regions is adopted.

### *4.1 The 'new' megacities: growth in low- and middle-income countries*

Between 1975 and 2000, the number of megacities in low- and middle-income countries rose from two to 13, with the global megacity population increasing more than seven-fold to 165 million (more if wider agglomeration boundaries are drawn) (Cohen 2006). Looking at Plate 1, of the seven new megacities anticipated by UN-HABITAT to emerge by 2025, all are in low- or middle-income countries with only four remaining in developed countries. Even more significantly, by 2011 three of these projections have already become reality, with Lagos, Shenzhen and Guangzhou already having passed ten million in population – and Jakarta is not far behind (UN-DESA 2012b). That two Chinese cities have done this reflects the 'staggeringly high' growth rates being experienced in China, which exceeded 10 per cent in 2008 (UN-HABITAT 2008).

These data indicate that generally, urban growth rates are higher in Asia, Africa and Latin America compared to the 'older' megacities in Europe, North America and Japan. Nevertheless despite the rapid growth of these 'new' megacities, recent reports are keen to dispel popular visions of a future led by 'explosive' urbanisation – a myth largely driven by megacities' high visibility and the media (Satterthwaite 2010). Whilst it is true that megacities have grown in size and number in the last few decades, this growth has been much slower than expected – particularly in high-income nations (Angel *et al.* 2011). Currently, 63 per cent of urban populations in developed countries reside in intermediate and small-sized cities, compared to 9 per cent being located in megacities (UN-HABITAT 2009). Some observations indicate trends of counter-urbanisation in large cities in Europe and North America since the 1970s, and the rate of growth of some low-income-country cities also slowed during the 1980s (McDonald 2000). Satterthwaite (2005) also observes that in many cities, movement outwards exceeds migration inwards, and few have reached anything close to the size that was predicted for them in the 1970s. However it is difficult to say the extent to which these trends are the product of the boundaries within which population data is collected, or changes to census methodologies. Furthermore, data suggesting declining urban growth rates in high-income countries may also reflect demographic factors such as the reduction of urban and rural fertility, and a game of diminishing returns in calculating net population change relative to a very large and increasing total.

Despite these complexities, overall it is clear that urban growth rates are much larger in less developed regions than in more developed regions. Plate 2 illustrates this, showing the fastest urban growth rates lie in low- and middle-income countries compared to low or even negative growth rates in high-income countries, where urban growth rates have been declining since 1950 (Montgomery 2008).

The most likely drivers of high rates of urbanisation in low- and middle-income countries are public perceptions about enhanced employment opportunities and better provision of basic services in cities. However, in many cases such assumptions about elevated living conditions in urban centres are highly inaccurate, and levels of urban poverty, inequality and social exclusion are very high in many megacities (Tacoli *et al.* 2008). Extreme urban poverty is fuelled by population growth and in-migration, stimulating rapid expansion of informal settlements in the absence of necessary improvements to public services and infrastructure (Cohen 2006). For example in Lagos, Nigeria, 40 per cent of the population live in over-crowded housing with low access to sanitation whilst the city continues to expand (UN-HABITAT 2010). Recent reports estimate urban densities in low-income countries to be twice as high as those in Europe and Japan (Angel *et al.* 2011) – high density combined with poverty and inadequate waste management and sanitation infrastructure can culminate in severe environmental health problems (see McGranahan *et al.* 1996, McMichael 2000). Aggressive gentrification policies, seeking to 'clean up' the city to make way for high-value developments by bulldozing informal settlements, are widespread, and have caused massive dislocation of the population, for example in Mumbai and Lagos (UN-HABITAT 2010). Unfortunately, attempts to curb megacity growth through the exclusion or neglect of migrants are just as likely to reinforce rather than counter these issues of inequality (Tacoli *et al.* 2008).

### 4.2 Declining urban densities

Arguably more significant than the apparent demographic slow-down in higher income nations is Angel *et al.*'s (2011) observation that urban *expansion* is occurring at a rate twice as fast as urban population growth, meaning the density of built-up areas must be declining. Angel *et al.* estimate that urban densities have decreased by 2 per cent every year between 1990 and 2000 and that if current rates continue global urban population will double within 43 years, whilst urban land cover will take only 19. 'Urban sprawl' is the most commonly used term to describe low-density urban expansion, and is a well-recognized phenomenon in cities literature (see for example Harvey and Clark 1965, and the more recent Yu and Ng 2007). UN-HABITAT (2009) state that urbanisation beyond metropolitan boundaries is a particularly significant trend in Asia. Sprawl is attributed with causing fragmentation of the open space surrounding urban centres, and there is a wide literature on its associated negative environmental impacts (Hasse and Lathrop 2003, UN-DESA 2011. In many low-income regions, sprawl is dominated by informal or 'slum' settlements, which grow on city outskirts where the land is often cheaper as a product of increased distance from the economic centre, lower quality land and higher exposure to risk. For example, in Rio de Janeiro, Brazil, favela settlements at the city edges extend up very steep slopes on highly unstable terrain (Ojima and Hogan 2009).

As cities expand through urban sprawl (in combination with other processes such as 'leap-frogging' (Korff and Rothfu 2009) and incorporation of outlying settlements), scale makes it difficult for megacities to perform as an organic whole. This can force a multiplication of loci for economic activity, industry, educational excellence, places of poverty, etc., creating pressures for geographical, social, administrative and political fragmentation and leading to a transition from uni-polarity to multi-polarity (Laquian 2011). Fragmentation may also arise where areas of new and old growth are not closely connected, either in terms of governance or transportation links. This mechanism of fragmentation may be particularly visible in low-income country megacities, where loose systems of urban governance can detract from planning that is informed by a holistic vision of the city as an organic whole. Nevertheless, fragmentation need not necessarily lead to decreased efficiency of the city en masse, as in many high-income countries (especially in the US and Australia) high motorization increases connectivity between outlying areas of the city (Huang and Sellers 2007). However, it can be problematic for these multiple centres to interact in planned ways that can benefit from scale economies at the city level.

Closely related to sprawl is peri-urbanisation, which refers to the semi-urban development of previously rural land. Peri-urbanisation is an ambiguous term (Allen 2003), defined by Martine (2011) as "the non-contiguous and patchwork form of urban expansion and leapfrog development which springs from land speculation, changing production modalities, and the spread of automobile transportation" (p. 13).

Allen (2003) argues that current planning systems and regulation are inadequate to deal with the unique development and environmental challenges of the peri-urban interface. These include increased need for transportation links, job creation, sanitation, and education and healthcare facilities, which are often lacking in peri-urban areas that have grown very rapidly (da Gama Torres 2011). Peri-urban areas almost always straddle more than one administrative jurisdiction, adding to the complexity of peri-urban governance. Allen (2003) notes the growing literature on peri-urbanisation as a legitimate land use, and calls for the development of planning regulations that consider the environmental, social and economic specificity of peri-urban areas.

### *4.3 The growth of megaregions, urban corridors and city regions*

As stated by UN-HABITAT's 'State of the world's cities: 2010/11' report (2010), the growth of mega-regions is a key trend in global urban development. This has been particularly significant along the coast, linked to the locational advantages of coastal areas for industry and trade. Megaregions form through the overlapping sprawl of multiple urban centres as agglomerations spread away from the coastal strip, facilitated by technological advances (Aguilar *et al.* 2003). Megaregions tend, therefore, to be polycentric. They may continue to draw on coast-based industries as a source of (amongst other things) trade, employment and energy, but as the model shifts from core-periphery to multi-core, economic dependency on the coast may become weaker or more indirect. Castells (1998) emphasizes that megaregions are a distinct entity from 'megacities', which may form part of a megaregion. In practice, however, the strong degree of interaction and interdependency between megacities and adjoining agglomerations can make them difficult to distinguish. Examples of coastal megaregions include the area extending from Accra, Ghana, to Lagos in Nigeria (through four countries), Osaka-Nagoya-Tokyo in Japan, and – the largest of all with 120 million people – the Hong Kong-Shenhzen-Guangzhou region in southern China (UN-HABITAT 2010).

As megacities and urban regions often result from the agglomeration of previously distinct urban (and rural) entities, different political and administrative institutions may begin to share authority. Megaregions may therefore retain much of the heterogeneity between its poly-centres that pre-existed their integration. However, this adds significantly to the challenge of cohesive urban governance, and may hinder attempts at coordinated development and environmental management, including adaptation to climate change and the management of other emerging pressures and risks, throughout the megaregion (Laquian 2011).

The growth of megaregions has implications at both regional and global scales. Cities in clusters, corridors and megaregions are described by UN-HABITAT as 'becoming the new engines of both global and regional economies' (2010 p. ix), by improving interconnectivity between cities and nations. Florida *et al.* (2008) identify 40 megaregions with a combined economy that exceeds $100 billion per annum – estimated to represent 66 per cent of global economic output and 85 per cent of innovation. However, the report also warns that megaregions represent both a symptom and a cause of the perpetuation of asymmetrical development, strengthening existing patterns of economic dependency rather than supporting 'diffused spatial development'. Reinforcement of 'status quo' patterns of economic development is of little benefit to populations living outside these economic 'powerhouses', a concern that is particularly salient to rural populations in low-income countries. This poses a significant challenge to development planners in national and local governments (UN-HABITAT 2010).

## 5 Risk, vulnerability and resilience in coastal megacities

The growth in the number and size of coastal megacities worldwide, particularly those in low- and middle-income countries, is simultaneously resulting in increasing pressures on the natural environment and changing the way that environmental risk is constructed and experienced. These goalposts are shifting further as a result of global environmental change, which is anticipated to bring changes in the behaviour and frequency of extreme climate events and cause additional stress to human and environmental systems. These systems are best viewed as being coupled, because disturbances in one can cause and reinforce disturbances in another, through complex feedback mechanisms. For example, urban expansion along the coast often results in concrete reinforcement of the coastline, which interrupts natural erosion cycles and tidal movements, which increases flood risk; also urban expansion leads to pollution and exploitation of resources, which harm ecosystems and ultimately reduce the availability and quality of ecosystem services and resources on which the urban population depends, with negative implications for human health, economies and well-being. In both cases, greater attention to finding sustainable forms of urban growth, in which human and environmental systems can co-exist in a balanced and symbiotic way, is necessary.

Vulnerability to hazards is defined as human sensitivity to hazard impacts and is the product of a wide range of factors (Wisner *et al.* 2004). Resilience, simply put, is the ability to withstand hazard impact, such that the system affected is able to continue to function, as well as learn from and adapt to new and unanticipated sources of hazard (Cutter *et al.* 2008). Whilst urban resilience is, in part, a product of a city's specific vulnerabilities and it's actions taken to mitigate these (Pelling 2011b). The most significant factors affecting vulnerabilities and resilience in coastal megacities are addressed in this section.

### *5.1 Drivers of risk in coastal megacities*

The hazards affecting any given megacity depend upon the geographical location and physical characteristics of the city in question. Flood risk, for example, is greatest in cities that are either in close proximity to major rivers or are susceptible to storm surges – and even more so if these coincide (Prasad *et al.* 2009, Dasgupta *et al.* 2009). Elevation is another strong predicator of coastal flood risk, and coastal plains may or may not be sufficiently raised above sea level to mitigate exposure to coastal flooding as well as near- or medium-term projections for sea level rise due to climate change. Deltaic settings tend to be particularly vulnerable for this reason also, although Manila and Mumbai are examples of non-deltaic but nonetheless very low-lying exposed cities. Some additional factors that contribute to heightened exposure to current and future risks in coastal megacities are outlined in Box 1.1.

### Box 1.1 Geographical characteristics increasing hazard exposure in coastal megacities

*Low elevation deltas* – Coastal cities tend to be highly exposed to coastal flooding, whether due to changes in sea level, tidal waves or the effects of cyclones or frontal systems. Moreover, many coastal cities are also located by rivers as this provides a source of fresh water and, historically, access points to interior lands. Given how floods propagate downstream through catchments, this also makes coastal cities vulnerable to flooding from upstream (for example the flooding of the Mississippi which affected New Orleans in May 2011).

*Topography* – Depending on their tectonic setting, coastal cities may be surrounded by mountains/topographic barriers that serve to enhance local precipitation as onshore, saturated winds are forced to rise. Weather systems, be these cyclones, frontal systems or simply air flowing on-land from features such as persistent trade winds, interact with these physical barriers resulting in heavy rainfall rates and runoff. Even in the absence of significant topography, changes in surface roughness can affect the movement of onshore storm systems.

*Land use* – Many large coastal cities are also ports and industrial processing areas, for example Shanghai is amongst the world's busiest ports. Some of the associated industrial activities, for example oil refineries, result in huge emissions with significant implications for local and regional air quality. The ratio of green space versus impermeable surface in a coastal megacity also has significant implications for flood risk, as it affects the infiltration rate of the land area.

*Sea/land breezes* – The proximity of water and land, and the differential heating of these two surfaces over the course of the day, generates day-time sea breezes (onshore flows of air) and land breezes (offshore flows of air). In summer, sea breezes are important in coastal cities in mitigating heat stress, though in some urban settings the density of buildings impedes the penetration of the sea breeze, reducing ventilation, with implications for thermal stress and air quality. Tokyo, for instance, is considering removing buildings to allow the sea breeze to penetrate and aerate the city. Other cities are also considering the orientation of the buildings in order to affect the inflow of sea air (e.g. Hong Kong, Singapore). This needs to take into account both wind flow and solar gain (e.g. Ng *et al.* 2011). Sea/land breezes also serve to concentrate and re-circulate pollutants across coastal cities with important implications, particularly at night when the urban boundary layer (and thus atmospheric mixing) is diminished.

*Population density and heat island effect* – Coastal megacities tend to be denser than megacities in other geographical settings, in part because of competing land uses and topography, and as a result some 13 of the 20 densest cities worldwide are in the coastal zone (McGranahan *et al.* 2007). Thus coastal cities tend to have a high proportion of impervious cover, which is a key determinant of energy flux partitioning – latent heat fluxes (evapotranspiration rates) are suppressed, while sensible and storage heat fluxes are enhanced. This results in greater heating of the air and substrate, culminating in the 'heat island effect'. High population densities (which often coincide with high poverty levels) experience heightened exposure to concentrated pollution, which exacerbates environmental health problems and enhances greenhouse gas emissions – contributing further to urban warming effects.

Source: Grimmond (2011)

The full range of hazards that coastal megacities are exposed to is outlined in detail in Chapter 3 of this volume. A selection of these are summarized in Plate 3.

The construction of risk, however, is more than simply exposure to hazard; risk is also the product of vulnerability (Wisner *et al.* 2004). Pelling (2011a) observes seven features of coastal megacities that influence their vulnerability to natural hazards, relative to other types of human settlement in other locations: (1) the concentration of physical assets, industries, energy installations and exposed populations; (2) a significant migrant population and cultural and socio-economic diversity; (3) their reach and dependency on coastal and interior networks of infrastructure; (4) the layering of coastal hazards – subsidence, salinization, liquefaction, sea-level rise, etc., which makes risk management more complex and raises the likelihood of risk-reduction policies with regards to one risk having adverse effects on vulnerability to another; (5) a capacity to trigger economic contagion at different scales through their strategic importance as finance and trade centres; (6) the possibilities of hotspot growth for new ecological assemblages as a result of degradation and interruption of ecosystems; (7) an extensive source of intervention in biophysical systems through extractive industry, water and air pollution, and others, implicated in the creation and reinforcement of urban hazards by reducing the sustainability and environmental quality of the city.

These features highlight that it is the unique juxtaposition between large-scale settlement and environmental processes that exists in coastal megacities that culminates in heightened risk in coastal megacities. It is where these systems interact that the potential for hazard arises. However, it is also in this space of overlap that the potential for adaptation and resilience arises, depending on cities' responses to the source of environmental pressure.

### 5.1 Possibilities for resilience and transformation

Despite the sources of hazard and vulnerability outlined above, some studies have highlighted the increased resilience of large urban regions compared to small towns and rural areas, due to greater capacity to respond and prepare for hazard impact (Cross 2001, Cutter *et al.* 2010). Megacities in particular may have advantages in scale and concentration of assets that can help find solutions to hazard and vulnerability driven by competition over land and social diversity (Pelling 2011a). They have the locational advantage of being able to benefit from coast-dependent industries, and access to international trade links. Furthermore, as hotbeds of economic prosperity, magnets for highly skilled migrants and having close connections with other centres of innovation, coastal megacities are uniquely placed for the development of innovative solutions to risk management and adaptation. This is facilitated through the emergence of formalized networks such as C40, or the UNISDR's global 'Making Cities Resilient' campaign (2010–2015).

A similar view is put forward by Klein *et al.* (2003), Parker (1995) and Handmer (1995) who argue that megacities in particular have more power, resources and the built-in complexities of economic, physical and social form and function that can confer capacity to adapt to hazards and reduce risk from mitigating climate change to air pollution control, and to lead such innovations that might then also benefit smaller cities and their hinterlands. Satterthwaite *et al.* (2009) argue that it is 'easy' to envisage a future where urban planning and development have embedded adaptation to natural hazard risks into their normal functions. They see this as an achievable goal so long as long-term policies are in place to ensure that the costs of adaptation are spread over time. Such measures could include directing urban growth away from high-risk areas, prioritizing risk-resistant infrastructure, or adjustments being made to building codes and standards. Klein *et al.* (2004) similarly observe that the survival of some cities through history (such as Istanbul, Baghdad and Cairo), whilst others have failed or been abandoned, indicates that there are some qualities that enable certain urban areas to resist and recover from external (and internal) shocks and continue to thrive.

Nevertheless, at the global scale, while urbanisation is rising in importance as an area for social and physical research and policy innovation, there remain important strategic gaps in understanding and knowledge. The lack of shared definitions for different types of human settlements and variation in the quality of national-level data that are relied upon for global assessment and modelling continues to be a concern – both for researchers and policy makers. This is a weakness not only of basic demographic data but also of data on populations exposed to or impacted by environmental, technological and disaster risk. Moreover, information on the depth and social distribution of poverty in cities is variable and often underestimated (Pelling 2003). The greater reliance on money economies and the absence of entrenched moral economies in urban places are two factors that make survival post-disaster and on the margins of the economy more difficult in urban places compared to rural. However, the erosion of social cohesion and continued incursion of the global economy in many rural areas are reducing this distinction, which is becoming associated more with governance than geography, demography or geology.

Goals such as urban sustainability and resilience of urban systems require a cohesive and responsive system of governance, with clear flows of information between research, policy and practice. Achieving these goals first requires improvements to the collection, management and collation of observations, modelling approaches and tools, as well as promoting knowledge exchange and understanding in tandem (see Table 1.2, Grimmond *et al.* 2010, NAS 2010).

Given that meteorological hazards occur at both ends of the frequency spectrum (i.e. high/low frequency/magnitude) to predict and inform those who are vulnerable requires a multi-agency approach (e.g. Tang 2006) and one that is embedded within systems of practice – including government, civil society and private sector actors.

*Table 1.2* High-priority recommendations from the World Climate Conference 3- Need for more sustainable cities: information for improved management and planning of cities

*Observations*

- Need for more operational *urban measurement station and networks*; this will require stations within the urban area and upwind. Station should to be sited and equipment exposed in conformity with WMO Urban Guidelines. Stations are especially needed in *rapidly developing cities in hot climates* and in their surroundings. Both simple and complex topographical settings should be represented.
- Where possible *vertical profiles* of physical and chemical variables should be sampled.
- *Long-term measurement stations* should be preserved or established in cities with different urban morphologies to determine universal flow and flux characteristics.

*Data*

- Need to establish an *international data archive* to aid translation of research findings into design tools and guidelines for different climate zones and urban land use. The archive should consist of high-quality data of use to a broad range of practitioners.
- The importance of fully documenting *urban station metadata* (e.g. description of instruments, site, data quality assurance and control, protocol) should be stressed.

*Understanding*

- Need to develop methods and frameworks to *analyse atmospheric data* measured above complex urban surfaces.
- Proposed actions to make cities sustainable need to be assessed to determine at what scale interventions are needed and are possible.
- Need methods to distinguish between signals attributable to urban climate change and those to regional and global change.

*Modelling*

- Need to improve short-range, high-resolution numerical prediction of weather, air quality and chemical dispersion in the urban areas through improved modelling of the biogeophysical features of the urban land surface and consequent exchange of heat, moisture, momentum and radiation with the atmospheric urban boundary layer.
- Need to improve or incorporate data assimilation from meteorological and biogeophysical observations from improved observing networks.

*Tools*

- Need to develop tools to allow models to be able to accommodate the wide differences in data availability (e.g. routine versus research intensive data) depending on the application from research to operational situation.
- Need to develop tools that allow probable impacts of proposed sustainable design measures to be assessed and ranked, including any unintended consequences of the proposed changes.

*Knowledge exchange*

- Need to ensure widespread education of the meteorological community (including National Meteorological and Hydrological Services, NMHS) in urban meteorology.
- Assist NMHS to appreciate the role of meteorology and hydrology in urban planning and management of more sustainable cities of all sizes.
- *Communication* across scientific disciplines and spatial and temporal scales must be encouraged.

Source: Grimmond et *al.* (2011), NAS (2010)

## 6 Discussion: Urban transitions, urban futures

At a global scale, the urban transition – a shift from a predominantly rural population to a predominantly urban population – has been accompanied by a demographic transition – a shift from high birth and death rates, to low birth and death rates (de Sherbinin and Martine 2007). Although the reasons for this are contested, this is thought to be due in part to improved employment opportunities for women and the elevated cost of living in the city, which both reduce fertility, and improved access (typically, although not always) to medical services, which reduces mortality and improves access to family planning (Caldwell 2005, de Sherbinin and Martine 2007). Also at the global scale, this urban metamorphosis has accompanied a transformation of society, culture, politics, economics and environmental management (de Sherbinin and Martine 2007). However, as outlined below, these associations are neither inherent nor universal.

The literature on the connection between urbanisation and economic development is wide, complex and contradictory. There is evidence for symbiotic relationships between various aspects of modern development, two facets of which are indeed economics and urbanisation (Tacoli *et al.* 2008). However, identifying which is the prime driving force is problematic. For example, UN-HABITAT identifies positive links between urbanisation and improvements to democratic governance and female empowerment, which are both viewed as characteristics of developed societies (UN-HABITAT 2008), and observes that "no country has ever achieved sustained economic growth or rapid social development without urbanising [sic]" (UN-HABITAT 2010). Henderson and Wang (2007) provide empirical evidence to suggest that urbanisation is an outcome of development, attributing urban growth to (1) technological advances and (2) democratization. Both, they argue, promote demographic urbanisation and urban expansion, but they affect differently sized cities in different ways. They find that whilst technology facilitates scale economies and management systems, and promotes growth in very large cities, increased democratization has the greatest growth impact on smaller cities because decentralized politics increases local autonomy.

However, this argument in itself accounts for neither the persistence of intensely asymmetrical rural–urban divides (Montgomery 2008) nor the high rates of urban poverty that exist in low-income countries (Tacoli *et al.* 2008). To understand these inequalities it is useful to distinguish different types of 'development', particularly human development and economic development. Human development is defined by the UNDP Human Development Reports as "enlarging people's choices . . . by expanding human capabilities and functionings", the prerequisites of which are 'for people to lead long and healthy lives, to be knowledgeable and to have a decent standard of living' (UNDP n.d.). Insofar as democracy can be understood as a means to express free choice, it is therefore a key facet of human development. It could therefore be posited that the growth of megacities with high poverty levels have succeeded in generating sufficient agglomeration of industry and labour to achieve economic development, but have failed to improve human development equitably. Ranis *et al.* (2000) argue that in the long-term it is preferable to promote human development in the first instance, because this can generate a virtuous cycle of mutual reinforcement between human and economic development, driven by a healthy and well-educated workforce. In making such a choice, governance is clearly central.

This logic supports the well-developed argument that urban governance is key to vulnerability and poverty reduction, and that 'good governance' is necessary to ensure that both human and economic development accompany urbanisation (Pelling 2003, Wisner *et al.* 2004, UN-HABITAT 2008, UN-HABITAT 2010). Good governance means having an accountable, responsive and well-structured governance system with well-designed policies and effective enforcement power. This is also key in maintaining a functioning land market, capable of revealing the value put on land by residents, developers and profit-making enterprises (a context that is important to good planning, and making interventions in the interest of environment, equity and public interests generally), and to ensure the benefits of urban growth reach the poor as well as the elite (Martine *et al.* 2008). However, there are strong forces preventing human development from being given priority over economic growth, and this view contrasts strongly with the claim that inequality is an almost inevitable consequence of development (World Bank 2009).

The impact of the urban transition on the environment varies according to many factors (see Chapter 2). Factors that affect the sustainability potential of megacities include the spatial extent, type of land, form of occupancy and ecological condition of urban residence (de Sherbinin and Martine 2007). The failure or damage of ecosystem services such as clean water, inadequate access to them, or disproportionate demand for available resources, are attributed to environmental health concerns in coastal megacities (McGranahan *et al.* 2006). Literature on the urban environmental transition states that the majority of environmental health problems are actually associated with large-scale poverty rather than urbanisation per se, and that increased wealth is associated with negative environmental impacts being located increasingly remotely from the point of consumption (McGranahan and Songsore 1994). In terms of future development, therefore, it seems certain that the management of urban growth in low-income megacities will dictate the proportion of the city that lives in poverty and has environmental health problems (Martine *et al.* 2008).

The following chapters develop in more detail the impacts of coastal megacities and city regions on the environment, and in turn the hazards and vulnerabilities experienced by these cities. The trade-offs and synergies between management options will be analyzed, outlining potential adaptive strategies now and in the future.

## References

Aguilar, A., Ward, P. and Smith Sr., C. (2003) Globalization, regional development, and mega-city expansion in Latin America: Analyzing Mexico City's peri-urban hinterland, *Cities*, 20 (1), pp. 3–21.

Allen, A. (2003) Environmental planning and management of the peri-urban interface: perspectives on an emerging field, *Environment and Urbanisation*, 15 (1), pp. 135–148.

Angel, S., Parent, J., Civco, D. and Blei, A. (2011) Making Room for a Planet of Cities, Policy focus Report Series, Cambridge MA: Lincoln Institute of Land Policy.

Beatley, T., Brower, D. and Schwab, A. K. (2002) *An introduction to coastal zone management*. Second edition, Washington, DC: Island Press.

Beguin, H. (1996) "Faut-il définir la ville?", in Derycke, P., Huriot, J. and Pumain, D. (eds.) *Penser la ville: Théories et modèles*, Paris: Anthropos, Collection Villes, pp. 301–320.

Caldwell, J. (2005) On net Intergenerational wealth flows: An update, *Population and Development Review*, 31 (4), pp. 721–740.

Caljouw, M., Nas, P. and Pratiwo (2005) Flooding in Jakarta: Towards a blue city with improved water management, *Bijdragen tot de Taal-, Land- en Volkenkunde*, 161 (4), pp. 454–484.

Castells, M. (1998) Why the Megacities Focus? Megacities in the new world disorder, The Megacities Project Paper MCP-018, first prepared for Mega-Cities Seventh Annual Coordinators Meeting in Jakarta, Indonesia, 1–7 August 1993. Available at http://mega-cities.net/pdf/publications_pdf_mcp018intro.pdf (last accessed 8 March 2012).

Center for International Earth Science Information Network (CIESIN)/Columbia University, International Food Policy Research Institute (IFPRI), The World Bank, and Centro Internacional de Agricultura Tropical (CIAT) (2011) Global Rural-Urban Mapping Project, Version 1 (GRUMPv1): Settlement Points. Palisades, NY: NASA Socioeconomic Data and Applications Center (SEDAC). Available from: www.sedac.ciesin.columbia.edu/data/set/grump-v1-settlement-points.

Champion, T. and Hugo, G. (2004) Introduction: moving beyond the urban–rural dichotomy, in Champion, T. and Hugo, G. (eds.), *New Forms of Urbanisation: beyond the urban–rural dichotomy*, Aldershot: Ashgate, pp. 3–24.

Cohen, B. (2006) Urbanization in developing countries: Current trends, future projections, and key challenges for sustainability, *Technology in Society*, 28, pp. 63–80.

Cross, J. (2001) Megacities and small towns: different perspectives on hazard vulnerability, *Global Environmental Change Part B: Environmental Hazards,* 3 (2), pp. 63–80.

Cutter, S., Barnes, L., Berry, B., Burton, C., Evans, E., Tate, E. and Webb, J. (2008) A place-based model for understanding community resilience to natural disasters, *Global Environmental Change,* 18 (4), pp. 598–606.

Cutter, S., Burton, C. and Emrich, C. (2010) Disaster Resilience Indicators for benchmarking Baseline conditions, *Journal of Homeland Security and Emergency Management,* 7 (1), Article 51.

da Gama Torres, H. (2011) Social and environmental aspects of peri-urban growth in Latin American mega-cities, in UN-DESA (eds) *Population Distribution, Urbanization, Internal Migration and Development: An International Perspective*, United Nations, Department of Economic and Social Affairs, Population Division, New York: Affairs, Population Division, New York: United Nations. pp. 123–143. Available at: www.un.org/esa/population/publications/PopDistribUrbanization/PopulationDistributionUrbanization.pdf (accessed 22 April 2013).

Dasgupta, S., Laplante, B., Murray, S. and Wheeler, D. (2009) Climate Change and the Future Impacts of Storm-Surge Disasters in Developing Countries. Centre for Global Development Working Paper 182.

de Sherbinin, A. and Martine, G. (2007) *Urban population, development and environment dynamics: situating PRIPODE*, Paris: Committee for International Cooperation in National Research in Demography (CICRED).

de Sherbinin, A., Schiller, A. and Pulsipher, A. (2007) The Vulnerability of Global Cities to Climate Hazards, *Environment & Urbanisation*, 19 (1), pp. 39–64.

Florida, R., Gulden, T. and Mellander, C. (2008) The Rise of the Mega-Region, *Cambridge Journal of Regions, Economy and Society*, 1, pp. 459–476.

Grimmond, C. S. B. (2011) Urban Climate in the Coastal Zone: Resilience to Weather Extremes, in Pelling, M., *Megacities and the Coast: Transformation for resilience. A preliminary review of knowledge, practice and future research*. June, LOICZ. Accessed from www.loicz.org/imperia/md/content/loicz/hotspots/urbanization/Megacities_and_the_coast_report_4_6_2011.pdf

Grimmond, C. S. B., Roth, M., Oke, T. R., Au, Y. C., Best, M., Betts, R., Carmichael, G., Cleugh, H., Dabberdt, W., Emmanuel, R., Freitas, E., Fortuniak, K., Hanna, S., Klein, P., Kalkstein, L. S., Liu, C. H., Nickson, A., Pearlmutter, D., Sailor, D. and Voogt, J. (2010) Climate & More Sustainable Cities: Climate Information for Improved Planning & Management of Cities (Producers/Capabilities Perspective), *Procedia Environmental Sciences*, 1, pp. 247–274.

Halfacree, K. (1993) Locality and Social Representation: space, discourse and alternative definitions of the rural, *Journal of Rural Studies,* 9 (1), pp. 23–37.

Handmer, J. W. (1995) Managing Vulnerability in Sydney: Planning or Providence? *GeoJournal*, 37 (3), pp. 355–368.

Harvey, R. and Clark, W. (1965) The Nature and Economics of Urban Sprawl, *Land Economics*, 41 (1), pp. 1–9.

Hasse, J. and Lathrop, R. (2003) Land Resource Impact Indicators of Urban Sprawl, *Applied Geography*, 23 (2-3), pp. 159–175.

Henderson, J. and Wang, H. (2007) Urbanization and city growth: The role of institutions, *Regional Science and Urban Economics*, 37 (3), pp. 283–313.

Huang, J., Lu, X. and Sellers, J. (2007) A global comparative analysis of urban form: Applying spatial metrics and remote sensing, *Landscape and Urban Planning*, 84 (4) pp. 184–197.

Kay, R. and Alder, J. (2005) Coastal Planning and Management, second edition, Abingdon: Taylor and Francis.

Klein, R. J. T., Nicholls, R. J. and Thomalla, F. (2003) The Resilience of Coastal Megacities to Weather-Related Hazards, in Kreimer, A., Arnold, M. and Carlin, A. (eds) *Building Safer Cities: The Future of Disaster Risk*. Disaster Risk Management Series No. 3. Washington, DC: World Bank Publications, pp. 101–120.

Klein, R., Nicholls, R. and Thomalla, F. (2004) Resilience to natural hazards: how useful is this concept?, EVA Working Paper No.9, DINAS-COAST Working Paper No.14, Potsdam Institute for Climate Impact Research, Potsdam, Germany.

Korff, R. and Rothfuß, E. (2009) Urban revolution as catastrophe or solution? Governance of megacities in the global South, *Die Erde*, 140 (4), pp. 355–370.

Laquian, A. (2011) The planning and governance of Asia's mega-urban regions, in UN-DESA (eds) *Population Distribution, Urbanization, Internal Migration and Development: An International Perspective*, United Nations Department of Economic and Social Affairs, Population Division, New York: United Nations , pp. 302–322. Available at: www.un.org/esa/population/publications/PopDistribUrbanization/PopulationDistributionUrbanization.pdf (accessed 22 April 2013).

Li, H. (2003) Management of coastal mega-cities – a new challenge in the 21st century, *Marine Policy*, 27, pp. 333–337.

Martine, G. (2011) Preparing for sustainable urban growth in developing areas, in UN-DESA (eds) *Population Distribution, Urbanization, Internal Migration and Development: An International Perspective*, United Nations, Department of Economic and Social Affairs, Population Division, New York: United Nations pp. 6–30. Available at: www.un.org/esa/population/publications/PopDistribUrbanization/PopulationDistributionUrbanization.pdf (accessed 22 April 2013)

Martine, G., McGranahan, G., Montgomery, M. and Fernandez-Castilla, R. (2008) Introduction: The new global frontier: cities, poverty and environment in the 21st century, in Martine, G., McGranahan, G., Montgomery, M. and Fernandez-Castilla, R. (eds) *The New Global Frontier: urbanization, poverty and environment in the 21st century*, London: Earthscan, pp. 1–13.

Martínez, M., Intralawan, A., Vázquez, G., Pérez-Maqueo, O., Sutton, P. and Landgrave, R. (2007) The coasts of our world: Ecological, economic and social importance, *Ecological Economics*, 63, pp. 254–272.

McDonald, R. (2000) Urban Places: The Challenges of Sustainable Living, *Geography Bulletin*, 32 (4), pp. 181–186.

McGranahan, G. and Songsore, J. (1994) Wealth, Health and the Urban Household, *Environment*, 36 (6), pp. 4–45.

McGranahan, G., Songsore, J. and Kjellén, M. (1996) Sustainability, Poverty and Urban Environmental Transitions, in Pugh, C. (eds) *Sustainability, the Environment and Urbanization*, London: Earthscan, pp. 103–133.

McGranahan, G., Balk, D. and Anderson, B. (2007) The rising tide: assessing the risks of climate change and human settlements in low-elevation coastal zones, *Environment & Urbanisation*, 19 (1), pp. 17–37.

McGranahan, G., Balk, D. and Anderson, B. (2007) Low Elevation Coastal Zone (LECZ) Urban-Rural Population Estimates, Global Rural-Urban Mapping Project (GRUMP), Alpha Version. Palisades, NY: NASA Socioeconomic Data and Applications Center (SEDAC). Available from: www.sedac.ciesin.columbia.edu/data/set/lecz-low-elevation-coastal-zone.

McGranahan, G., Marcotullio, P., Bai, X., Balk, D., Braga, T., Douglas, I., Elmqvist, T., Rees, W., Satterthwaite, D., Songsore, J., Zlotnik, H., Eades, J., Ezcurra, E. (2006) Chapter 27: Urban Systems, in *Ecosystems and Human Well-being: Current State and Trends,* Millennium Ecosystem Assessment, Washington, DC: Island Press, pp. 795–825.

McMichael, A. (2000) The urban environment and health in a world of increasing globalization: issues for developing countries, *Bulletin of the World Health Organisation,* 78 (9), pp. 1117–1126.

Montgomery, M. (2008) The demography of the urban transition: what we know and don't know, in Martine, G., McGranahan, G., Montgomery, M. and Fernandez-Castilla, R. (eds) *The New Global Frontier: urbanization, poverty and environment in the 21st century,* London: Earthscan, pp. 17–35.

NAS (National Academy of Sciences) (2010) *When Weather Matters: Science and Service to Meet Critical Societal Needs*, Washington, DC: National Academies Press.

Ng E, C Yuan, L Chen, C Ren, JCH Fung (2011) Improving the wind environment in high-density cities by understanding urban morphology and surface roughness: A study in Hong Kong, Landscape and Urban Planning, 101, pp. 59-74

Nicholls, R. J. (1995) Coastal Megacities and Climate Change, *GeoJournal* , 37 (3), pp. 369–379.

Nur, Y., Fazi, S., Wirjoatmodjo, N. and Han, Q. (2001) Towards wise coastal management practice in a tropical megacity – Jakarta, *Ocean & Coastal Management,* 44, pp. 335–353.

Ojima, R. and Hogan, D. (2009) The demographic composition of urban sprawl: local and regional challenges concerning global environmental change in Brazilian metropolitan areas, XXVI IUSSP International Population Conference, 27 September – 2 October, Morocco.

Parker, D. (1995) Hazard Transformation and Hazard Management Issues in the London Megacity, *GeoJournal*, 37 (3), pp. 313–328.

Pelling, M. (2003) *The Vulnerability of Cities: Natural Disasters and Social Resilience*, London: Earthscan.

Pelling, M. (2011a) Introduction, in Pelling, M., Megacities and the Coast: Transformation for resilience. A preliminary review of knowledge, practice and future research. June, LOICZ, www.loicz.org/science/hotspot/Urbanization/News_Reports/index.html.en (last accessed 30 October 12).

Pelling, M. (2001b) *Adaptation to Climate Change: From resilience to transformation*, London: Routledge.

Pickett, S., Cadenasso, M., Grove, J., Nilon, C., Pouyat, R., Zipperer, W. and Costanza, R. (2001) Urban Ecological Systems: Linking Terrestrial Ecological, Physical, and Socioeconomic Components of Metropolitan Areas, *Annual Review of Ecology and Systematics*, 32, pp. 127–157.

Pumain, D. and Robic, M. (1996) Théoriser la ville, in Derycke, P., Huriot, J. and Pumain, D. (eds) *Penser la ville, Théories et modèles*, Paris: Anthropos, Collection Villes, pp. 107–161.

Prasad, N., Ranghieri, F., Shah, F. Trohanis, Z., Kessler, E. and Sinha, R. (2009) *Climate Resilient Cities: A Primer on Reducing Vulnerability to Disasters*, Washington, DC: World Bank.

Ranis, G., Stewart, F. and Ramirez, A. (2000) Economic Growth and Human Development, *World Development,* 28 (2), pp. 197–219.

Satterthwaite, D. (2005) *The Scale of Urban Change Worldwide 1950–2000 and Its Underpinnings*, Human Settlements Discussion Paper Series, Theme: Urban change – 1, London: International Institute for Environment and Development.

Satterthwaite, D. (2010) Urban Myths and the Mis-use of Data that Underpin them, UNU-WIDER Working Paper No.2010/28.

Satterthwaite, D., Huq, S., Reid, H., Pelling, M. and Lankao, P. R. (2009) Adapting to Climate Change in Urban Areas: the possibilities and constraints in low- and middle-income nations,in Bicknell, J., Dodman, D. and Satterthwaite, D. (eds) *Adapting Cities to Climate Change*, London: Earthscan.

Tacoli, C., McGranahan, G. and Satterthwaite, D. (2008) Ubanization, Poverty and Inequity: is rural–urban migration a poverty problem, or part of the solution?, in Martine, G., McGranahan, G., Montgomery, M. and Fernandez-Castilla, R. (eds) *The New Global Frontier: urbanization, poverty and environment in the 21st century*, London: Earthscan, pp. 37–54.

Tang, X. (2006) Managing disaster risk in a mega-city, *Bulletin of the World Meteorological Organization*, 55 (4), pp. 268–273.

*The Jakarta Post* (2011) With a daytime population of 12 million, Jakarta is overpopulated, Andreas D. Arditya, 10 September 2011, www.thejakartapost.com/news/2011/09/10/with-a-daytime-population-12-million-jakarta-overpopulated.html (last accessed 30 October 2012).

UN-DESA (2009) World Urbanization Prospects: The 2009 Revision, United Nations, Department of Economic and Social Affairs, Population Division, New York: United Nations.

UN-DESA (2011) Population Distribution, Urbanization, Internal Migration and Development: An International Perspective, United Nations Department of Economic and Social Affairs Population Division, New York: United Nations (CD-ROM Edition).

UN-DESA (2012a) World Urbanization Prospects: The 2011 Revision – Highlights, United Nations Department of Economic and Social Affairs, Population Division, New York: United Nations.

UN-DESA (2012b) World Urbanization Prospects: The 2011 Revision, United Nations, Department of Economic and Social Affairs, Population Division, New York: United Nations, CD-ROM Edition.

UNDP (n.d.) Human Development Reports Glossary of Terms, http://hdr.undp.org/en/reports/glossary/ (accessed 6 July 2012).

UNEP (2005) Assessing coastal vulnerability: developing a global index for measuring risk, report prepared by Singh, A., Pathirana, S. and Shi, H., Division of Early Warning and Assessment, UNEP, Nairobi.

UN-HABITAT (2008) *State of the World's Cities 2008/2009: Harmonious Cities*, United Nations Human Settlements Programme, London: Earthscan.

UN-HABITAT (2009) *Planning Sustainable Cities: Global Report on Human Settlements 2009*, London: Earthscan.

UN-HABITAT (2010) *State of the World's Cities 2010/2011: Bridging the Urban Divide*, United Nations Human Settlements Programme, London: Earthscan.

Uzzell, D. (1979) Conceptual Fallacies in the Rural–Urban Dichotomy, *Urban Anthropology*, 8 (3/4) Urbanization in Latin America, pp. 333–350.

von Glasow, R., Jickells, T., Baklanov, A., Carmichael, G., Church, T., Gallardo, L., Hughes, C., Kanakidou, M., Liss, P., Mee, L., Raine, R., Ramachandran, P., Ramesh, R., Sundseth, K., Tsunogai, U., Uematsu, M. and Zhu, T. (2012) Megacities and Large Urban Agglomerations in the Coastal Zone: Interactions Between Atmosphere, Land, and Marine Ecosystems, *Ambio*, 42 (1), pp. 13–28.

Vörösmarty, C., Syvitski, J., Day, J., de Sherbinin, A., Giosan, L. and Paola, C. (2009) Battling to save the world's river deltas, *Bulletin of the Atomic Scientists*, 65 (2), pp. 31–43.

Ward, P., Marfai, M., Yulianto, F., Hizbaron, D. and Aerts, J. (2011) Coastal inundation and damage exposure estimation: a case study for Jakarta, *Natural Hazards*, 56, pp. 899–916.

Wisner, B., Blaikie, P., Cannon, T. and Davis, I. (2004) *At Risk: Natural hazards, people's vulnerability and disasters*, second edition, London: Routledge.

World Bank (2009) *World Development Report 2009: Reshaping economic geography*, New York: World Bank.

Yu, XJ. and Ng, CN. (2007) Spatial and temporal dynamics of urban sprawl along two urban–rural transects: A case study of Guangzhou, China, *Landscape and Urban Planning*, 79, pp. 96–109.

Chapter 2

# The environmental impacts of megacities on the coast

Lead author: *Antje Bruns (Humboldt Universität zu Berlin, Germany)*

Contributing authors: *Sophie Blackburn (King's College London, UK), Carlos Castellanos Perez Bolde (LOICZ, Germany), Katrin Burkart (Humboldt Universität zu Berlin, Germany), Rob Francis (King's College London, UK), Dietmar Kraft (Carl von Ossietzky Universität Oldenburg), Yangfan Li (Nanjing University, China), Qingshui Lu (Chinese Academy of Sciences, China), Mohsen Makki (Humboldt Universität zu Berlin, Germany), Akeem Olaniyi (Universiti Putry Malaysia), Xin Zhou (Ocean University of China, China)*

*With support from Claudia Hosch (Humboldt Universität zu Berlin), Germany), Jessica Jache (Humboldt Universität zu Berlin, Germany)*

## 1 Introduction

Humans have been an element of coastal ecosystems for thousands of years (Agardy and Alder 2005). According to McGranahan *et al.* (2007) the Low Elevation Coastal Zone (LECZ) "covers 2 per cent of the world's land area but contains 10 per cent of the world's population and 13 per cent of the world's urban population" (ibid., 2007: 17). This coastal zone comprises a unique set of ecosystems (Burke *et al.*, 2001; Crossland *et al.*, 2005) such as mangroves, tidal wetlands, sea-grass beds, estuaries or coral reefs (see Table 2.1). About 71 per cent of the coastal population lived within 50 kilometres of an estuary (CIESN, 1995). Mangrove forests, sea grass and coral reef systems are attractive for humans due to their high productivity (Vörösmarty *et al.*, 2005). Urbanisation, however, is a major cause for the loss of coastal wetlands; Burke *et al.* (2001, see Plate 4) states that up to 50 per cent of the world's mangrove forests are lost; the Millennium Ecosystem Assessment refers to 35 per cent (Agardy and Alder, 2005).

Wilson *et al.* (2003) conclude that costal zones provide 53 per cent of the global ecosystem services (ESS) with an economic value of US$17 trillion per year, they are among the most productive systems providing a range of services to human well-being (Costanza *et al.*, 1997). Depending on the ecosystem subtype, coastal areas offer a wide range of services: estuaries, marshes and lagoons are filtering water as well as mixing nutrients from tidal and upstream sources, thus maintaining the hydrological balance and creating a highly fertile coastal environment (Agardy and Alder, 2005). For example, mangroves not only act as a habitat for many species and provide timber and fuel-wood. They also offer crucial protection against soil erosion and tropical storms. Consequently, coastal ecosystems' resources and services support urbanisation along coastal zones and are one reason that about 40 per cent of the world's population lives within 100 kilometres of the coast (Agardy and Alder, 2005).

*Table 2.1* Coastal environment subtypes

| | |
|---|---|
| Terrestrial | Dunes, cliffs, rocky and sandy shores |
| Intertidal | Estuaries, deltas, lagoons, mangrove forests, mudflats, salt marshes, other coastal wetlands |
| Benthic environments | Kelp forests, sea grass beds, coral reefs |

Source: Compiled from Burke *et al.* (2001)

Moreover, the attractiveness of coastal locations is historically rooted in trade (McGranahan *et al.*, 2007). As economic globalisation brought an increase in shipping and commerce, migration towards the coast increased as well, leading to uneven spatial development (Seto, 2011: 101). It also reinforced the importance of coastal cities for national economies and even far outperformed entire regions in terms of GDP.

The continuing urbanisation of coastal zones puts multiple pressures on coastal subsystems and these are outlined and discussed in this chapter. Our aim is to give an overview on (1) human activities related to coastal urbanisation and (2) their ensuing consequences on the lithosphere, atmosphere, hydrosphere, biosphere and anthroposphere. We then describe (3) ecosystem functions and their services for urban settlements and the spatial and temporal patterns of environmental change.

The results presented here draw on case studies from a wide range of megacity contexts that differ significantly in levels of economic and social development and in governance systems. Though habitat degradation, changes in the freshwater inflow and sediment discharges are worldwide problems, it is impossible to draw universally valid conclusions on how urbanisation affects socio-ecological urban systems on the coast (Tibbetts, 2002; Newton *et al.*, 2011). In order to understand important differences in the environmental performance of cities it is therefore useful to employ urban environmental transition theory to highlight valuable findings (McGranahan and Songsore, 1994; McGranahan, 2007). It forms the basis for our review of the environmental impacts of megacities on the coast.

## 2 Human activities in coastal megacities

Human activities generate pressures on coastal environmental systems. Equally, they trigger the responses to direct and indirect anthropogenic impacts. Apart from a clear dependence on both physical and environmental characteristics of place, human activities are shaped by the respective capacities and cultural, socioeconomic, political, historical and ecological values a society holds (GESAMP, 1996; McFadden, 2007). However, a number of key requirements can be identified such as food, water, shelter and the promise of safety, affection and entertainment. Such needs exert common pressures across localities: land degradation and habitat loss, exploitation of resources, land use change, waste and emissions (Maslow, 1970; Kristensen, 2004). Urban development on the coast has thus a significant environmental impact both on its immediate environment and globally.

According to Roy and Alsayyad (2004), Varis *et al.* (2006) and Sorensen and Okata (2011), 'rich' megacities progress as global market beneficiaries, while 'poor' ones

**Box 2.1 Economic sectors in coastal cities as drivers for environmental impacts**

*Primary sector: raw materials.* This sector makes direct use of natural resources and is usually more important in developing countries (Krueger, 2008). Primary sector activities in coastal megacities include fishing and aquaculture; urban agriculture – both subsistence and commercial – farming and forestry, mining, and oil and gas extraction. Their impacts include an overexploitation and depletion of natural resources (Baran and Hambrey, 1998; Hodgson, 1999; Sekovski *et al.*, 2011), the input of nutrients (nitrogen and phosphorous) and chemical and waste emissions to water and soil as point and diffuse sources due to agriculture and husbandry (Euan-Avila *et al.*, 2005; UN-DESA, 2007; Newton *et al.*, 2011; Shaw, 2011; Zhang and Huang, 2011). They endanger ecosystems, e.g. trawling that destroys the sea bottom or cutting mangroves to develop aquacultures (Baran and Hambrey, 1998; Klein *et al.*, 2003; Primavera, 2006; Newton *et al.*, 2011) and lead to a reduction of natural defences against phenomena such as coastal erosion, storm surges and hurricanes by mining sand and gravel or the removal of coastal dune vegetation, sea-grasses and mangroves (Sato and Mimura, 1997; Arthurton, 1998; Klein *et al.*, 2003; UN-DESA, 2007) and increase the risk of massive oil spills (Newton *et al.*, 2011).

*Secondary or industrial sector: manufacturing.* This sector creates finished goods and encompasses all manufacturing, processing and construction, including energy and textile production as well as the chemical, engineering and desalination industries (Krueger, 2008), all of which are present or being seriously considered in megacities (Joubert *et al.*, 2003; Boutkan and Stikker, 2004; Gurjar *et al.*, 2004; Molina and Molina, 2004; Varis *et al.*, 2006; Favez *et al.*, 2008; Schuetze *et al.*, 2009; Srinivasan *et al.*, 2010). Here, related pressures may include chemical waste and noise emissions to air, water and soil (Phiri *et al.*, 2005; UN-DESA, 2007; Newton *et al.*, 2011); changes in land use, particularly loss to urbanisation (Sekovski *et al.*, 2011; Shaw, 2011) and by-products with immense impact potential such as salt from desalination processes, which not only has very low market potential, but is expensive to confine and could disrupt an aquatic ecosystem or turn fertile soil into a desert by changing its salinity (UN-DESA, 2007). Reduction or obstruction of sediment flows, e.g. by damming rivers for hydroelectric generation, radiation and waste heat transmitted to ecosystems via cooling water are present across the manufacturing sector, particularly in the energy producing industry: coal-fired and nuclear power plants pose fatal threats for entire ecosystems such as coral reefs bleaching, soil acidification and the altering of river water quality (Arthurton, 1998; Syvitski *et al.*, 2005; UN-DESA, 2007).

*Tertiary sector: services.* Tertiary sector activities include gas and electricity distribution, sanitary services such as waste disposal, water treatment and landfills; healthcare, public utilities, transportation, retail and wholesale trade, dredging, beach nourishment, real estate, banking and the entertainment, restaurant and tourism industries (Krueger, 2008). Their potential for negative impacts often depends on the nature of regulative systems and their enforcement capacity, e.g. in healthcare waste or the shipping industry where it often depends on goodwill not to wash fuel tanks on high seas and readiness to prevent and contain the effects of spills and accidents can be limited (Carlin, 2001; Hampton *et al.*, 2003; Sekovski *et al.*, 2011). Transportation pressures also include emissions into air and water bodies, e.g. combustion and greenhouse effect gases (Streets *et al.*, 1997; Newton *et al.*, 2011), and the risk of spreading overnight infectious diseases by travellers or cargo in aircrafts and ships. Dredging (Nakanishi *et al.*, 1989; Brown *et al.*, 1990; Arthurton, 1998; Barbosa and de Almeida, 2001) and beach nourishment (Greene, 2002; Pezzuto, 2006; Defeo *et al.*, 2009) may restore contaminated sediments into the water column, soils or onto human-frequented places affecting them, habitats and fauna both directly and indirectly. Ports are land-consuming infrastructure (often) within the city centre risking air and water pollution (Wolanski, 2006), while tourism is a strong driver for land use change of coasts and mangroves, the removal of coastal vegetation for aesthetic purposes and the construction of 'hard' coastal structures (Sato and Mimura, 1997; Goessling, 2002; Klein *et al.*, 2003; Bulleri and Chapman, 2010).

*Quaternary and quinary sectors.* The quaternary sector consists of intellectual activities such as education, scientific and technology research and development as well as information and culture-related activities. Largely requiring expensive inputs and a highly educated workforce, it is most developed in industrialised countries (Farvacque and McAuslan, 1992; Chaudhury *et al.*, 2006). The quinary sector here refers to the highest levels of decision making, that is, top government, industry, higher education and civil society bodies and influencers in culture and media.

attract rural migration predestined to urban poverty (even though living conditions in the rural hinterland may even be worse). In these places, precarious living conditions contribute to internal displacement, urban expansion and sprawl including into ecologically fragile areas such as coastal wetlands and mangroves, and sealing and degradation of fertile soils. They are characterised by low wage labour, large informal housing and economic activity with often weak public accountability systems and underdeveloped public services. Rising social and economic disparity exerts pressures sometimes leading to urban unrest (Kraas, 2007). Yet, the United Nations Population Fund (UNFPA, 2007: 1) recognises urbanisation as a precondition for significant national economic growth in cities that "represent the best hope of escaping it [poverty]". Further, Seto *et al.* (2010) amongst others indicate that cities present opportunities for sustainability due to the observed benefits of innovation from large agglomerations and per capita energy savings through economies of scale.

The framework of Fisher and Clark's structural change theory (Clark, 1940), which divides the economy in sectors according to patterns of development, is used to show how socio-economic activities in coastal megacities determine their environmental impacts. The model's additional quaternary and quinary sectors allow the same process for the area of education and governance.

Coastal megacities draw their economic strength from several sectoral activities. Of greatest importance to regulate and govern those activities is the quaternary and quinary sector as it is their influence and leverage that extends not only onto the first three sectors but also onto the societal processes of social capital generation, awareness raising and decision making. The extent to influence urban planning and development is closely related to the respective governance system. Governance, clearly not being a synonym of government, deals with those interactions in decision-making processes between government, markets, social organisations and citizens (Olsen *et al.*, 2006; McFadden, 2007; Bruns and Gee, 2009; Bruns, 2010).

Among the multiple governance-related causes for inactivity or poor effectiveness of coastal zone management strategies are conflicting authorities, unclear or lack of responsibilities allocation, poor stakeholders' involvement/commitment and lack of scientific knowledge (Bruns, 2010). Good governance of coastal megacities requires balancing the interests of multiple stakeholders across those scales and sectors as mentioned in Box 2.1 above. Megacities in particular present distinct challenges due to their size, global connectedness and complex administrative structures.

## 3 Urbanisation on the coast and local environmental burdens

> [S]ome of the greatest challenges posed by urbanisation and economic growth have been environmental.
>
> (Marcotuillio and McGranahan, 2007: 2)

Environmental impacts of urbanisation on the coast range in their spatial dimension from households and neighbourhoods up to regional and global challenges (Turner II *et al.*, 1990; Grimm *et al.*, 2008). In the following chapters we concentrate on local causes and consequences of urban environmental burdens. Here, the local scale covers the urban built-up area, its associated urban fringe as well as the immediate coastal region.

A large body of literature exists that explores the spatial dimensions of urban environmental burdens and their relation to development and economic performance. According to McGranahan (2007) an environmental burden "can be defined as a threat to human well-being, resulting from human activities that damage the environment" (ibid., 2007: 21). When assessing and comparing the environmental impacts of cities it is necessary to take key variables into account: cities differ widely in terms of population size, population density, economic conditions and environmental governance. In reaction to the environmental Kuznets Curve literature, McGranahan and Songsore (1994) have shown that household and neighbourhood environmental problems are highest in low-income settlements and decrease with increasing affluence. Many of those problems pose direct health risks originating from poor sanitation, inadequate access to water and lack of waste management. While this is valid for most of the cities in low-income countries independent from their proximity to the coast, coastal cities face additionally specific challenges (see Table 2.2). Furthermore, for middle-income countries city-wide burdens may first increase and then decline, such as in the case of ambient air pollution. Global impacts like carbon dioxide emissions, however, are associated with urban affluence in addition to local and regional burdens (McGranahan *et al.*, 2001).

It is clear, however, that in rapidly developing countries, urban growth differs from the stylised model of urban transitions and the effects of wealth that characterised much of European following the industrial revolution. For example, Marcotullio

*Table 2.2* Urban environmental burdens in coastal (mega)cities

| *Environmental indicator* | *Household and neighbourhood* | *City-regional* | *Coastal city characteristics* |
|---|---|---|---|
| *Water* | Inadequate access to freshwater | River water pollution | Intrusion of saltwater |
| | Inadequate sanitation | | Pollution/eutrophication of coastal waters |
| *Waste* | Unsafe waste handling | Inadequate waste treatment | Marine litter |
| | | Soil pollution | |
| *Air* | Indoor air pollution | Ambient air pollution | Ship emissions |
| | | | Concentration of emission intensive industry |

Source: Based on McGranahan *et al.* (2001)

(2003) shows that the rate of urbanisation in Asian developing countries exceeds that experienced in the era of industrialisation, resulting in stronger overlaps of different environmental burdens (local, regional and global). Warren-Rhodes and Koenig (2001) analysed the urban metabolism of Hong Kong in the period of 1971–97 and found that resource consumption and waste generation has risen immensely due to increases in the levels of wealth. This resulted in a "systemic overload of land, atmospheric and water systems" (ibid., 2001: 429).

Urban environmental transition theory (UET) draws attention to the fact that urban environmental challenges and, subsequently, governance responses shift in scale from the household to the city-regional to the global level. While affluence, i.e. GDP per capita, may be one independent variable conducive for predicting broad environmental trends, other factors weigh in heavily and thus make it difficult to predict urban trajectories. We will show later that the urban poor are most vulnerable to those burdens (Pelling, 2003; McGranahan, 2007). Urbanisation brings opportunities – city dwellers generally are of better health than their rural counterparts – but also challenges regarding the inequalities and health differences within cities (WHO, 2002; WHO, 2008), though there is limited analysis on the health inequalities in megacities that focus on gender, ethnicity or geographical location (Krämer *et al.*, 2011). Urban environmental burdens, closely linked to the concept of urban environmental transition, call for an integrative perspective. Thus we conceptualise (coastal) megacities as coupled social–ecological systems: cities are seen as engines of the economy and innovators for social and political processes. At the same time urban development causes significant environmental problems affecting human systems. We address those issues in the following chapters.

## 4 Impacts on the lithosphere

In this section we summarise selected impacts of large urban areas on underground and land-surface processes and elements and especially the interaction and alteration of soil, substrate and water. The human-induced burdens and intensified processes on these systems range from soil-sealing, changes of the coastal morphology and fostered coastal erosion to the problem of solid and fluid waste caused by waste dump. Particularly land subsidence has become a widespread urban problem linked to excessive groundwater use.

### 4.1 Urban coastal morphology

Geomorphological processes dictate the shape and hydrology of rivers, estuaries and the coast. Coastal areas have been strongly degraded or altered as a consequence of human activities (see Table 2.3). Dredging, land reclamation, engineering works as well as fishing practices led to the destruction of coastal forests, wetlands or coral reefs (Agardy and Alder, 2005). Many coastal (mega) cities are located in deltas; their dense population and economic growth have noticeably altered the coastal morphology. The extraction of oil, gas and groundwater from underlying deposited solid material resulted in sediment compaction (Syvitski *et al.*, 2009). For example, Miao *et al.* (2010) outline how population growth, reclamation and deforestation have led to soil erosion in the Yellow River catchment and that sediment transport of the river increased significantly. Paradoxically, the flux of sediment actually reaching the coasts is reduced by its retention in man-made reservoirs (Syvitski *et al.*, 2005). Agardy and Alder (2005) estimate that the sediment flows increased worldwide by about 20 per cent due to human activities but that at the same time 30 per cent are being held back by engineered works. Thus, about a 10 per cent high net reduction of sediments reaches the coasts, resulting in the sinking of highly populated delta areas below mean sea level, as the trapped upstream sediment cannot maintain current heights. The submergence in combination with the rising global sea level is one contributing factor for severe flooding during the past decade, endangering 500 million people living close to the shoreline and deltas (Syvitski *et al.*, 2009).

*Table 2.3* Deltas and coastal lowlands under threat from human activities

| *Observed changes* | *Human activities* |
|---|---|
| Changes in discharge | Rivers are engineered (dams, diversions) |
| | Distributary channels are reduced and often replaced by irrigation channels |
| Changes in sediment delivery | Reduced sediment input from large dams built |
| Changes in relative sea level | Accelerated subsidence (extracted groundwater/petroleum, peat oxidation) |
| | Reduced aggradation |
| Changes in land use | Deforestation for agribusiness crops and aquaculture |
| | Rampant nitrogen pollution |
| | Loss of mangroves and other wetlands accelerating shoreline retreat |

Source: Compiled after Syvitski, 2011 (presentation on the LOICZ Open Science Conference in Yantai, China).

Further, opencast sand mining in coastal areas – used as raw material for the construction sector – has adversarial effects on coastal cities, particularly in countries with weak environmental regulations (Borges *et al.*, 2002; Mensah, 1997). Mensah (1997) showed for Ghana that the process of sand mining along with dam construction and other engineering activities had enhanced coastal environmental degradation at an alarming rate. As a result millions of dollars are spent to combat sea erosion, requiring sophisticated and expensive resource management in Accra alone. Addo (2011) estimates that anthropogenic activities account for 70–90 per cent of the coastal erosion problems in the densely populated coastal Ghana. Affected are mainly the urban poor who settle directly along the shoreline and thereby increase their vulnerability (Baker, 2008).

Another example is the southwest coast of India, which is equally under pressure due to indiscriminate sand extraction. The area is host to one of the fastest developing industrial centres. As a result of indiscriminate sand mining one observes for the Vembanad lake catchments that the riverbed in the storage zone has lowered significantly during the last two decades. This imposes severe damage to the physical and biological environments of the river systems (Padmalal *et al.*, 2008). Similarly, the mining of coral reefs for the construction industry off the Jakarta coastline caused severe morphological damage (see the case study on Jakarta). Land-based pollution and destructive fishery practices put reefs further at risk (UNESCO, 2000) though the introduction of regulatory controls a decade ago led to a decline in heavy metal contamination in marine sediments (Hosono *et al.*, 2011).

Coastal areas with the greatest population densities are those with the most altered shorelines, mostly for coastal protection purposes. The east coast of the United States with New York City and other major centres as well as the coastlines along the Mediterranean Sea and much of Asia show the highest proportion of altered shorelines. In contrast, sparsely populated coastal zones have relatively natural coastlines (Burke *et al.*, 2001; Harrison *et al.*, 2001).

Coastal erosion is – like aggradation – a natural phenomenon. Paradoxically, urban growth and the need to protect valuable infrastructures "has turned coastal erosion from a natural phenomenon into a problem of growing intensity" (Eurosion, 2004). There is high evidence that human-induced factors now surpass natural coastal erosion; amongst those are engineering of the waterfront, harbour development, river water regulation, dredging for construction and nourishment purposes, as well as land reclamation (ibid., 2004).

These single examples illustrate some of the impacts cities have on the urban coastal morphology. The combined effect of dams and coastal structures on sediment fluxes, artificial shorelines and reef destruction are human-induced activities, which severely change natural processes.

### *4.2 Urban soils in coastal cities*

Soils are part of many ecological processes and essential for ecosystem services such as flood mitigation or waste recycling (Dominati *et al.*, 2010). They are a core component of nutrient and hydrologic cycles and serve as a location for human existence. Clearly, urban soils are highly affected by human activities, with significant spatial and temporal variability on the observed outcomes (Effland and Pouyat,

1997). A wide variety of urban processes is altering the structure of soils. Soil destruction occurs through erosion, salinisation, desiccation and compaction as well as through contamination by hazardous substances (Burghardt, 1994; Pavao-Zuckerman, 2008; Craul, 1999). The recent rapid urbanisation trends in Bangkok for instance have led to an increased volume of landform transformation thereby reducing the soil's capacity to store flood waters (Hara *et al.*, 2005).

Urban growth goes hand in hand with urban area expansion and thus implies soil modification and degradation. Since 1970, China (with large domestic variations), India and parts of sub-Saharan Africa have the highest rates in urban land expansion though North America experienced the largest change in total urban extent (Seto *et al.*, 2011). In their meta-analysis of 326 studies mapping urban land conversion, Seto *et al.* (2011) show that one-third of the cities are within 10 metres of low elevation coastal zones indicating above-average growth rates.

Even though this trend partially reflects the high urbanisation rates in OECD countries (OECD, 2008) it also suggests that "urban land expansion rates are higher than or equal to urban population growth rates, suggesting that urban growth is becoming more expansive than compact" (Seto *et al.*, 2011: x). Angel *et al.* (2005) warn that the expansion of built-up areas will triple by 2030 while the total population doubles, provided that the average densities continue to decline at the annual rate of 1.7 per cent. That means that cities, i.e. rapidly growing coastal areas, would consume more land relative to their population. Due to the fact that coastal cities spread along the coast rather than growing in concentric circles, further shoreline degradation may be the result (see also previous chapter).

In general, urban soil types range from relatively undisturbed to completely man-made soils. These often arise from relocated substrates like sand, slag or rubble. Many large cities are built on land that has been extremely modified by humans. This made land is identified by a high percentage of anthropogenic materials, more precisely, solid waste. Wood, masonry, glass, plastic, metal, asphalt and organic garbage do not only mix with the soil profiles, but also change their biochemical and physical properties (Craul, 1985). As a result urban soils are generally characterised by a high spatial variability of soil types. Further, urban soil formations described in one city may not apply to other cities as urban areas often expand into former agricultural, industrial or mining areas (Pavao-Zuckermann, 2008). While we can hypothesise that soil in coastal cities can develop out of harbour mud or excavated (riverine, coastal or marine) sediments, not much is known about their impact.

We do know, however, that urban soils show a "modified soil reaction (in most cases higher pH values), low organic matter content, restricted aeration and water drainage, high content of anthropic materials, modified soil organism population and activity" (Bretzel and Calderisi, 2006: 219). Point source pollution as well as non-point pollution causes severe soil quality degradation in the coastal urban space (Xu *et al.*, 2008). Bretzel and Calderisi (2006) show that urban soils are often polluted by metals deriving from human activities in their study on Tuscany (Italy) where high heavy metal concentrations were found in urban areas close to the coast. The source was attributed to the transportation sector and road network use. Li *et al.* (2004) carried out a study to assess the anthropogenic pollution of soils by metals in Hong Kong. They identified several hotspots for metal contamination, such as in old industrial sites and venues close to road junctions.

Furthermore, urban soils are characterised by sealing and compaction. The share of sealed surface in comparison to the total area differs significantly within and across cities (Scalenghe and Marsan, 2009). In Europe alone, one observes diverse soil sealing patterns among cities (EEA, 2011). With the rapid expansion of cities, soil compaction accelerates, bringing forth decreasing infiltration rates as its porosity decreases. This process increases the risk for water runoff after storms or flash floods (Gregory *et al.*, 2006). Coastal erosion in combination with those modified soils can have negative effects on water quality as well as cause strong feedbacks with hydrological fluxes, as described in Section 6 on the hydrosphere. High soil sealing also increases vulnerability to heat waves impacts, i.e. in the Northern hemisphere, and further exacerbates the heat island effect outlined in more detail below.

In general, most research on urban soils consists of a wide geographic range of case studies while the general focus of soil governance, an emerging policy issue, remains predominantly rural (ISRIC, 2012). The importance of urban soils for ecosystem functions and services remains to a certain extent neglected, particularly in urban planning. An exception is the US where several government agencies recognise its importance for urban planning (NRCS, 2011). So far, urban soils at the coast are mainly addressed in order to quantify hazardous substances or soil pollution.

### 4.3 Coastal megacities and the subsurface environment

In studies of the urban environment, the importance of the subsurface has often been overlooked in favour of more visible processes, partly due to the difficulty of evaluation (Jago-on *et al.*, 2009). However, the use of subsurface environments has increased in recent decades following accelerated urbanisation processes stemming from industrialisation, particularly in Asia (Taniguchi *et al.*, 2009a; Jago-on *et al.*, 2009).

Cities are using subsurface space for subterranean water supply, sanitation and drainage infrastructures but also for the construction of subways, roadways, underground stations, underground shopping areas and subterranean railways (Jago-on *et al.*, 2009). This necessitates renewed consideration of the various lithospheric alterations associated with urban development. The scale of subsurface change is related to the stage of urbanisation and local hydrogeology (Yoshikoshi *et al.*, 2009, Taniguchi *et al.*, 2009c). The most significant issues are subsidence due to excessive pumping of groundwater, groundwater contamination and seepage, and subsurface thermal anomalies (Taniguchi *et al.*, 2009a). These will be briefly addressed below.

First, groundwater has become a very attractive raw material "without adequate understanding of the balance between its occurrence (in space and time) and replenishment (through recharge) and impact on the groundwater environment" (Pandey *et al.*, 2011). Excessive groundwater use causes land subsidence, due to the removal of pore-water leading to compaction of sediments (Taniguchi *et al.*, 2009b). It varies according to the physical characteristics of the aquifer (subsidence is magnified in urbanised delta areas), the drawdown level and duration of groundwater pumping. Subsidence causes damage to infrastructure both above and below ground and leads to flooding, which has become a major hazard in most large Asian coastal cities (Taniguchi *et al.*, 2009a). Sinking deltas such as the Pearl River in China and Mekong in Vietnam are particularly at risk as these are already below sea level and possess limited coastal protection.

Salinity intrusion in coastal aquifers has equally become a major concern in coastal urban centres (Konikow and Kendy, 2005). Further, urbanisation processes impact negatively upon groundwater quality and climate change impacts exert additional pressures as surface stores become less reliable, as observed in Taiwan (Taniguchi *et al.*, 2009b; Shreshta and Kataoka, 2008).

In most cases land subsidence is a symptom of much wider urban processes, such as population growth and infrastructure planning, exerting pressure on groundwater levels in combination with certain geographical and hydrological factors (Kaneko and Toyota, 2011). Subsidence in Manila and Jakarta has been described as resulting from "unaddressed overdevelopment" (Kaneko and Toyota, 2011). Increasingly, however, cities seek to respond to subsidence issues by introducing regulations on groundwater extraction. In Tokyo, for example, subsidence of 17–18 centimetres per year in the 1950s led to groundwater regulation, which reduced subsidence rates to 2 centimetres per year in 2006 (Jago-on *et al.*, 2009). However, there is often a time lag between the point of maximum groundwater pumping and subsidence affected by the level of urbanisation (Taniguchi *et al.*, 2009a). This could be associated with excessive building practices combined with loose ground material, as illustrated in Jakarta by maximum subsidence occurring in central urban areas (Delinom *et al.*, 2009). An additional problem is that following such regulation, new rise in groundwater may cause further damage to subsurface infrastructures constructed during periods of high drainage (the 'drawdown period') (Taniguchi *et al.*, 2009a).

Second, recent research suggests that groundwater discharge is a direct and highly significant source of water and material shift from land to ocean (Taniguchi *et al.*, 2009a). Groundwater containing high concentrations of nutrients, organic carbon and metals compared to both surface water and ocean waters are now 'seeping' out along many stretches of coastline. This causes widespread environmental problems including coastal water pollution, coastal hypoxia, nuisance algal blooms, ecosystem damage and eutrophication (Taniguchi *et al.*, 2009a; Lee *et al.*, 2009). Sources of contamination include broken sewers, aqueducts, drains and landfills (Jago-on *et al.*, 2009) containing both domestic and industrial waste (Delinom *et al.*, 2009). For example, the Tokyo Metropolitan Government has reported that leakage from drains and aqueducts in Tokyo averages $44x10^4$ $m^3$ per day (Jago-on *et al.*, 2009).

Third, subsurface temperature anomalies have been associated with climate change and the anthropogenic 'heat island effect' using borehole data. This is of global concern because temperature elevation alters the chemical and microbiological balance of the groundwater system via geochemical and geo-biological reactions, leading to associated degradation of groundwater reserves (Green *et al.*, 2007).

## 5 Impacts on the atmosphere

Urban climates usually differ from the atmospheric environments of their surroundings with regard to physical and chemical characteristics. Due to their significance in terms of energy and water conservation, fresh air ventilation, air pollution dispersion or human health and well-being, these particular climates have received much attention in both science and planning (Taha *et al.*, 1998; Basu and Samet, 2002; Akbari and Konopacki, 2005; Kikegawa *et al.*, 2006; Hancock *et al.*, 2007; Lai and Cheng, 2009; Agarwal and Tandon, 2010; Burkart *et al.*, 2011). The most commonly known anthropogenic modification of the urban microclimate is the so-called urban heat island (UHI), a phenomenon referring to urban excess temperatures (Howard, 1833; Landsberg, 1981; Oke, 1987; Kuttler, 1998). Most coastal megacities face UHI issues including London (Kolokotroni *et al.*, 2005), Seoul (Kim and Baik, 2005), Sao Paolo (Freitas *et al.*, 2007), Shanghai and Tokyo (Zhu *et al.*, 2010) and Bangkok (Sintusingha, 2011), just to mention a few.

The formation of the urban climate and atmospheric environment is a highly complex process resulting from the interplay of a multitude of different drivers. The urban structure and geometry, anthropogenic heat generation and the emission of pollutants all shape the urban atmospheric environment (Oke, 1973, Landsberg, 1981, Oke, 1987, Kuttler, 1998, Marsh and Grossa, 2002, Tran *et al.*, 2006, Weng and Yang, 2006). Further, urban areas exhibit differences in other meteorological variables as well. The urban atmosphere displays a lower level of specific humidity as a consequence of the high amount of sealed surface and sparse transpiring vegetation as well as lower relative humidity levels associated with reduced water disposability coupled with higher ambient temperatures. A modified airflow and rainfall pattern is usually ascribed to the particular urban structures and the physicochemical characteristics of the atmosphere (Changnon *et al.*, 1977; Changnon, 1981; Landsberg, 1981; Huff, 1986; Oke, 1987; Kuttler, 1998).

In addition to these physical characteristics, the urban atmosphere usually involves higher levels of air pollution from a number of sources (Marsh and Grossa, 2002; Weng and Yang, 2006). Urban–rural differences are usually pronounced under *autochthonous* weather situations with calm and clear conditions and tend to diminish in cloudy and windy *allochthonous* weather situations (Landsberg, 1981; Oke, 1987; Kuttler, 1998; Lai and Cheng, 2009). Consequently, the meso-scale urban climate is more pronounced in regions and during seasons with stable and dry weather conditions. The UHI shows diurnal variations, which are basically determined by the physical characteristic of building materials (Kuttler, 2004a, Kuttler, 2004b, Burkart and Endlicher, 2011). In mid-latitude countries, urban excess temperatures are usually highest during night time due to the high heat conductivity and capacity of urban material, which is associated with decelerated heating during daytime and decelerated cooling during night time (Lai and Cheng, 2009). Nonetheless, with changing physical properties of the urban body such patterns are quite likely to vary (Burkart and Endlicher, 2011).

### *5.1 The coastal urban climate*

Coastal climates display distinctive features. In general, the ocean exerts a mitigating influence displayed in a narrowed diurnal and seasonal temperature distribution (Schönwiese, 2008; Weischet, 2008). In addition, coastal cities are usually dominated by a convective circulation system generated by the temperature and pressure difference between the ocean and the land, called the land/sea breeze (Scorer, 1978; Kraus and Businger, 1994). Given the stronger warming of landmasses during daytime, a thermally induced pressure system triggers the onset of the so-called sea breeze, a wind coming from the sea in an onshore direction. During night time, landmasses cool down faster than the sea and the resulting change in the pressure constellation generates a so-called land breeze, a wind coming from the land in an offshore direction (Scorer,

1978; Kraus and Businger, 1994). Despite the oceanic influence and mitigating effect of the sea, urban excess temperatures have been observed in coastal cities of different climatic zones (Gedzelman *et al.*, 2003; Streutker, 2003; Alcoforado and Andrade, 2006; Tran *et al.*, 2006; Sasaki *et al.*, 2008; Kolokotsa *et al.*, 2009). Moreover, a number of investigations have established pronounced and ongoing urban warming trends for several coastal Asian megacities, outweighing regional warming (Sakakibara and Owa, 2005; Kataoka *et al.*, 2009). However, the amount of research into the urban climate of coastal cities, and particularly into this of coastal megacities, remains rather limited. Considering the complexity of the phenomenon, general conclusions applying to the coastal UHI remain difficult to draw. Nevertheless, some insightful observation made throughout several case studies shall be outlined below.

In essence, the research findings so far demonstrate competing mechanisms shaping the urban mesoclimate. First, several studies highlight the influence of the urban structure and characteristics such as the sky-view factor, building height, the percentage of built-up areas as well as the importance of geographic factors such as altitude, geology and topography, or land use. Beyond that, a number of studies demonstrate the effects of the urban form, air circulation and the distance from water bodies, lagoons or the ocean (Pinho and Orgaz, 2000; Saaroni *et al.*, 2000; Alcoforado and Andrade, 2006; Charabi and Bakhit, 2011). For Asian megacities it was found that the diurnal variations of the surface UHIs were "different for the cities depending on topography, rural surroundings and geographic locations of the cities in relation to the sea" (Tran *et al.*, 2006: 34). Moreover, the role of façade material and colour is highlighted in the literature (Priyadarsini *et al.*, 2008; Rizwan *et al.*, 2008; Mirzaei and Haghighat, 2010).

In addition, several studies underline the importance of the urban structure and geometry for micro- and mesoscale wind systems and air flow patterns thus indicating potential mitigation measures through the strategic placing of buildings and planning (Priyadarsini *et al.*, 2008, Wong *et al.*, 2010, Luo and Li, 2011). Moreover, Tran *et al.* (2006) show that the spatial shape of the night-time UHIs is significantly affected by the land/sea breeze circulations as the coastal sides exhibit considerably higher temperatures than those inland. Moreover, the authors find that night-time urban–rural differences are less pronounced for the inland cities in comparison to coastal cities. Gedzelsman *et al.* (2003) conclude that the spatial occurrence of the UHI in New York City is crucially influenced by the strength of the sea breezes.

So far, only a few studies cover urban climates in the lower latitudes; this gap is especially pronounced with regard to investigations into the impact of urbanisation on tropical coastal regions (Arnfield, 2003; González *et al.*, 2005; Roth, 2007). Nonetheless, a number of studies have demonstrated the existence of urban excess temperatures in the tropics and in tropical coastal cities (González *et al.*, 2005; Velazquez-Lozada *et al.*, 2006; Roth, 2007; Burkart and Endlicher, 2011). Considering that a substantial share of future urban growth will take place in cities located in low latitudes as well as the already extreme climatic conditions prevailing in these regions, the urban climate of tropical (coastal) cities certainly merits a closer investigation. The characteristic climate and unique features of tropical cities with regard to urban structure and the building materials used might well produce quite distinctive warming and cooling patterns.

## *5.2 Air pollution in coastal cities*

Air pollution is a common and often severe ecological and epidemiological problem in urban areas. Generally, elevated urban temperatures are likely to accelerate a number of chemical reactions leading for example to increased ozone pollution (DeWitt and Brennan, 2001; Lai and Cheng, 2009). In addition to the typical sources of urban air pollution, related to traffic and domestic activities, in coastal cities, emissions from industrial plants and ships are furthermore critical (Sekovski *et al.*, 2011; Weng and Yang, 2006; Chan and Yao, 2008). Due to their infrastructural advantages and proximity to ports, coastal cities attract various sorts of trade and industrial activities. Several recent studies highlight the importance of emissions from ships (Agrawal *et al.*, 2008; Moldanová *et al.*, 2009; González *et al.*, 2011). In addition, the impact of sea spray on particle chemistry and formation adds a further dimension of complexity to air pollution in coastal cities (Pilinis *et al.*, 2000; Knipping and Dabdub, 2002; Pryor *et al.*, 2008). For instance, a study by Knipping and Dabdub (2002) suggests an increase in ozone predictions in the South Coast Air Basin of California through the inclusion of sea salt-derived chlorine chemistry. Another study, modelling the impact of sea spray on particle concentration in the coastal city of Vancouver, Canada, showed that reactions in or on sea spray particles affect the entire particle ensemble and particularly the size distribution of particle nitrate (Pryor *et al.*, 2008). Generally, the particle size distribution is relevant from an epidemiological perspective as it fundamentally influences particle penetration and depositing in the human respiratory tract (Wichmann *et al.*, 2000; Kan *et al.*, 2007). Nonetheless, the various epidemiologically relevant determinants such as size, shape, distribution, number, volume, and chemical composition complicate evaluating the role of marine effects for human and public health and certainly constitute an area in which further research is needed.

Meteorological conditions have a critical influence on the dispersion and deposition of air pollution. In addition to the mesoscale weather conditions prevalent in coastal cities, the land/sea breeze circulation system exerts a profound effect on levels of air pollution (Tirabassi *et al.*, 1990; Lai and Cheng, 2009). While several studies found a negative correlation between wind speed and air pollution due to an enhanced dispersion of pollutants, a common feature of coastal cites is the recirculation of air pollution due to the diurnal variations of the land/sea breeze circulation (Yimin and Lyons, 2003). A tracer experiment conducted by Shair *et al.* (1982) showed that material released into a land breeze at night was transported back during the day's sea breeze. Furthermore, a study by Yoshikado and Tsuchida (1996) found high levels of winter air pollution in the shore of Tokyo, which the authors ascribed to the influence of the UHI. They suggested that the coupling of the sea breeze with the UHI leads to a convergence zone over the urbanised areas, which in turn results in increased air pollution concentrations (Yoshikado and Tsuchida, 1996).

When discussing air pollution in coastal urban areas, it is important to understand that the land/sea-breeze circulation interacts with other circulation systems and is strongly modified by the morphology and topography of the (urban) landscape (Sarrat *et al.*, 2006; Weng and Yang, 2006). For instance, the urban geometry could disrupt the air flow pattern, a phenomenon observed in the coastal city of Trabzon, Turkey (Cuhadaroglu and Demirci, 1997). One study of the Los Angeles area identified complex interactions between the mountains and sea breeze, producing some kind of vertical transport leading to the formation of elevated pollution layers (Lu and Turco, 1994).

The distinct physical and chemical characteristics of urban climates in coastal cities play a key role in atmospheric research. Not only is the urban coastal atmospheric environment shaped by the multitude of determinants common to all urban areas, but above that, the oceanic water body exerts influence on short- to long-term seasonal scales by means of the land/sea breeze circulation system. In turn, this mesoscale wind system interacts with the regional and local morphology, topology, the urban spatial structure and geometry and the thermally induced pressure constellation to produce a highly complex system of competing mechanisms.

## 6 Impacts on the hydrosphere

Water plays an essential role for humans and ecosystems. The hydrosphere is closely linked to all other earth systems including the anthroposphere and the rapid growth of (coastal) cities is clearly associated with multiple impacts on the hydro-systems. Impacts range from increased demand for freshwater (where the smallest part is for drinking) to significant pressures on riverine and coastal ecosystems caused by habitat degradation or pollution of water bodies from point or diffuse sources (Meybeck, 2003, Vörösmarty *et al.*, 2005; Newton *et al.*, 2011). Hydro-systems are changed to provide adequate water services for domestic use and industrial production – which still is a key challenge particularly in the rapidly growing cities of low-income countries – as well as to protect from floods.

Given this context we focus in this section particularly on water provision and pollution sources, such as sewage treatment. Other issues related to aquatic ecosystems – such as habitat loss or degradation – are tackled elsewhere in the chapter.

### *6.1 Megacities and the freshwater-coast continuum*

From a geographical perspective the impact of urban areas on the water system can be assessed by using a scaling approach (for a similar approach see McGranahan, 2005: 815): (i) impact assessment of the water quality and quantity within the urban area, (ii) impact assessment of the urban area on upstream water systems (e.g. dams for energy generation or impounding reservoirs) and (iii) impact assessment of the urban area on downstream water systems (e.g. eutrophication, pollution, water temperature rise). Coastal cities are characterised by the particularity that they can directly discharge into coastal waters. At the same time they are often located downstream of large rivers and thus affected by their pollutants. Generally the catchment to coast continuum is a core issue in the scholarly work of LOICZ (Crossland *et al.*, 2005). So far, various natural and anthropogenic fluxes from land to sea and vice versa with a direct impact on the water quality (see Table 2.4) were assessed.

Given our focus on the assessment of urbanisation on the local water system, it is essential to point out that the physical structure of a city, such as buildings and sealed surfaces, modifies the urban water cycle (see also Section 4 on the lithosphere). Due to soil sealing and other impervious surfaces rainwater cannot completely infiltrate, thus precipitation contributes to a higher proportion of the surface runoff; however,

*Table 2.4* Fluxes from land to sea and from sea to land, natural and anthropocentric factors

| | *Natural* | *Anthropogenic* |
|---|---|---|
| *Land to sea fluxes* | | |
| | river discharge | |
| | sediment | sediment (increase from land use, decrease from dams) |
| | nutrients and minerals | nutrients and organic matter (from agriculture and sewage) |
| | humics and organics | coliform bacteria |
| | groundwater | herbicides and pesticides |
| | storm debris | heavy metals |
| | | oil and chemicals |
| | | solid waste |
| *Sea to land fluxes* | | |
| | cold water/nutrients from upwelling | oil and chemical spills |
| | wave action | chronic input of oil and chemicals |
| | salt and salt aerosols | sewage from ships |
| | sand | ballast water with exotic organisms |
| | nutrients through carcasses, guano | debris from ships |
| | | brackish infiltrations of groundwater reservoirs by water extraction |
| | | pharmaceuticals |

Source: Modified from Kjerfve *et al.*, 2002; Agardy and Alder 2005

this does not necessarily mean that the groundwater recharge is reduced as there can be leakages (Mitchell, 2001). Furthermore, as vegetation cover is limited, evaporation is reduced (Kuttler, 1998), which significantly affects the urban water cycle; generally, urban areas have very low evaporation values (Liu *et al.*, 2010). With the rapid pace of urban sprawl, evaporation tends to change, albeit unevenly: while evaporation is likely to decrease in China's coastal areas of Haihe River (Tang *et al.*, 2011) and Yangtze River (Xu *et al.*, 2006), there is evidence for an increase in South Florida (Abtew *et al.*, 2011).

## 6.2 Water supply of coastal cities

Adequate water supply is the necessary condition to maintain life and economic activities in coastal cities. Global trends in water withdrawals show a doubling between 1960 and 2000 (Vörösmarty *et al.*, 2005) – a development that is closely linked to population growth and rising water demands of the economic sector. The predicted increase in human population and urban development will be the dominant pressure on running water ecosystems (Malmqvist and Rundle, 2002) in two respects: the total water demand increases while the available water is reduced due to pollution. City growth is often correlated to "the area required to provide water and nutrients to its inhabitants" (Swaney *et al.*, 2011: 1). Changes in demand may necessitate that watershed boundaries are artificially changed to cover the water need, possibly causing land-use conflicts. Vörösmarty *et al.* (2005) stress that water is also of greatest importance to maintain ecosystem functions. Particularly in drought-prone areas water stress can function as a constraint not only to human well-being but also to economic development.

The supply of treated surface water, the predominant source of urban water (in part a response to excessive extraction of ground water and consequent urban subsidence), places water security at risk from water pollution and ecosystem degradation, as well as from changing precipitation patterns due to climate change (Vörösmarty *et al.*, 2000; Meehl *et al.*, 2007). Groundwater, from renewable or fossil sources, is another major source for drinking water. If water supply usage from renewable groundwater sources surpasses the natural recharge of the aquifer the water table tends to fall (Vörösmarty *et al.*, 2005). In coastal city areas of the North China Plain, persistent groundwater overexploitation has caused an average water table decline of 0.425 metres a year on average (Liu and Yu, 2001). As discussed above, such groundwater pumping in coastal cities can lead to seawater intrusion, which is a globally known phenomenon. However, local environmental factors can enhance seawater intrusion, including "geological characteristics of the subsoil, the hydraulic gradient, the topography and morphology of the territory, the existence of [other] subsidence phenomena and low lying areas, and the land use of the region" (Di Sipio, 2011: 1387). Continuous overexploitation of freshwater in urban areas of the United Kingdom led to sea water intrusion and water table decline during the industrial era – being reversed in later years due to industrial decline (Lerner and Barrett, 1996). Significant seawater intrusion is documented in Tunisia (Kouzana *et al.*, 2010) – ranging three kilometres inland – the Green Mountain region of Libya (Abdulmonem, 2007), in Australia, for instance in Cairns (Werner, 2010; Bohnet and Pert, 2010) and large parts of South Asia including the coastal megacities of Dhaka, Kolkata, Mumbai, Chennai and Karachi (ADB, 2012; Mohapatra *et al.*, 2011).

### 6.3 Water quality

Whereas surface water quality has improved in most of the developed countries, due to higher progress in reduction of point source pollutants, the water quality in developing countries continues to pose serious health risks (Burke *et al.*, 2001; UN-Water 2011). An estimated 780 million people lacked safe drinking water in 2010 (WHO, 2012), despite progress being made over the past decade. Urbanisation, which generally leads to an increase of water demand, also creates more wastewater and eventually environmental impacts (Swaney *et al.*, 2011). Every day "two million tons of sewage and industrial and agricultural waste are discharged into the world's water" (UN-WWP, 2003). Especially within urban and industrial areas water pollution is high and the capacity of hydro-ecosystems for self-cleaning and recovery is often limited due to ecosystem degradation and overexploitation. Many aquatic ecosystems are degraded as a result of inadequately treated waste. "Untreated wastewater contributes pathogens, toxic contaminants, suspended solids, organic wastes and dissolved nutrients" (Tibbetts, 2002; A678).

While functioning and sufficient infrastructure systems are lacking in many places, some cities have started to respond to the severe environmental problems stemming from water pollution. Bangkok authorities invested heavily in expanding and modernising wastewater plants through joint public–private partnerships; equally Manila is set to expand its wastewater management – an important indicator for urban pollution – as currently only 17 per cent of wastewater is treated before being discharged into the sea (Worldbank, 2012). Sao Paolo currently treats 70 per cent of its sewage water (Jordan *et al.*, 2010). For Europe, there are remarkable spatial disparities: while 65 per cent of the total population is connected to wastewater treatment, only 40 per cent are connected in South Eastern Europe (Turkey, Bulgaria and Romania) (EEA, 2010). Yin *et al.* (2005) indicated for Shanghai a strong spatial interrelation between land cover, population density and water quality due to a lack of wastewater treatment and non-point pollution. Interestingly the authors detected that the most problematic water quality was found in the urban fringe where water treatment capacities are particularly low (ibid., 2005). Close to Shanghai is Taihu Lake, the third-largest freshwater lake in China, which provides a typical example of water quality degradation caused by urban clusters (Shao *et al.*, 2006). Taihu Lake region is not only experiencing rapid urban growth and becoming the most developed region in China (Su *et al.*, 2010), the lake is a crucial irrigation source for its adjoining cotton and grain industry. Further, it hosts a large fisheries sector (Guo, 2007). However, the water quality has deteriorated greatly during the last 30 years, largely as a result of this rapid economic growth (Shao *et al.*, 2006). The case study of Jakarta Bay (Chapter 7) highlights further aspects of water quality in the context of urban sprawl in coastal regions.

Monitoring data on water quality in over 100 countries is provided by the United Nations Global Environment Monitoring System (GEMStat). There is evidence that coastal megacities release more dissolved inorganic P to global surface water (Harrison *et al.*, 2010), however, phosphorous contamination has been reduced considerably over the last decades (Vörösmarty *et al.*, 2005). Nitrogen from urban sewage is projected to increase, especially in Asia (OECD, 2008). Færge *et al.* (2001) carried out one of the first studies on urban nutrient flows, studying Bangkok, and highlight the process of eutrophication, as a result of mismanaged waste management as a key driver for declining water quality (ibid.: 63). An examination of the net-Nitrogen (imported by food) into New York City shows its contribution "for over 70% nitrogen input to the region" (Swaney *et al.*, 2011: 16). The role of nitrogen in- and output for coastal megacities and its links to the wider framework of urban metabolism deserves further research.

With regard to biological water status, Malmqvist and Rundle (2002) find an increased transfer of exotic species between lotic (river) systems. Mollusc invasion in several large river systems have changed complex ecosystem balances, resulting for example in altered phytoplankton populations. Generally port cities are seen as the entry point for exotic species, as described in more detail in the following section on the biosphere.

### 6.4 Political ecology of urban water

While some regions experience well-documented water shortages such as the MENA (Middle East and North Africa) countries (World Bank, 2008), discussions on water rights and the social and political dimension of water in urban areas have gained prominence in both research and local politics. In view of conflicts over entitlements, the management of water and its "uneven access or control . . . is invariably the outcome of combined geographical conditions, technical choices and politico-legal arrangements" (Swyngedouw, 2009: 58). The lack of clean water and sanitation services remains a major health threat throughout the developing world (WHO, 2008); 2.6 billion people lack basic water sanitation worldwide (WHO, 2012), while access to water is limited for just over one billion people, using less than 19 litres per day (in comparison, the average American uses about 575 litres of water daily) (Circle of Blue, 2009; World Bank, 2008). Around one-third of the urban population in developing countries are living in informal settlements with no or limited public service provision, often linked to issues of land rights, rural–urban migration patterns and the consequence of cheap labour demand (Varis *et al.*, 2006).

Clean water provision, storm water disposal and wastewater collection have become serious challenges for most megacities impacting the functioning of the city as a whole (Lundqvist *et al.*, 2005). It highlights how closely intertwined local environmental burdens in megacities are: the lack of urban infrastructure for waste and sewage treatment results in water contamination and contributes to ill-health (Aylward *et al.*, 2005). It further increases the likelihood of floods when waste blocks the water channels, soils are largely sealed and rivers channelled. Local impacts on the urban water system need to be coupled with the global water system to also address large-scale interrelations (Alcamo *et al.*, 2008). Multi-level environmental governance therefore plays a decisive role in shaping the political ecology of urban areas (Varis *et al.*, 2006).

Further, water is a crucial factor in meeting energy requirements of megacities as they use large quantities of energy as well, up to five to ten times more than the national averages (Circle of Blue, 2012). An often-overlooked fact here is that all large-scale electricity generation needs substantial quantities of water, be it hydropower or for cooling, as well as for bioenergy production.

## 7 Impacts of megacities on the biosphere

Coastal megacities influence the biosphere through three different but interlinked processes: (1) changes to the physical environment or ecosystems, (2) changes in community composition and (3) changes in biotic interactions. Most research on these is on the basis of case studies, further, the coastal aspect is often neglected. Where possible we note this throughout the section.

### *7.1 Impacts on ecosystems within the urban region*

As with all large urban regions, megacities have direct impacts on pre-urban ecosystems via land use change to impervious land cover with low habitat facility (Arnold and Gibbons, 1996). Within the city and the wider urban region, pre-urban ecosystems are highly fragmented by infrastructure development (particularly roads) leading to reduced patch sizes and population/biodiversity decline. Following well-established species/area relationships, the patterns are complex and sometimes contradictory for different taxa (Crooks *et al.*, 2004; Yamaguchi, 2005; Pauchard *et al.*, 2006; Forman,

2008; McKinney, 2008; Wu *et al.*, 2011). Cities are composed of a heterogeneous mix of impervious cover with patches and corridors of woodland, brownfield sites, waterways, gardens, etc. This mix, along with new species introductions (Sukopp, 1998), can sometimes lead to increased biodiversity in less developed urban places compared to exurban areas, though this varies with location and taxa (e.g. McKinney, 2008). The position of coastal megacities in or near estuaries or coastal ecosystems clearly has a disproportional impact on such systems, often displaying high levels of biodiversity, productivity or rarity (MEA, 2005). This is illustrated in the number of biodiversity hotspots that are found in coastal areas (see Chapter 4). Coastal wetlands (including salt marshes, mudflats, dunes and other inter-tidal zones) have experienced substantial decline globally (e.g. 35 per cent of mangrove areas and 20 per cent of coral reefs; MEA, 2005), and it is likely that coastal urbanisation is one of the key drivers. As coastal urban expansion follows the coastline as well as stretches inland unevenly (rather than the usual "concentric rings" pattern of expansion found in most inland cities), quantifying impact globally remains difficult (Forman, 2008). Certainly wetland and estuarine areas are rapidly lost during urbanisation, especially where they are seen as unproductive (Pauchard *et al.*, 2006); in a survey of major urban regions Forman (2008) found that cities with populations >8 million had no major wetlands and wetland decline is directly linked to population size.

Urban coastal habitat degradation in particular occurs due to the conversion of such areas for aquaculture, which has become a substantial industry in some coastal cities. The construction of docks, turning basins, housing and industry, and construction of coastal or estuarine flood defence walls and groynes all contribute to degradation processes (Folke *et al.*, 1998; Francis *et al.*, 2008; Chapman and Underwood, 2011). An example can be seen in the reduction of coral cover and species composition in reefs adjacent to Jakarta (Cleary *et al.*, 2006). Most impacts relate to the delivery of aquatic pollutants to coastal systems, which results in changes in community composition (see below). At the same time, the destruction of ecosystems may be accompanied by the creation of artificial 'habitat analogues' that may have some value for certain organisms (Lundholm and Richardson, 2010). Seawalls and tidal river walls can support a range of plants and invertebrates (Lam *et al.*, 2009; Browne and Chapman, 2011; Francis and Hoggart, in press), and further ecological engineering may produce good examples for the "reconciliation of ecology", that is, encouraging biodiversity in human-dominated systems (Francis, 2009; Browne and Chapman, 2011).

### *7.2 Changes in community composition*

Changes to the physical environment invariably influence species populations and community composition, and urban ecosystems are widely populated by generalists with flexible responses to novel their environments (Francis and Chadwick, 2012). Most work on changes to distributions have focused on urban land use/gradients in relation to plants, insects and birds, including that for coastal cities (e.g. Miyashita *et al.*, 1998; Morimoto *et al.*, 2006; Vallejo *et al.*, 2009; Duncan *et al.*, 2011). The main influence specific to coastal cities is the delivery of pollutants to coastal ecosystems, which range from excessive nutrients causing eutrophication to heavy metals and plastics often associated with sewage, shipping waste or industry (e.g. Brown *et al.*, 1998; Liu and Kueh, 2005; Peng *et al.*, 2009). These may have direct effects on biodiversity, including algal or phytoplankton blooms that can prove toxic to many marine organisms in response to eutrophication (Nixon, 1995 and New York, Chapter 7), direct health/mortality impacts on marine invertebrates, fishes and birds from ingestion of plastics (Moore *et al.*, 2011), or mortality from heavy metals or other industrial pollutants (He and Morrison, 2001).

Further impacts may result from non-native species introductions. All megacities are notable points of accidental and intentional introduction of non-natives, but coastal cities are quite unique in that they are international shipping hubs that represent a major vector for aquatic species introductions via hull fouling and ballast (Mack, 2003; Drake and Lodge, 2004). During any 24-hour period it is estimated that 7.000–10.000 species are being transported in cargo ships, and notable aquatic invaders that have probably originated from major ports include the Chinese mitten crab (*Eriocheir sinensis*; Bentley, in press), naval shipworm (*Teredo navalis*; Pimental *et al.*, 2005) and Asian clam (*Potamocorbula amurensis*; Bax *et al.*, 2003), among many other examples. Such species can change physical habitats (e.g. via sediment erosion) or biotic interactions, disrupt food webs, mating, competition or predator/prey interactions, ultimately resulting in detrimental changes at the community level and a loss of biodiversity. Whether megacities are more vulnerable or have in fact a better capacity to act as a node in a network of global spread is uncertain (see Drake and Lodge, 2004); it certainly constitutes an important consideration with regard to potential impacts. Bax *et al.* (2003: 313) noted "[n]ew estuarine and marine species have been establishing once every 32 (San Francisco Bay) to 85 weeks in six studied ports in the US, Australia and NZ, and the rate of establishment appears to be increasing."

### *7.3 Changes to species interactions*

All of the impacts described above cause changes in interactions both between biota (the total number of organisms within a distinct geographical place or temporal scale), and between biota and abiota. Urban environments create situations with resources and productivity at an artificially high level, along with a consequent stabilisation of species populations (mainly of generalist species), reduction of seasonal fluctuations and reduced predation/mortality of many species; with resulting changes in food-web dynamics (Faeth *et al.*, 2005). Although there is little research in this area, it is likely that coastal food webs are affected by the impacts of megacities noted above; certainly evidence exists that, for example, species invasions can significantly change costal food webs in general (Byrnes *et al.*, 2007). Behavioural or morphological changes are also common (and to a lesser extent hybridisation); this has mainly been reported for terrestrial taxa such as birds and mammals (Francis and Chadwick, 2012) and has sometimes been linked to processes of natural selection and microevolution (e.g. Badyaev *et al.*, 2008). Again, such responses have not been investigated in coastal urban scenarios, but it is possible that estuarine/coastal species exposed to urban influences may exhibit similar responses and potentially some level of genetic differentiation between populations.

## 8 Urbanisation on the coast in the Anthropocene

In recent decades theoretical thinking and empirical research on human–environment relations, seeking to overcome the dominant dichotomy between culture and nature, has become a key issue of socio-ecological research (Heynen *et al.*, 2006). Socio-ecological approaches (and that subset which draws directly from social–ecological systems theory) are based on the assumption that there are no natural systems without any human interventions (Folke, 2006, Kareiva *et al.*, 2007). We have shown in the previous sections that human activities cause manifold changes on the coastal morphology and sediment discharges, the urban atmosphere, hydro-systems and the biosphere. Thus urbanisation here engages with local environments and drives land use changes within and well beyond city boundaries up to the global level (see Chapter 4). However, this is not a one-sided process but, rather, anthroposphere and geo-biosphere are interrelated and develop in parallel across time and space. Thus, cities are part of the bio-geophysical system (Elmqvist *et al.*, 2008). Kareiva *et al.* (2007) depict cities as the most domesticated landscape on the globe. Here, cities are socio-ecological systems that create distinct habitats with benefits and costs – not only for humans. According to Grove (2009: 285) such a view

> opens the way toward understanding feedbacks among the biophysical and human components of the system, toward placing them in their spatial and temporal contexts, and toward examining their effects on ecosystem inputs and outputs at various social scales, including individuals, households, neighbourhoods, municipalities, and regions.

In the following sections we outline the framework of ecosystem functions, goods and services and its relevance for urban landscapes.

### *8.1 Coastal cities as social–ecological systems*

An urban perspective on ecosystems complements the focus on biophysical structures and functions with socioeconomic and demographic dynamics that form a distinct complex system (Machlis *et al.*, 1997, cited by Grove, 2009; Picket *et al.*, 2001). Urban features of socio-ecological systems (SES) such as the built environment, physical diversity and social and economic heterogeneity clearly differ from rural areas. They contribute to a distinct urban ecology of cities being "open complex adaptive systems that can be characterized in terms of vulnerability and resilience" (Grove, 2009: 283). Further, the elements making up the urban SES may be less coupled than within other systems as human behaviour heavily influences the environment and in ways that are hard to predict, thus being non-linear and difficult to model (Marcotullio *et al.*, 2003). Key features of megacities such as rapid environmental change, market and industry dynamics, constant urban physical alterations and changing norms and values all add complexity: a variety of drivers operates at different scales and with a myriad of inter-relationships.

One possible way to strengthen the conceptual link between ecosystems and human societies is the ecosystem services (ESS) approach that gained scholarly interest through the work of the Millennium Ecosystem Assessment (MEA) in the past decade (MEA,

2005). It builds on a series of interdisciplinary efforts seeking to define a framework that recognises system dynamics such as resistance, persistence, resilience and variability (Marcotullio *et al.*, 2003). Central is the concept of human well-being, which depends on a variety of services provided by ecosystems. If ecosystems are degraded or altered in such ways that they no longer provide services, human well-being is at risk as well (Daily, 1997).

The value of the ecosystem approach is its systemic take on these urban challenges; it also highlights the fact that urban ecosystems to a large extent operate remotely in that they sustain themselves on the 'carrying capacity' of global commons elsewhere (Rees, 2003). The well-known example on land-use change in the Amazon for livestock to meet surging demand for beef elsewhere is a neat illustration for the lack of key ecosystem components in cities.

The ESS on offer are likely to be more indirect and subject to highly fluctuant spatial and temporal dynamics (Pickett *et al.*, 2006). For example, the livelihood of the urban poor depends to a large extent on local ecosystem services (Glavovic and Boonzaier, 2007), whereas more affluent inhabitants can afford to import ecosystem services from outside the city (McGranahan *et al.*, 2005). Urban gardening to sustain livelihoods is a prominent example of this linkage.

The MEA describes four loose categories of services by the environment that are of benefit to people. First, *supporting* services are those processes necessary to produce all other kinds of provisioning, regulating and cultural services. These include soil formation, oxygen production or habitat provision (MEA, 2005). Human impact on these was described in the previous sections, such as observed changes in urban soils. Second, *provisioning* services relate to the products of ESS, such as food production, storm water or fuel. Ecosystem goods here may be vegetables or hydropower (through water). Third, the goods and benefits deriving from *regulating* ecosystem functions can be linked to climate, water or subsidence control, among others (Lundy and Wade, 2011). Fourth, *cultural* services are those holding recreational or aesthetic value, such as parks or other green spaces or waterways. Their restorative benefits contribute to human well-being.

De Groot *et al.* (2010: 260) observe:

> [T]he ecosystem service approach and ecosystem service valuation efforts have changed the terms of discussion on nature conservation, natural resource management, and other areas of public policy. . . . Investments in conservation, restoration and sustainable ecosystem use are increasingly seen as a "win-win situation" which generates substantial ecological, social and economic benefits.

For example, the Shanghai region increased its public green spaces from 761 to 30,000 hectares over the past three decades (Breuste and Qureshi, 2011). Yet, urban ESS as a planning tool for policymaking is still in its infancy and McGranahan *et al.* (2005) highlight that a large part of literature regards human actions more as a nuisance and less as a dynamic feature of a socio-ecological system.

## *8.2 Ecosystems functions, goods and services in urban areas*

Humans are dependent on biogeochemical cycles and other fundamental natural processes provided by ecosystems (Daily, 1997). ESS as a concept assesses their ecological character and state, the relevance of their components and scope of their functions from an anthropocentric (i.e. economic) point of view. In the past decade the concept of ESS has received considerable scientific attention as it seeks to bridge ecology, economy and social science (Burkhard *et al.*, 2010). Not only discussions on the global consequences of land use emerged (Foley *et al.*, 2005) but also more robust concepts through which to value ecosystem functions and services. Discourse centred around ecosystem appropriation of cities (Folke *et al.*, 1997) or on the consequences of biotic invasions – for which cities are a hotspot (Mack *et al.*, 2000). More broadly, the phenomenon of changing biodiversity (Chapin III *et al.*, 2000) and its impact on ecosystem functioning (Hooper *et al.*, 2005) were examined. One prominent approach is the ecological footprint analysis to represent the amount of biologically productive land and sea area necessary to supply the resources a human population consumes as well as to mitigate associated waste (see Box 2.2). More recently, the ESS approach was applied extensively especially in urban areas as well as coastal zones, and is further being developed (e.g. Abdulmonem 2007; Batker *et al.*, 2010; Butler *et al.*, 2011, Yang *et al.*, 2011). Barkmann *et al.* (2008) demonstrate its usefulness by identifying crucial hydrological ecosystem functions for stakeholders in Indonesia. According to McGranahan *et al.* (2007) it is useful to distinguish between various linkages that exist between cities and ESS. It is clear that not only the net flow of ESS into urban areas has increased over the past centuries but also their average spatial distance has taken on global dimensions, particularly for more affluent cities.

Within urban ecology, important research and policy agendas address the shaping of landscapes and ecosystems for human welfare (Kareiva *et al.*, 2007), the environmental performance of land cover types (Pauleit and Duhme, 2000) and the heterogeneity of interactions of socio-systems and environment (Luck *et al.*, 2001). A corresponding chapter in the MEA report on coastal systems highlights the importance of coasts, the ESS they provide and the need for and challenges of an adequate management (Agardy and Alder, 2005).

### *8.2.1 Ecosystem services and coastal cities*

Given that the human population is concentrated near coasts, the demand for ESS within coastal zones is quite significant: Costanza *et al.* (1997) estimate that 63 per cent of ESS are provided by marine ecosystems, Wilson *et al.* (2003) conclude that costal zones provide 53 per cent of global ESS with an economic value of US$17 trillion per year. Coastal habitats are host to some of the most productive ecosystems of the world, yet, as highlighted in the previous sections, diverse pressures have altered coastlines and threaten coastal marine environments (MA, 2005; Halpern *et al.*, 2007).

Further, as outlined in detail in Chapter 3, urban vulnerability to natural disasters such as storm surges has increased, often as a result of embankments and manmade changes of the natural morphological dynamics of shores and estuaries (UN-HABITAT, 2009; De Sherbinin *et al.*, 2007). Batker *et al.* (2010) assessed the value of restoring the Mississippi River Delta to reduce potential damages caused by hurricanes. Following the motto "economics need nature" they evaluated the main types of ESS like waste treatment, water flow regulation and storm protection – over all 18 types of services aggregated in 11 land cover types – calculating a total value of up to US$47 billion. A sustainable restoration of deltaic processes and maintaining land expansion would outweigh the value by about 30 per cent.

### *8.3 Resilience and complex adaptive systems*

Humans are dependent on ESS as they provide the conditions for a decent, healthy and secure life (MEA, 2005). This seems to be especially the case for megacities (Butsch *et al.*, 2008) and coastal areas (Agardy and Alder, 2005). Cities have a long history of continuous inhabitation and significance, while they are constantly adapting to changing conditions. Obviously these cities required a specific ability to manage with and regain from external stresses. Analogous to economic as well as ecological concepts, this ability may be regarded as the specific resilience of large cities (Klein *et al.*, 2003). Resilience is commonly seen as a desirable systems characteristic in order to absorb changes and still persist, thus understood as a highly dynamic process (Holling, 1973). Particularly urban domains constantly "balance between stabilizing feedbacks that buffer the system against stresses and disturbance and innovation that creates opportunities for change" (Grove, 2009: 290). Various concepts highlight resilience opposition to vulnerability (Klein *et al.*, 2003), though it can also be viewed as an independent property (Pelling, 2011) emphasising learning and experimentation (O'Brien *et al.*, 2012) and operationalised through social infrastructure (Adger, 2003). This line of thinking embraces multiple scales and identifies feedbacks between the local, regional and global taking place in various dynamics (Bahadur *et al.*, 2010). That is, changes in international trade, global climate or regional institutions impact upon local processes, often leading to unexpected changes that question linear thinking. Similarly, lack of water and sanitation infrastructures in large slum areas may threaten the well-being of more affluent areas. Resilience concepts are now widely used in studies concerned with understanding complex human–nature systems such as megacities, including disaster risk reduction and climate change. Yet, the warning of Klein *et al.* (2003: 112), that the popularity of resilience "shows that what was once a straightforward concept used only in mechanics is now a complex multi-interpretable concept with contested definitions and even relevance," is more pertinent than ever.

Despite the growing interest in ESS research on coastal megacities several gaps remain. One is the prevailing focus on single aspects and selected goods and services, presumably due to the high complexity inherent to such ecosystems. Consequently, data collection and use necessary to quantify the value of ESS is scattered, incomplete and difficult to use (Villa *et al.*, 2002). The same applies for dynamic models that would allow the projection of policy changes in the future (ibid.). Further, the outcomes of research modelling on ecosystem functions and services need to be linked to policy and include a discussion on possible urban management scenarios (Barkmann *et al.*, 2008). Resilience frameworks that acknowledge the importance of governance, i.e. institutions, and social capital so central for the structure of urban conglomerates are useful tools.

## 9 Spatio-temporal scales of environmental impacts on the coast

In the following sections we summarise cross-sectoral issues of contemporary urbanisation on the coast. Discourse centres on the temporal and spatial scales of urban and economic growth and its sustainability.

### 9.1 *Timelines*

Many of the problems faced by megacities derive from urbanisation rates far exceeding the planning and management capacities necessary for dealing efficiently with the size of the increments. In difference to urbanisation and growth experienced by cities like London and New York during the Industrial Revolution in the nineteenth century, contemporary urbanisation faces a number of new challenges (Varis *et al.*, 2006). Primary service provision such as water and sanitation, power supply and waste disposal differs in the megacities of the developing countries particularly with regard to their capacity in addressing local environmental burdens and the task of providing basic services to rapidly growing and low-income populations, often over geographically widespread areas and covering different administrations (Biswas *et al.*, 2004; Turok, 2009; Chatterjee, 2010; Mavropoulos, 2010).

Where earlier urban expansion in Europe went hand in hand with the development of institutions, governance capacity and infrastructure development, coastal megacities such as Shanghai, Dhaka, Mumbai, Istanbul and Lagos have had explosive urbanisation rates mostly after 1960 (Biswas *et al.*, 2004; Varis *et al.*, 2006). In addition, megacities in developed countries and their economies have grown simultaneously, providing the required jobs, housing, and financial and human resources (Scott and Storper, 2003; Wheeler, 2009). Further, while they often benefited from economies of scale in providing socioeconomic services (Hardoy *et al.*, 2001; Rodriguez-Pose, 2008), basic services like water and sanitation remain a pressing challenge in much of the developing world (Wisner, 2003; Varis *et al.*, 2006; UNFPA, 2007).

The expansion rates of coastal megacities often exceed the rates of national population growth (Sekovski *et al.*, 2011) and jeopardises urban and environmental services. Land reclamation from the sea is a common response to spatial growth in coastal megacities, either for more arable land, ports, tourism or residence purposes (Tibbetts, 2002). Additionally, many grow rapidly across river deltas, usually at the expense of prime agricultural areas and wetlands, e.g. Dhaka (Griffiths *et al.*, 2010), Jakarta (Nur *et al.*, 2001), Rio de Janeiro (De Sherbinin *et al.*, 2007), Manila (Jacinto *et al.*, 2006) and Lagos (Adelekan, 2009), just to mention a few. Moreover, with continued growth cities that were originally established inland may eventually reach the coast, as experienced by Lima (Devine, 1992). According to Haase (2009), the cumulative impact of land use change and surface sealing is likely to impair the urban water balance favouring surface flooding and runoff pollution and reducing rainwater absorption and aquifer replenishment, as outlined in the beginning of this chapter. Although most coastal megacities were fostered by ports from the very beginning, as port–city relationships evolved, significant differences arose between developed and developing countries (Lee *et al.*, 2008). In general, port cities in developed countries have a low, stable port–city interdependence index, while port cities in developing nations are "forced" to follow the urban hierarchy; however, such strong interdependence has dramatically decreased (Ducruet, 2006; Sorensen and Okata, 2011).

According to the urban environmental transition hypothesis, the environmental challenges cities face may follow a hierarchy of priority associated with economic development stage: most pressing for poorer cities are the 'brown' issues – clean water supply and waste management, followed in middle-income cities by the 'grey' issues – air and water pollution, both typical for low-income countries. Green issues including sustainable ecosystems and life-support systems are prioritised by richer cities (McGranahan *et al.*, 1999; Marcotullio, 2007, in Sorensen and Okata, 2011). Closely related to these challenges of urban economic growth is the loss of biodiversity, of which 50 per cent to 80 per cent is concentrated in countries such as Indonesia, Malaysia, Brazil, China, Thailand and India (Harris, 2004). Most urbanisation worldwide is projected to happen in these countries, where most coastal megacities are both established and emerging (UN, 2010).

Derived from all brown, grey and green issues, the ecological footprint measures the impact on nature in terms of the required land to sustain the consumption and waste disposal of a person in a specific place (Rees, 1996; Chambers *et al.*, 2001). Consequently coastal cities need an agenda that addresses, on the one hand, local and city-regional burdens and threats to socio-ecological systems at a regional and global level, on the other.

### *9.2 Urban forms and spatial patterns*

Dense urban areas with a large proportion of sealed surface have other impacts than "rurban" megacities with low-density and/or discontinuous but extensive commuter, suburban and peri-urban fringe and satellite communities. Angel *et al.* (2010) found that urban densities in developing countries are often more than four times higher than those of the United States with its well-known urban sprawl. Generally the form of urban expansion (i.e. the size and shape of built-up areas) is of most importance (Angel *et al.*, 2005). Soil sealing by impervious materials changes the capacity of ecosystems to provide services and can thus "strongly affect the environmental quality of urban areas" (Pauleit and Duhme, 2000: 7). The amount of sealed surfaces is related to the "urban heat island" introduced in previous sections.

Large cities in North America (as well as London) are characterised by low densities, urban sprawl and long-distance commuting (Angel *et al.*, 2010). Those elsewhere not only experience much poorer informal settlements but also largely unregulated urban growth resulting in the mentioned environmental and health burdens associated with social segregation (Sanders, 1986; Gilbert, 1989; Kim *et al.*, 1993; McGranahan and Songsore, 1994; Scalenghe and Marsan, 2009; Sorensen, 2011; Sorensen and Okata, 2011). Examples of degrading water and higher concentration of faecal coliforms in coastal megacities such as Istanbul (Geymen and Baz, 2008) or Sao Paulo (Groppo *et al.*, in press) have also been related to denser urban land uses.

However, sheer population growth is certainly not the only driver of urban expansion along the coast. In Cairns, Australia, located close to the Great Barrier Reef, tourism is the main driver of land consumption and degradation of coastal habitats

### *Box 2.2* The ecological footprint of cities

To sustain human well-being and quality of life urban dwellers rely on ecosystem services that go well beyond the city's boundaries (Grimm *et al.*, 2008). Formally the ecological footprint of a specified population can be defined as the area of productive land required, on a continuous basis, to produce the resources that the population consumes, and to assimilate the wastes that the population produces, wherever on Earth the relevant land/water is located (Rees, 2001). The quantitative analysis of eco-footprints counts the population of a city, their material standard of living and couples this with an analysis of land use, fishing grounds and the amount of CO2 emissions from burning fossil fuels. Recent data clearly shows that the footprint of high-capita urbanised countries dwarfs that of low- and middle-income countries (McGranahan *et al.*, 2005; WWF, 2012). While urban slum dwellers in a developing country such as India require 0.8 hectares per capita and urban residents in Australia require seven to eight to support their lifestyles, the global average is two hectares per capita (Hamilton and Denniss, 2005; Beeton, 2006; Newton 2006, 2007; in Newton and Bai, 2008). Moreover, the WWF (2012) observes that the footprint of some very poor countries has in fact declined by 60 percent and that, increasingly, rich countries import resources from elsewhere: for example, the increased "demand for beef . . . has transformed New World tropical rainforests into grazing land" (Grimm *et al.*, 2008: 756). McGranahan *et al.* (2005: 817) point out that no urban region could presently be self-sustainable given that "a city may represent as little as 0.1% of the area of the host ecosystems that sustain it." However, this analysis can be misleading in that it ignores the issue of efficiency, that is, how much area (occupied plus footprint area) is needed to sustain a certain group of people. One could also argue that urbanisation "is critical to delivering a more ecologically sustainable and resource-efficient world because the per-person impact of city-dwellers is generally lower than people in the countryside" (Grove, 2009: 283). Further, the footprint analysis does not account for temporal dynamics in both ecosystems and markets, i.e. innovations in resource efficiency and the impact urbanisation has on fertility, land-use patterns or waste production, for example.

(Bohnet *et al.*, 2007). Hence, changes in city functions or sectoral activities – such as port development, coastal tourism or changing lifestyle patterns – may lead to urban expansion (Scalenghe, 2009). The ongoing urban expansion along the coast characteristic to so many urban settlements finds its expression in highly diverse spatial patterns. It depends on the planning and control ability of environmental governance systems to effectively regulate these processes. Land-use policies (i.e., zoning, growth plans) can influence spatial patterns, though these are not always successful as low-density sprawl may continue just beyond set boundaries (Grimm *et al.*, 2008). The urban fringe induces changes in soils, markets and informal housing, not to mention additional burden through increased traffic routes and pollution, which all exert pressure on particular ecosystems.

## 10 Key uncertainties and future research priorities

In this review chapter we assessed a significant part of the peer-reviewed literature as well as grey literature on environmental impacts of urbanisation processes on the local scale. The information we gathered clearly shows the growing influence of urbanisation on coastal systems. While the cited impacts were assessed explicitly for the local level, in their cumulative consequence they may as well qualify as global syndromes as much as services, consumption and the diversity inherent to large cities make up the "global city" (Sassen, 2004; Grimm *et al.*, 2008; Newton *et al.*, 2011).

The literature we found illustrates that the research field is very heterogeneous. Research is selective in terms of methodology, case-study site, spatial and temporal scales as well as with regard to the drivers and pressures in a given locality. However, there seems to be a guiding paradigm: most case studies conceptualise human activities as disturbances (i.e. negative impact) on otherwise functioning earth subsystems (see Haberl *et al.*, 2006).

Although we observe that the human impacts on hydrological, geomorphological and ecological processes are reasonably well understood, there remain major knowledge gaps with regard to their interrelations and feedback loops. In particular, we observe a lack of quantitative data; similarly, data monitoring – for instance time series of water quality – is patchy and dependent on environmental regulations and standards (Vörösmarty *et al.*, 2005). Despite the progress being made in recent years – not least due to the efforts of LOICZ – long-term monitoring of environmental trends on the coast that can be linked to urban expansion rates still need to be improved (especially in developing countries). To name a few examples: the importance of soils for urban social–ecological systems and their services is rarely considered despite their centrality for human well-being in terms of food security, health and overall resilience (Brabec *et al.*, 2002). Even though there is high evidence that urbanisation is a major driver for wetland loss, there is no consistent global monitoring programme addressing this issue (Burke *et al.*, 2001; Turner *et al.*, 2000), although progress is being made especially in the area of remote sensing (see Bartsch *et al.*, 2009). Yet the coastal urban climate receives more scholarly attention (and funding). Modelling the complex interplays between mesoscale wind systems and thermally induced pressure constellation, on the one hand, and the urban spatial structure and geometry, on the other, could provide tailor-made solutions for urban planning and reduce heat stress. However, there are significant knowledge gaps on process interactions across temporal and spatial scales, crucial for our understanding of regional and global changes in the operating mechanisms of coastal ecosystems (Crossland *et al.*, 2005; Nicholls *et al.*, 2007). Further, we attest a general lack of spatially explicit assessments with regard to urban ecosystem services and their dynamics at larger scales (Daily and Matson, 2008).

In general, while the link between urban expansion, environmental impacts and their feedback on human well-being are gaining prominence, their application into policy such as urban planning strategies remains weak. Urban governance strategies such as coastal zone management clearly have a proactive role to play in determining future urbanisation pathways on the coast.

The ecosystem approach is useful in its ability to consider complex sets of interrelationships across scales and different disciplines. Its comprehensive perspective can be a good starting point to develop indicators necessary for a global assessment of resilient megacities. So far, however, most research on ecosystem services still tends to focus on separate aspects and remains local, a weak point scholars criticised for a long time (Bolund and Hunhammar, 1999).

The ongoing intense mega-urbanisation, particularly in Asia and Africa, with the high vulnerabilities of urban dwellers to environmental burdens and numerous hazard risks clearly calls for an improved understanding of socio-ecological system dynamics (Folke *et al.*, 2004). These vulnerabilities coupled with declining values of ecosystem services will not only impact negatively at the local level; they also act as pressures at the regional and global level. Agardy and Alder (2005) call for improved coastal management to optimise the use of ecosystem services but recognise the remaining gaps in scientific knowledge and the lack of science-policy cooperation.

Based on these observations future research on costal urbanisation processes may further develop integrative system perspectives to examine the particular drivers of urban growth along the coast and the resulting pressures on social–ecological coastal systems. Further needed are comparative city-regional case studies that examine key parameters such as socioeconomic patterns, infrastructure development, forms of urban expansion, governance regimes).

## References

Abdulmonem, E. (2007) Sea water intrusion in Derna located in the Green Mountain region, Libya – a threatening recurrent phenomenon calling for desalination, *Desalination,* 220, 189–193.

Abtew, W., Obeysekera, J. and Iricanin, N. (2011) Pan evaporation and potential evapotranspiration trends in South Florida, *Hydrological Processes,* 25, 6, 958–969.

Addo, K.A. (2011) Changing morphology of Ghana's Accra coast, *Journal of Coastal Conservation*, 15, 433–443.

Adelekan, I.O. (2009) Vulnerability of Poor Urban Coastal Communities to Climate Change in Lagos, Nigeria, Paper Presented at the Fifth Urban Research Symposium, Marseille, June.

Agardy, T. and Alder, J. (2005) Coastal Systems, in Hassan, R., Scholes, R. and Ash, N. (eds) *Ecosystems and Human Well-Being: Current Status and Trends*, Washington, DC: Island Press, 513–549.

Agarwal, M. and Tandon, A. (2010) Modeling of the urban heat island in the form of mesoscale wind and of its effect on air pollution dispersal, *Applied Mathematical Modelling*, 34, 2520–2530.

Agrawal, H., Malloy, Q.G.J., Welch, W.A., Wayne Miller, J. and Cocker, I.D.R. (2008) In-use gaseous and particulate matter emissions from a modern ocean going container vessel, *Atmospheric Environment*, 42, 5504–5510.

Akbari, H. and Konopacki, S. (2005) Calculating energy-saving potentials of heat-island reduction strategies, *Energy Policy*, 33, 721–756.

Alcamo, J., Henrichs, T. (2002) Critical regions: A model-based estimation of world water resources sensitive to global changes, *Aquatic Science*, 64, 352–362.

Alcamo, J.M., Vörösmarty, C.J., Naiman, R.J., Lettenmaier, D.P. and Pahl-Wostl, C. (2008) A grand challenge for freshwater research: understanding the global water system, *Environmental Research Letters*, 3, 010202.

Alcoforado, M.J. and Andrade, H. (2006) Nocturnal urban heat island in Lisbon (Portugal): main features and modelling attempts, *Theoretical and Applied Climatology*, 84, 151–159.

Andrulis, D.P. (1997) The Urban Health Penalty: New Dimensions and Directions in Inner City Health Care, *Inner City Health Care*, 126, 6, 485–490.

Angel, S., Sheppard, S.C. and Civco D.L. (2005) *The Dynamics of Global Urban Expansion*, Washington, DC: Transport and Urban Development Department – The World Bank.

Angel, S., Parent, J., Civco, D., Blei, A. and Potere, D. (2010) *A Planet of Cities: Urban Land Cover Estimates and Projections for All Countries, 2000–2050*, Cambridge, MA: Lincoln Institute of Land Policy.

Arnfield, A.J. (2003) Two decades of urban climate research: a review of turbulence, exchanges of energy and water, and the urban heat island, *International Journal of Climatology*, 23, 1–26.

Arnold, C.L. and Gibbons, C.J. (1996) Impervious surface coverage: the emergence of a key environmental indicator, *Journal of the American Planning Association*, 62, 243–258.

Arthurton, R.S. (1998) Marine-Related Physical Natural Hazards Affecting Coastal Megacities of the Asia-Pacific Region: Awareness and Mitigation, *Ocean and Coastal Management*, 40, 65–85.

Aylward, B., Bandyopadhyay, J., Belausteguigotia, J.-C., Börkey, P., Cassar, A., Meadors, L., Saade, L., Siebentritt, M., Stein, R., Tognetti, S., Tortajada, C., Allan, T., Bauer, C., Bruch, C., Guimaraes-Pereira, A., Kendall, M., Kiersch, B., Landry, C., Rodriguez, E.M., Meinzen-Dick, R., SuzanneMoellendorf, Pagiola, S., Porras, I., Ratner, B., Shea, A., Swallow, B., Thomich, T., Voutchkov, N. (2005) Freshwater ecosystem services, in: Hassan, R., Scholes, R. and Ash, N. (eds.), *Millennium Ecosystem Assessment Series*, Washington, DC: Island Press, 213–255.

Badyaev, A.V., Young, R.L., Oh, K.P. and Addison, C. (2008) Evolution on a local scale: Developmental, functional, and genetic bases of divergence in bill form and associated changes in song structure between adjacent habitats, *Evolution*, 62, 1951–1964.

Bai, X. and Imura, H. (2000) A Comparative Study of Urban Environment in East Asia: Stage Model of Urban Environmental Evolution, *International Review for Environmental Strategies*, 1, 1, 135–158.

Baker, J. (2008) Urban poverty: A global view, *World Bank Urban Papers UP-5*, Washington, DC: World Bank.

Balvanera, P., Pfisterer, A.B., Buchmann, N., He, J.-S., Nakashizuka, T., Raffaelli, D. and Schmid, B. (2006) Quantifying the evidence for biodiversity effects on ecosystem functioning and services, *Ecology letters*, 9, 10, 1146–56.

Baran, E. and Hambrey, J. (1998) Mangrove Conservation and Coastal Management in Southeast Asia: What Impact on Fishery Resources, *Marine Pollution Bulletin*, 37, 431–440.

Barbosa, M.C. and de Almeida, M.S.S. (2001) Dredging and disposal of fine sediments in the state of Rio de Janeiro, Brazil, *Journal of Hazardous Materials*, 85, 1–2, 15–38.

Barkmann, J., Glenk, K., Keil, A. and Leemhuis, C. (2008) Confronting unfamiliarity with ecosystem functions: The case for an ecosystem service approach to environmental valuation with stated preference methods, *Ecological Economics*, 65, 48–62.

Bartsch, A., Wagner, W., Scipal, K., Pathe, C., Sabel, D. and Wolski, P. (2009) Global monitoring of wetlands – the value of ENVISAT ASAR Global mode, *Journal of Environmental Management*, 90, 7, 2226–2233.

Basel Convention (2012) *Where are WEee in Africa? Findings from the Basel Convention E-Waste Africa Programme.*

Basu, S.R. and Main, H.A.C. (2001) Calcutta's water supply: demand, governance and environmental change, *Applied Geography*, 21, 23–44.

Basu, R. and Samet, J.M. (2002) Relation between Elevated Ambient Temperature and Mortality: A Review of the Epidemiologic Evidence, *Epidemiologic Reviews*, 24, 190–202.

Batker, D., Torre, I.D., Costanza, R., Swedeen, P., Day, J., Boumans, R. and Bagstad, K. (2010) Gaining ground: Wetlands, hurricanes, and the economy: The value of restoring the Mississippi River Delta, *Environmental Law Institute*, 40, 11106–11110.

Bax, N., Williamson, A., Aguero, M., Gonzalez, E. and Geeves, W. (2003) Marine invasive alien species: a threat to global biodiversity, *Marine Policy*, 27, 313–323.

Beeton, R.J., Buckley, K.I., Jones, G.J., Morgan, D., Reichelt, R.E. and Trewin, D. (2006) *Australian State of the Environment 2006*, Canberra: Department of the Environment and Heritage.

Bentley, M.G. (in press) Eriocheirsinensis H. Milne-Edwards (Chinese mitten crab), in: Francis, R.A. (ed.) *A handbook of global freshwater invasive species*, London: Routledge.

Biswas, A., Lundqvist, J., Tortajada, C. and Varis, O. (2004) Water management for megacities, *Stockholm Water*, 2, 12–13.

Bohle, H.-G. and Warner, K. (ed.) (2008) *Megacities: Resilience and Social Vulnerability*, Bonn: Publications Series of UNU-EHS, 10.

Bohnet, I.C. and Pert, P.L. (2010) Patterns, drivers and impacts of urban growth – A study from Cairns, Queensland, Australia from 1952 to 2031, *Landscape and Urban Planning*, 97, 4, 239–248.

Bolund, P. and Hunhammar, S. (1999) Ecosystem services in urban areas, *Ecological Economics*, 29, 293–301.

Borges, P., Andrade, C. and Freitas, M.C. (2002) Dune, Bluff and Beach Erosion due to Exhaustive Sand Mining – the Case of Santa Barbara Beach, São Miguel (Azores, Portugal), *Journal of Coastal Research*, 95, 36, 89–95.

Boutkan, E. and Stikker, A. (2004) Enhanced water resource base for sustainable integrated water resource management, *Natural Resources Forum*, 28, 150–158.

Brabec, E., Schulte, S. and Richards, P.L. (2002) Impervious surfaces and water quality: A review of current literature and its implications for watershed planning, *Journal of Planning Literature*, 16, 4, 499–514.

Bretzel, F. and Calderisi, M. (2006) Metal contamination in urban soils of coastal Tuscany (Italy), *Environmental monitoring and assessment*, 118, 1–3, 319–335.

Breuste, J.H. and Qureshi, S. (2011) Urban sustainability, urban ecology, and Society for Urban Ecology (SURE), *Urban Ecosystems*, 14, 3, 313–317.

Brigden, K., Labunska, I., Santillo, D. and Johnston, P. (2008) *Chemical Contamination at e-waste recycling and disposal sites in Accra and Korforidua, Ghana*, Greenpeace Research Laboratories, Technical note 10/2008.

Brown, B.E., Le Tissier, M.D.A., Scoffin, T.P. and Tudhope, A.W. (1990) Evaluation of the environmental impact of dredging on intertidal coral reefs at Ko Phuket, Thailand, using ecological and physiological parameters, *Marine ecology progress series*, 65, 273–281.

Brown, D.W., McCain, B.B., Horness, B.H., Sloan, C.A., Tilbury, K.L., Pierce, S.M., Burrows, D.G., Chan, S.L., Landahl, J.T. and Krahn, M.M. (1998) Status, correlations and temporal trends of chemical contaminants in fish and sediment from selected sites on the Pacific coast of the USA, *Marine Pollution Bulletin*, 37, 67–85.

Browne, M.A. and Chapman, M.G. (2011) Ecologically informed engineering reduces loss of intertidal biodiversity on artificial shorelines, *Environmental Science and Technology*, 45, 8204–8207.

Bruns, A. (2010) *Governance im Küstenraum. Europäische Umweltpolitik im Wandel – Die Umsetzung des Integrierten Küstenzonenmanagements und der Wasserrahmenrichtlinie an der Westküste Schleswig-Holsteins*. Dissertation, Mathematisch-Naturwissenschaftliche Fakultät der Christian-Albrecht-Universität zu Kiel.

Bruns, A. and Gee, K. (2009) From State-Centered Decision-Making to Participatory Governance, *GAIA Ecological Perspectives for Science and Society*, 2, 150–157.

Bulleri, F. and Chapman, M.G. (2010) The introduction of coastal infrastructure as a driver of change in marine environments, *Journal of Applied Ecology*, 47, 26–35.

Burghardt, W. (1994) Soils in urban industrial environments, *Zeitschrift für Pflanzenernähung und Bodenkunde*, 157, 205–214.

Burkart, K. and Endlicher, W. (2011) Human Bioclimate and Thermal Stress in the Megacity of Dhaka, Bangladesh – A Climatological Approach to Health Relevance Assessment, in: Krämer, A., Khan, M. and Kraas, F. (eds) *Health in Megacities and Urban Areas*, Heidelberg: Springer.

Burkart, K., Schneider, A., Breitner, S., Khan, M.H., Krämer, A. and Endlicher, W. (2011) The effect of atmospheric thermal conditions and urban thermal pollution on all-cause and cardiovascular mortality in Bangladesh, *Environmental Pollution*, 159, 2035–2043.

Burke L., Kura Y., Kassem K., Revenga C., Spalding M., McAllister D., (2001) *Pilot Analysis of Global Ecosystems: Coastal Ecosystems*, Washington, DC: World Resource Institute.

Burkhard, B., Petrosillo, I. and Costanza, R. (2010) Ecosystem services – Bridging ecology, economy and social sciences, *Ecological Complexity*, 7, 257–259.

Butler, J.R.A., Wong, G.Y., Metcalfe, D.J., Honzák, M., Pert, P.L., Rao, N., van Grieken, M.E., Lawson, T., Bruce, C., Kroon, F.J. and Brodie, J.E. (2011) An analysis of trade-offs between multiple ecosystem services and stakeholders linked to land use and water quality management in the Great Barrier Reef, Australia, *Agriculture, Ecosystems & Environment*, 876–893.

Butsch, C., Etzold, B. and Sakdapolrak, P. (2008) The Megacity Resilience Framework, in: Bohleand, H.-G. and Warner, K. (eds) *Megacities: Resilience and Social Vulnerability*, Bonn: Publication Series of the United Nations University – Institute for Environment and Human Security (UNU-EHS), 10.

Byrnes, J.E., Reynolds, P.L. and Stachowicz, J.J. (2007) Invasions and extinctions reshape coastal marine food webs, PLoS ONE 2(3): e295. doi:10.1371/journal.pone.0000295.

Carlin, E.M. (2001) Oil Pollution from Ships at Sea: The Ability of Nations to Protect a Blue Planet, in: Miles, E.L., Underdal, A., Andresen, S., Wettestad, J. Skjaerseth, J.B. and Carlin, E.M. (eds) *Environmental Regime Effectiveness. Confronting Theory with Evidence*, Cambridge, MA: The MIT Press, 331–356.

CEPA, Centre for Policy Analysis (2007) Ghana: The Energy Crisis and Growth Performance of the Economy, CEPA Issues Paper, *Ghana – Selected Economic Issues*, 15, 1–19, Accra.

Cesar, H., Burke, L. and Pet-Soede, L., (2003) *The Economics of Worldwide Coral Reef Degradation*, WWF-Report: Zeist.

Chambers, N., Simmons, C. and Wackernagel, M. (2001) *Sharing nature's interest: Ecological footprints as an indicator of sustainability*, London: Earthscan Publications.

Chan, C.K. and Yao, X. (2008) Air pollution in mega cities in China, *Atmospheric Environment*, 42, 1–42.

Changnon, S.A. (1981) *METROMEX: A Review and Summary*, American Meteorological Society.

Changnon, S.A., Huff, F.A., Schickendanz, P.T. and Vogel, J.L. (1977) Summary of METROMEX, 1: Weather anomalies and impacts, *Illinois State Water Survey Bulletin*, 62.

Chapin III, F.S., Zavaleta, E.S., Eviner, V.T., Naylor, R.L., Vitousek, P.M., Reynolds, H.L., Hooper, D.U., Lavorel, S., Sala, O.E., Hobbie, S.E., Mack, M.C. and Díaz, S. (2000) Consequences of changing biodiversity, *Nature*, 405, 6783, 234–242.

Chapman, M.G. and Underwood, A.J. (2011) Evaluation of ecological engineering of 'armoured' shorelines to improve their value as habitat, *Journal of Experimental Marine Biology and Ecology*, 400, 302–313.

Charabi, Y. and Bakhit, A. (2011) Assessment of the canopy urban heat island of a coastal arid tropical city: The case of Muscat, Oman, *Atmospheric Research*, 101, 215–227.

Chatterjee, M. (2010) Slum dwellers response to flooding events in the megacities of India, *Mitigation and Adaptation Strategies for Global Change*, 15, 4, 337–353.

Chaudhury, N., Hammer, J., Kremer, M., Muralidharan, K. and Rogers, F.H. (2006) Missing in Action: Teacher and Health Worker Absence in Developing Countries, *Journal of Economic Perspectives*, 20, 1, 91–116.

CIESIN, Center for International Earth Science Information Network (2006) *Where the Poor Are: An Atlas of Poverty*, Columbia University, Available at: http://www.ciesin.columbia.edu/povmap

Clark, C. (1940) *The conditions of economic progress*, London: Macmillan.

Cleary, D.F.R., Suharsono and Hoeksema B.W. (2006) Coral diversity across a disturbance gradient in the Pulau Seribu reef complex off Jakarta, Indonesia, *Biodiversity and Conservation*, 15, 3653–3674.

Costanza, R., d'Arge, R., de Groot, R., Farber, S., Grasso, M., Hannon, B., Limburg, K., Naeem, S., O'Neill, R.V., Paruelo, J., Raskin, R.G., Sutton, P. and van der Belt, M. (1997) The value of world's Ecosystem Services and natural capital, *Nature*, 387, 6630, 253–260.

Craul, P.J. (1985) Urban Soils, *METRIA 5: Selecting and Preparing Sites for Urban Trees*, Proceedings of the Fifth Conference of the Metropolitan Tree Improvement Alliance, Pennsylvania, 45–61.

Crooks, K.R., Suarez, A.V. and Bolger, D.T. (2004) Avian assemblages along a gradient of urbanisation in a highly fragmented landscape, *Biological Conservation*, 115, 451–462.

Crossland, C.J., Kremer, H.H. and Lindeboom, H.J., (2005) *Coastal fluxes in the Anthropocene: The land-ocean interactions in the coastal zone project of the International Geosphere-Biosphere Programme*, Berlin Heidelberg: Springer.

Crutzen, P.J. (2002) Geology of mankind, *Nature*, 415, 6867, p. 23.

Cuhadaroglu, B. and Demirci, E. (1997) Influence of some meteorological factors on air pollution in Trabzon city, *Energy and Buildings*, 25, 179–184.

Daily, G.C. (1997) *Nature's services: societal dependence on natural ecosystems*, Washington, DC: Island Press.

Daily, G.C. and Matson, P.A. (2008) Ecosystem services: from theory to implementation, *Proceedings of the National Academy of Sciences of the United States of America*, 105, 9455–9456.

de Groot, R., Alkemade, R., Braat, L., Hein, L. and Willemen, L. (2010) Challenges in integrating the concept of ecosystem services and values in landscape planning, management and decision making, *Ecological Complexity*, 7, 260–272.

Defeo, O., McLachlan, A., Schoeman, D.S., Schlacher, T.A., Dugan, J., Jones, A., Lastra, M. and Scapini, F. (2009) Threats to sandy beach ecosystems: A review, *Estuarine, Coastal and Shelf Science*, 81, 1–12.

Delinom, R., Assegaf, A., Abidin, H., Taniguchi, M., Suherman, D., Lubis, R. and Yulianto, E. (2009) The contribution of human activities to subsurface environment degradation in Greater Jakarta Area, Indonesia, *Science of the Total Environment (Special Issue)*, 407, 3129–3141.

De Sherbinin, A., Schiller, A. and Pulsipher, A. (2007) The vulnerability of global cities to climate hazards, *Environment and Urbanization*, 19, 1, 39–64.

Devine, N.P. (1992) *Urban vulnerability to sea-level rise in the Third World*, MS Thesis, Rutgers, New Brunswick, NJ: The State University of New Jersey.

DeWitt, J. and Brennan, M. (2001) Taking the Heat, *Imaging Notes*, 16, 20–23.

Diaz, R.J., Solan, M. and Valente, R.M., (2004) A review of approaches for classifying benthic habitats and evaluating habitat quality, *Journal of environmental management*, 73, 3, 165-81.

Douglas, I. (1983) *The Urban Environment*, London: Arnold.

Drake, J.M. and Lodge, D.M. (2004) Global hot spots of biological invasions: evaluating options for ballast-water management, *Proceedings of the Royal Society of London B: Biological Sciences*, 271, 575–580.

Ducruet, C. (2006) Port-city relationships in Europe and Asia, *Journal of International Logistics and Trade*, 4, 2, 13–35.

Duncan, R.P., Clemants, S.E., Corlett, R.T., Hahs, A.K., McCarthy, M.A., McDonnell, M.J., Schwartz, M.W., Thompson, K., Vesk, P.A. and Williams, N.S.G. (2011) Plant traits and extinction in urban areas: a meta-analysis of 11 cities, *Global Ecology and Biogeography*, 20, 509–519.

Durfor, N.C. and Becker, E. (1962) *Public water supplies of the 100 largest cities in the united states*, Geological Survey Water Supply Paper, Washington: U.S. Government.

EEA, European Environmental Agency (2010) *Urban waste water treatment* (CSI 024) – Assessment published December.

Effland, W.R. and Pouyat, R.V. (1997) The genesis, classification, and mapping of soils in urban areas, *Urban Ecosystems*, 1, 4, 217–228.

Elhassadi, A. (2008), Sea water intrusion in Derna located in the Green Mountain region, Libya – a threatening recurrent phenomenon calling for desalination, *Desalination*, 220, 189–193.

Elmqvist, T., Alfsen, C. and Colding, J. (2008) Urban systems, in: Jorgensen, S.E. and Fath, B.D. (eds) *Encyclopedia of Ecology*, 5, 3665–3672.

Euan-Avila, J.I., Liceaga-Correa, M.A. and Rodriguez-Sanchez, H. (2005) GIS for assessing land-based activities that pollute coastal environments, in: Bartlett, D. and Smith, J. (eds) *GIS for coastal zone management*, Boca Raton, FL: CRC Press, 229–238.

Eurosion (2004) *Living with coastal erosion in Europe: Sediment and Space for Sustainability*, Part IV: A guide to coastal erosion management in Europe, Lessons learned, Europe.

Færge J., Magid J. and de Vries F. (2001) Urban nutrient balance for Bangkok, *Ecological Modelling*, 139, 63–74.

Faeth, S.H., Warren, P.S., Shochat, E. and Marussich, W.A. (2005) Trophic dynamics in urban communities, *BioScience*, 55, 399–407.

FAO (2010) *The State of World Fisheries and Aquaculture*, Rome: FAO.

Farvacque, C. and McAuslan, P. (1992) *Reforming urban land policies and institutions in developing countries*, Urban Management Program, Washington, DC: The World Bank.

Favez, O., Cachier, H., Sciare, J., Alfaro, S.C., El-Araby, T.M., Harhash, M.A. and Abdelwahab, M.M. (2008) Seasonality of major aerosol species and their transformations in Cairo megacity, *Atmospheric Environment*, 42, 7, 1503–1516.

Foley, J.A., DeFries, R., Asner, G.P., Barford, C., Bonan, G., Carpenter, S.R., Chapin, F.S., Coe, M.T., Daily, G.C., Gibbs, H.K., Helkowski, J.H., Holloway, T., Howard, E.A., Kucharik, C.J., Monfreda, C., Patz, J.A., Prentice, I.C., Ramankutty, N. and Snyder, P.K. (2005) Global consequences of land use, *Science*, 309, 5734, 570–574.

Folke, C. (2006) Resilience: The Emergence of a Perspective for Social-Ecological Systems Analyses, *Global Environmental Change*, 16, 253–267.

Folke, C., Jansson, A., Larsson, J. and Costanza, R. (1997) Ecosystem appropriation by cities, *Ambio*, 26, 3, 167–172.

Folke, C., Kautsky, N., Berg, H., Jansson, A. and Troell, M. (1998) The ecological footprint concept for sustainable seafood production: A review, *Ecological Applications*, 8, S63–S71.

Folke, C. Carpenter, S., Walker, B., Scheffer, M., Elmqvist, T., Gunderson, L. and Holling, C.S. (2004) Regime shifts, resilience, and biodiversity in ecosystem management, *Annual Review of Ecology, Evolution, and Systematics*, 35, 557–581.

Forman, R.T.T. (2008) *Urban regions: ecology and planning beyond the city*, Cambridge: Cambridge University Press.

Francis, R.A. (2009) Perspectives on the potential for reconciliation ecology in urban riverscapes, *CAB Reviews: Perspectives in Agriculture, Veterinary Science, Nutrition and Natural Resources*, 4, 73, 20.

Francis, R.A. and Chadwick, M.A. (2012) What makes a species synurbic?, *Applied Geography*, 32, 514–521.

Francis, R.A. and Hoggart, S.P.G. (in press) The flora of urban river wallscapes, *River Research and Applications*.

Francis, R.A., Hoggart, S.P.G., Gurnell, A.M. and Coode, C. (2008) Meeting the challenges of urban river habitat restoration: developing a methodology for the River Thames through central London, *Area*, 40, 435–445.

Freitas, E.D., Rozoff, C.M., Cotton, W.R. and Dias, P.L.S. (2007) Interactions of an urban heat island and sea-breeze circulations during winter over the metropolitan area of São Paulo, Brazil, *Boundary-Layer Meteorology*, 122, 43–65.

Gandy, M. (2004) Rethinking urban metabolism: water, space and the modern city, *City*, 8, 363–379.

Gandy, M. (2006) Urban nature and the ecological imaginary, in Heynen, N., Kaika, M. and Swyngedouw, E. (ed.) *In the Nature of Cities. Urban Political Ecology and the Politics of Urban Metabolism*, Routledge, London, 63–74.

Gedzelman, S.D., Austin, S., Cermak, R., Stefano, N., Partridge, S. and Quesenberry, S., (2003) Mesoscale aspects of the Urban Heat Island around New York City, *Theoretical and Applied Climatology*, 75, 29–42.

GESAMP (IMO/FAO/UNESCO-IOC/WMO/WHO/IAEA/UN/UNEP Joint Group of Experts on the Scientific Aspects of Marine Environmental Protection) (1996) *The Contributions of Science to Integrated Coastal Management*, Reports and Studies 61, Rome: GESAMP.

Geymen, A. and Baz, I. (2008) Monitoring urban growth and detecting land-cover changes on the Istanbul metropolitan area, *Environmental Monitoring and Assessment*, 136, 1–3, 449–459.

Gilbert, O.L. (1989) *The Ecology of Urban Habitats*, London: Chapman and Hall.

Gill, S.E., Handley, J.F., Ennos, A.R. and Pauleit, S. (2007) Adapting Cities for Climate Change: The Role of the Green Infrastructure, *Built Environment*, 33, 1, 115–133.

Glavovic, B. and Boonzaier, S. (2007) Confronting coastal poverty: Building sustainable coastal livelihoods in South Africa, *Ocean & Coastal Management*, 50(1–2), 1–23.

Goessling, S. (2002) Global environmental consequences of tourism, *Global Environmental Change*, 12, 283–302.

González, J.E., Luvall, J.C., Rickman, D., Comarazamy, D., Picón, A., Harmsen, E.W., Parsiani, H., Ramírez, N., Vásquez, R.E., Williams, R., Waide, R.B. and Tepley, C.A. (2005) Urban heat islands developing in coastal tropical cities, *EOS Trans. AGU*, 86, 397–399.

González, Y., Rodríguez, S., Guerra García, J.C., Trujillo, J.L. and García, R. (2011) Ultrafine particles pollution in urban coastal air due to ship emissions, *Atmospheric Environment*, 45, 4907–4914.

Graves, H., Watkins, R., Westbury, P. and Littlefair, P. (2001) *Cooling Buildings in London: Overcoming the Heat Island*, London: BRE and DETR.

Greene, K. (2002) *Beach Nourishment: A Review of the Biological and Physical Impacts*, ASMFC Habitat Management Series No. 7, Washington, DC: Atlantic States Marine Fisheries Commission.

Gregory, J.H., Dukes, M.D., Jones, P.H. and Miller, G.L. (2006) Effect of urban soil compaction on infiltration rate, *Journal of Soil and Water Conservation*, 61, 3, 117–124.

Griffiths, P., Hostert, P., Gruebner, O. and van der Linden, S. (2010) Mapping megacity growth with multi-sensor data, *Remote Sensing of Environment*, 114, 2, 426–439.

Grimm, N.B., Faeth, S.H., Golubiewski, N.E., Redman, C.L., Wu, J.G., Bai, X.M. and Briggs, J.M. (2008) Global change and the ecology of cities, *Science*, 319, 756–760.

Groppo, J.D., De Moraes, J.M., Beduschi, C.E., Genovez, A.M. and Martinelli, L.A. (2008) Trend analysis of water quality in some rivers with different degrees of development within the Sao Paulo State, Brazil, *River Research and Applications*, 24, 8, 1056–1067.

Grübner, O., Khan, M.M.H., Lautenbach, S., Mueller, D., Kraemer, A., Lakes, T. and Hostert, P. (2011) 'A spatial epidemiological analysis of self-rated mental health in the slums of Dhaka', *International Journal of Health Geographics*, 10, 36, 15 pp.

Guo, L. (2007) Ecology – Doing battle with the green monster of Taihu Lake, *Science*, 317, 1166–1166.

Gurjar, B.R., van Aardenne, J.A., Lelievelda, J. and Mohan, M. (2004) Emission estimates and trends (1990–2000) for megacity Delhi and implications, *Atmospheric Environment*, 38, 5663–5681.

Haase, D. (2009) Effects of urbanisation on the water balance – a long-term trajectory, *Environmental Impact Assessment Review*, 29, 211–219.

Haberl, H., Winiwarter, V., Andersson, K., Ayres, R.U., Boone, C., Castillo, A., Cunfer, G., Fischer-Kowalski, M., Freudenburg, W.R. and Furman, E. (2006) From LTER to LTSER: Conceptualizing the Socioeconomic Dimension of Long-term Socioecological Research, *Ecology and Society*, 11, 2, 1–34.

Hallegraeff, G.M. (1993) A review of harmful algal blooms and their apparent global increase, *Phycologia*, 32, 2, 79–99.

Hamilton, C. and Denniss, R. (2005) *Affluenza: When too much is never enough*, Sydney: Allen and Unwin.

Hampton, S., Kelly, P.R. and Carter, H.R. (2003) Tank vessel operations, seabirds and chronic oil pollution in California, *Marine Ornithology*, 31, 29–34.

Hancock, P.A., Ross, J.M. and Szalma, J.L. (2007) A Meta-Analysis of Performance Response Under Thermal Stressors, *Human Factors: The Journal of the Human Factors and Ergonomics Society*, 49, 851–877.

Hara, Y., Takeuchi, K. and Okubo, S. (2005) Urbanisation linked with past agricultural landuse patterns in the urban fringe of a deltaic Asian mega-city: a case study in Bangkok, *Landscape and Urban Planning*, 73, 1, 16–28.

Hardoy, J., Mitlin, D. and Satterthwaite, D. (2001) *Environmental problems in an Urbanizing world: Finding solutions for cities in Africa, Asia and Latin America*, London: Earthscan.

Harris, F. (2004) Conserving biodiversity resources, in: Harris, F. (ed.) *Global Environmental Issues*, London: John Wiley and Sons, 95–97.

Harrison, J.A., Bouwman, A.F., Mayorga, E. and Sybil, S. (2010) Magnitudes and sources of dissolved inorganic phosphorus inputs to surface fresh waters and the coastal zone: A new global model, *Global Biogeochemical Cycles*, 24, 1003.

Harrison, P. and Pearce, F. (2001) *AAAS Atlas of Population and Environment 2001*, American Association for the Advancement of Science, Berkley: University of California Press.

He, Z.J. and Morrison, R.J. (2001) Changes in the marine environment of Port Kembla Harbour, NSW, Australia, 1975–1995: A review, *Marine Pollution Bulletin*, 42, 193–201.

Henry, J.A. and Dicks, S.E. (1987) Association of urban temperatures with land use and surface materials, *Landscape Urban Plan*, 14, 21–29.

Heynen, N., Kaika, M. and Swyngedouw, E. (2006) Urban Political Ecology, in: Heynen, N., Kaika, M. and Swyngedouw, E. (ed.) *In the Nature of Cities. Urban Political Ecology and the Politics of Urban Metabolism*, Abingdon: Routledge.

Hien, W.N. (2003) *Urban heat island effect: Sinking the heat*, Innovations, 3, 2, Singapore: The National University of Singapore and World Scientific Publishing Co.

Hodgson, G. (1999) A Global Assessment of Human Effects on Coral Reefs, *Marine Pollution Bulletin*, 38, 5, 345–355.

Holling, C.S. (1973) Resilience and stability of ecological systems, *Annual Review of Ecology and Systematics*, 4, 1–23.

Hooper, D.U., Chapin III, F.S., Ewel, J.J., Hector, A., Inchausti, P., Lavorel, S., Lawton, J.H., Lodge, D.M., Loreau, M., Naeem, S., Schmid, B., Setälä, H., Symstad, A.J., Vandermeer, J., and Wardle, D.A. (2005) Effects of biodiversity on ecosystem functioning: A consensus of current knowledge, *Ecological Monographs*, 75, 1, 3–35.

Howard, L. (1833) *The climate of London*, London: n.p.

Huff, F.A. (1986) Urban hydrological review, *Bulletin of the American Meteorological Society*, 67, 703–712.

Hughes, T.P., Bellwood, D.R., Folke, C., Steneck, R.S. and Wilson, J. (2005) New paradigms for supporting the resilience of marine ecosystems, *Trends in ecology and evolution*, 20, 7, 380–386.

Jacinto, G.S., Azanza, R.V., Velasquez, I.B. and Siringan, F.P. (2006) Manila Bay: Environmental Challenges and Opportunities, in: Wolanski, E. (ed.) *The Environment in Asia Pacific Harbour*, Dordrecht: Springer, 309–328.

Jago-on, K.A.B., Kaneko, S., Fujikura, R., Fujiwara, A., Imai, T., Matsumoto, T., Zhang, J., Tanikawa, H., Tanaka, K., Lee, B. and Taniguchi, M. (2009) Urbanization and subsurface environmental issues: An attempt at DPSIR model application in Asian cities, *Science of the Total Environment (Special Issue)*, 407, 3089–3104.

Jiang, Y., Kirkman, H., Hua, A. (2001) Megacity development: managing impacts on marine environments, *Ocean & Coastal Management*, 44, 293–318.

Joubert, A., Stewart, T.J. and Eberhard, R. (2003) Evaluation of water supply augmentation and water demand management options for the City of Cape Town, *Journal of Multi-Criteria Decision Analysis (Special Issue): 16th International Conference on Multiple Criteria Decision Making*, 12, 1, 17–25.

Kan, H., London, S.J., Chen, G., Zhang, Y., Song, G., Zhao, N., Jiang, L. and Chen, B. (2007) Differentiating the effects of fine and coarse particles on daily mortality in Shanghai, China, *Environment International*, 33, 376–384.

Kaneko S. and Toyota, T. (2011) Long-term urbanization and land subsidence in Asian megacities: an indicators system approach, in: Taniguchi, M. (ed.) *Groundwater and Subsurface Environments: Human Impacts in Asian Coastal Cities*, London: Springer, 249–270.

Kareiva, P., Watts, S., McDonald, R. and Boucher, T. (2007) Domesticated nature: shaping landscapes and ecosystems for human welfare, *Science*, 316, 5833, 1866–1869.

Kataoka, K., Matsumoto, F., Ichinose, T. and Taniguchi, M. (2009) Urban warming trends in several large Asian cities over the last 100 years, *Science of the Total Environment*, 407, 3112–3119.

Kikegawa, Y., Genchi, Y., Kondo, H. and Hanaki, K. (2006) Impacts of city-block-scale countermeasures against urban heat-island phenomena upon a building's energy-consumption for air-conditioning, *Applied Energy*, 83, 649–668.

Kim, K., Ventura, S.J., Harris, P.M., Thum, P.G. and Prey, J. (1993) Urban non-point-source pollution assessment using a geographical information system, *Journal of Environmental Management*, 39, 157–170.

Kim, Y.H. and Baik, J.J. (2005) Spatial and temporal structure of the urban heat island in Seoul, *Journal of Applied Meteorology and Climatology*, 44, 591–605.

Kjerfve, B., Wiebe, W.J., Kremer, H., Salomons, W., Marshall Crossland, J.I.C., Morcom, N., Harvey, N. and Marshall Crossland, J.I.O. (2002) *Caribbean Basins: LOICZ Global Change Assessment and Synthesis of River Catchment/Island-Coastal Sea Interactions and Human Dimensions; with a desktop study of Oceania Basins*, Texel: LOICZ-IPO.

Klein, R.J.T., Nicholls, R.J. and Thomalla, F. (2003a) Resilience to natural hazards: How useful is this concept?, *Environmental Hazards*, 5, 1, 26.

Klein, R.J.T., Nicholls, R.J. and Thomalla, F. (2003b) The resilience of coastal megacities to weather-related hazards, in: Kreimer, A., Arnold, M. and Carlin, A. (eds) *Building Safer Cities: The Future of Disaster Risk*, Disaster Risk Management Series No. 3, Washington, DC: World Bank Publications, 101–120.

Knipping, E.M. and Dabdub, D. (2002) Impact of Chlorine Emissions from Sea-Salt Aerosol on Coastal Urban Ozone, *Environmental Science and Technology*, 37, 275–284.

Kolokotroni, M., Zhang, Y. and Watkins, R. (2005) *The London heat island and building cooling design*. International Conference 'Passive and Low Energy Cooling for the Built Environment', Santorini, Greece, pp. 743–748.

Kolokotsa, D., Psomas, A. and Karapidakis, E. (2009) Urban heat island in southern Europe: The case study of Hania, Crete, *Solar Energy*, 83, 1871–1883.

Kouzana, L., Benassib, R., Abdallah Ben Mammoua, A. and Sfar Felfoulc, M. (2010) Geophysical and hydrochemical study of the seawater intrusion in Mediterranean semi arid zones. Case of the Korba coastal aquifer (Cap-Bon, Tunisia), *Journal of African Earth Sciences*, 58, 2, 242–254.

Kraas, F. (2007) *Megacities and Global Change in East, Southeast and South Asia*, ASIEN, 103 (April), 9–22.

Krämer, A., Khan, M.H. and Kraas, F. (2011) *Health in Megacities and Urban Areas*, London: Springer.

Kraus, E.B. and Businger, J.A. (1994) *Atmosphere-Ocean Interaction*, Oxford: Oxford University Press.

Kristensen, P. (2004) *The DPSIR Framework*, Workshop on a comprehensive/detailed assessment of the vulnerability of water resources to environmental change in Africa using river basin approach, UNEP Headquarters, Nairobi, Kenya, 27–29 September.

Krueger, J.J. (2008) Productivity and structural change: A review of the literature, *Journal of Economic Surveys*, 22, 2, 330–363.

Kuttler, W. (1998) Stadtklima, in: Sukopp, H. and Wittig, R. (ed.) *Stadtökologie. Ein Fachbuch für Studium und Praxis*, Stuttgart: Fischer Verlag.

Kuttler, W. (2004a) *Stadtklima. Teil 1: Grundzüge und Ursachen*, Beitragsserie: Klimaänderung und Klimaschutz, Umweltwissenschaften und Schadstoff-Forschung – Zeitschrift für Umweltchemie und Ökotoxikologie, 16, 187–199.

Kuttler, W. (2004b) *Stadtklima. Teil 2: Phänomene und Wirkungen*, Beitragsserie: Klimaänderung und Klimaschutz, Umweltwissenschaften und Schadstoff-Forschung – Zeitschrift für Umweltchemie und Ökotoxikologie, 16, 263–274.

Lai, L.-W. and Cheng, W.-L. (2009) Air quality influenced by urban heat island coupled with synoptic weather patterns, *Science of the Total Environment*, 407, 2724–2733.

Lam, N.W.Y., Huang, R. and Chan, B.K.K. (2009) Variations in intertidal assemblages and zonation patterns between vertical artificial seawalls and natural rocky shores: a case study from Victoria Harbour, Hong Kong, *Zoological Studies*, 48, 184–195.

Landsberg, H.E. (1981) *The Urban Climate*, Maryland: Academic Press.

Lee, S.W., Song, D.W. and Ducruet, C. (2008) A tale of Asia's world ports: the spatial evolution in global hub port cities, *GeoForum*, 39, 1, 372–385.

Lee, Y.-W., Hwang, D.-W., Kim, G., Lee, W.-C. and Oh, H.-T. (2009) Nutrient inputs from submarine groundwater discharge (SGD) in Masan Bay, an embayment surrounded by heavily industrialized cities, Korea, *Science of the Total Environment (Special Issue)*, 407, 3181–3188.

Lerner, D.N. and Barrett, M.H. (1996) Urban groundwater issues in the UK, *Hydrogeology Journal*, 4(1), 80–89.

Li, W., Zhang, Y., Liu, Z., Cai, J., Zhang, X. & Cheng, S. (2009) Outline for establishment of the Taihu-Lake Basin early warning system, Ecotoxicology, 18, 768–771.

Li, X., Lee, S.-I., Wong, S., Shi, W. and Thornton, I. (2004) The study of metal contamination in urban soils of Hong Kong using a GIS-based approach, *Environmental Pollution*, 129, 113–124.

Li, Y. and Zhu, X., (2010) Urbanization and Global Environmental Change and Its Implications for China's Future Urban Socio-Economic Development Strategy, *UGEC Viewpoints*, 3, 23–26.

Li, Y., Zhu, X., Sun X. and Wang, F. (2010) Landscape effects of environmental impact on bay-area wetlands under rapid urban expansion and development policy: A case study of Lianyungang, China, *Landscape and Urban Planning*, 94, 218–227.

Liu, C., Yu, J. and Kendy, E. (2001) Groundwater exploitation and its impact on the environment in the North China Plain, *Water International*, 26, 2, 265–272.

Liu, J.H. and Kueh, C.S.W. (2005) Biomonitoring of heavy metals and trace organics using the intertidal mussel Pernaviridis in Hong Kong coastal waters, *Marine Pollution Bulletin*, 51, 857–875.

Liu, W., Hong, Y. and Khan, S.I. (2010) Actual evapotranspiration estimation for different land use and land cover in urban regions using Landsat 5 data, *Journal Of Applied Remote Sensing*, 4, 041873.

Low, B., Costanza, R., Ostrom, E., Wilson, J. and Simon, C.P. (1999) Human–ecosystem interactions: a dynamic integrated model, *Ecological Economics*, 31, 227–242.

Lu, R. and Turco, R.P. (1994) Air Pollutant Transport in a Coastal Environment. Part I: Two-Dimensional Simulations of Sea-Breeze and Mountain Effects, *Journal of Atmospheric Sciences*, 51, 2285–2308.

Luck, M.A., Jenerette, G.D., Wu, J. and Grimm, N.B. (2001) The Urban Funnel Model and the Spatially Heterogeneous Ecological Footprint, *Ecosystems*, 4, 8, 782–796.

Lundholm, J.T. and Richardson, P.J. (2010) Habitat analogues for reconciliation ecology in urban and industrial environments, *Journal of Applied Ecology*, 47, 966–975.

Lundqvist, J., Tortajada, C., Varis, O. and Biswas, A. (2005) Water management in megacities, *Ambio*, 34, 269–270.

Luo, Z. and Li, Y. (2011) Passive urban ventilation by combined buoyancy-driven slope flow and wall flow: Parametric CFD studies on idealized city models, *Atmospheric Environment*, 45, 5946–5956.

Mack, R.N. (2003) Global plant dispersal, naturalization, and invasion: Pathways, modes and circumstances, in: Ruiz, G.M. and Carlton, J.T. (eds) *Invasive Species: Vectors and Management Strategies*, Washington, DC: Island Press, 3–30.

Mack, R.N., Chair, Simberloff, D., Lonsdale, W.M., Evans, H., Clout, M. and Bazzaz, F. (2000) Biotic invasions: Causes, epidemiology, global consequences, and control, *Ecological Applications*, 10, 3, 689–710.

Machlis, G.E., Force, J.E. and Burch, W.R. Jr. (1997) Human Ecosystem Part I: the Human Ecosystem as an Organization Concept in Ecosystem Management, *Society and Natural Resources*, 10, 347–367.

Malmqvist, B. and Rundle, S. (2002) Threats to the running water ecosystems of the world, *Environmental Conservation*, 29, 2, 134–153.

Mansfield, B. (2011) '"Modern" industrial fisheries and the crisis of overfishing', in Peet, R., Robbins, P. and Watts, M. (ed.) *Global Political Ecology*. London: Routledge, pp. 84–99.

Marcotullio, P.J. (2003) Globalisation, Urban Form and Environmental Conditions in Asia-Pacific Cities, *Urban Studies*, 40, 219–247.

Marcotullio, P.J. (2007) Variations of urban environmental transitions: the experiences of rapidly developing Asia-Pacific cities, in: Marcotullio, P.J. and McGranahan, G. (ed.) *Scaling Urban Environmental Challenges. From Local to Global and Back*, London: Earthscan, 45–68.

Marcotullio, P.J. and Gordon McGranahan (2007) Scaling the urban environmental challenge, in: Marcotullio, P.J. and McGranahan, G. (eds) *Scaling Urban Environmental Challenges. From Local to Global and Back*, London: Earthscan, 1–17.

Marsh, W.M. and Grossa, J.M.J. (2002) *Environmental Geography: Science, Land Use, and Earth Systems*, second edition, New York: John Wiley and Sons.

Maslow, A. (1970) *Motivation and personality*, second edition, New York: Harper and Row.

Mavropoulos, A. (2010) *Megacities Sustainable Development and Waste Management in the 21st Century*, Hamburg: ISWA Conference.

McFadden, L. (2007) Governing Coastal Spaces: The Case of Disappearing Science in Integrated Coastal Zone Management, *Coastal Management*, 35, 429–443.

McGranahan, G. (2007) Urban Transitions and the Spatial Displacement of Environmental Burdens, in: Marcotullio, P.J. and McGranahan, G. (ed.) *Scaling Urban Environmental Challenges. From Local to Global and Back*, London: Earthscan, 1–17.

McGranahan, G. and Songsore, J. (1994) Wealth, health, and the urban household: Weighing environmental burdens in Accra, Jakarta, and São Paulo, *Environment*, 36, 6, 4–11, 40–45.

McGranahan, G., Songsore, J. and Kjellen, M. (1999) Sustainability, poverty and urban environmental transitions, in: Satterthwaite, D. (eds) *Sustainable Cities*, London: Earthscan, 107–130.

McGranahan, G., Jacobi, P., Songsore, J., Surjadi, C. and Kjellen, M. (2001) *The Citizens at Risk: From Urban Sanitation to Sustainable Cities*, London: Earthscan.

McGranahan, G., Marcotullio, P., Bai, X., Balk, D., Braga, T., Douglas, I., Elmqvist, T., Rees, W., Satterthwaite, D., Songsore, J. and Zlotnik, H. (2005) Urban Systems, in: Hassan, R., Scholes, R. and Ash, N. (eds) *Ecosystems and Human Well-being: Current State and Trends*, The Millennium Ecosystem Assessment Series, Washington, DC: Island Press, 795–825.

McKinney, M.L. (2008) Effects of urbanization on species richness: A review of plants and animals, *Urban Ecosystems*, 11, 161–176.

McMichael, A.J. (2000) The urban environment and health in a world of increasing globalization: Issues for developing countries, *Bulletin of the World Health Organization*, 78, 9, 1117–1126.

Meehl, G.A., Stocker, T.F., Collins, W.D., Friedlingstein, P., Gaye, A.T., Gregory, J.M., Kitoh, A., Knutti, R., Murphy, J.M., Noda, A., Raper, S.C.B., Watterson, I.G., Weaver A.J. and Zhao, Z.-C. (2007) Global Climate Projections, in: *Climate Change 2007: The Physical Science Basis*, Contribution of Working Group I to the Fourth Assessment Report of the Intergovernmental Panel on Climate Change, Cambridge: Cambridge University Press.

Mensah, J.V. (1997) Causes and Effects of coastal sand mining in Ghana, *Journal of Tropical Geography*, 18, 1, 69–88.

Meybeck, M. (2003) Global analysis of river systems: from Earth system controls to Anthropocene syndromes, *Philosophical transactions of the Royal Society of London*, Series B, Biological Sciences, 358, 1935–1955.

Miao, C., Ni, J. and Borthwick, A.G.L. (2010) Recent changes of water discharge and sediment load in the Yellow River basin, China, *Progress in Physical Geography*, 34, 4, 541–561.

Millennium Ecosystem Assessment, MEA (2005) *Ecosystems and human well-being: wetlands and water*, Washington, DC: World Resources Institute.

Minchin, D. and Gollasch, S. (2003) Fouling and Ships' Hulls: How Changing Circumstances and Spawning Events may Result in the Spread of Exotic Species, *Biofouling*, 19, 1, 111–122.

Mirzaei, P.A. and Haghighat, F. (2010) Approaches to study Urban Heat Island – Abilities and limitations, *Building and Environment*, 45, 2192–2201.

Mitchell, V.G., Mein, R.G. and McMahon, T.A. (2001) Modelling the urban water cycle, *Environmental Modelling and Software*, 16, 615–629.

Miyashita, T., Shinkai, A. and Chida, T. (1998) The effects of forest fragmentation on web spider communities in urban areas, *Biological Conservation*, 86, 357–364.

Mohapatra, P.K., Vijay, R., Pujari, P.R., Sundaray, S.K. and Mohanty, B.P. (2011) Determination of processes affecting groundwater quality in the coastal aquifer beneath Puri city, India: a multivariate statistical approach, *Water Science and Technology*, 64, 4, 809–817.

Moldanová, J., Fridell, E., Popovicheva, O., Demirdjian, B., Tishkova, V., Faccinetto, A. and Focsa, C. (2009) Characterisation of particulate matter and gaseous emissions from a large ship diesel engine, *Atmospheric Environment*, 43, 2632–2641.

Molina, M. and Molina, L.T. (2004) Megacities and Atmospheric Pollution, *Journal of the Air and Waste Management Association*, 54, 644–680.

Moore, C.J., Lattin, G.L. and Zellers, A.F. (2011) Quantity and type of plastic debris flowing from two urban rivers to coastal waters and beaches of Southern California, *Journal of Integrated Coastal Zone Management*, 11, 65–73.

Morimoto, T., Katoh, K., Yamaura, Y. and Watanabe, S. (2006) Can surrounding land cover influence the avifauna in urban/suburban woodlands in Japan?, *Landscape and Urban Planning*, 75, 143–154.

Myers, N., Mittermeier, R.A., Mittermeier, C.G., da Fonseca, G.A.B. and Kent, J. (2000) Biodiversity hotspots for conservation priorities, *Nature,* 403, 853–858.

Nakanishi, H., Ukita, M., Masayuki, S.M. and Sadaaki Murakami, S. (1989) Mercury pollution in Tokuyama Bay, *Hydrobiologia*, 176/177, 1, 197–211.

Newton, A., Carruthers, T.J.B. and Icely, J. (2011) The coastal syndromes and hotspots on the coast, *Estuarine, Coastal and Shelf Science.*

Newton, P.W. (2006) *Australia state of the environment. Human settlements: theme report*, Canberra: Department of Environment and Heritage.

Newton, P.W. (2007) Horizon 3 planning: meshing liveability with sustainability, *Environment and Planning B: Planning and Design*, 34, 571–575.

Newton, P.W. and Bai, X. (2008) Transitioning to Sustainable Urban Development Urbanisation, landuse, and water quality in Shanghai 1947–1996, *Environment International*, 29, 5, 649–659.

Nicholls, R.J., Wong, P.P., Burkett, V.R., Codignotto, J.O., Hay, J.E., McLean, R.F., Ragoonaden, S. and Woodroffe, C.D. (2007) Coastal systems and low-lying areas. Climate Change 2007: Impacts, Adaptation and Vulnerability, in: Parry, M.L., Canziani, O.F., Palutikof, J.P., van der Linden, P.J. and Hanson, C.E. (eds) *Working Group II to the Fourth Assessment Report of the Intergovernmental Panel on Climate Change*, Cambridge: Cambridge University Press, 315–356.

Niu, Z, Zhang, H. and Gong, P. (2011) More protection for China's wetlands, *Nature*, 471, 7338, 305.

Nixon, S.W., Granger, S.L. and Nowicki, B.L. (1995) An assessment of the annual mass balance of carbon, nitrogen, and phosphorus in Narragansett Bay, *Biogeochemistry*, 31, 15–61.

Nur, Y., Fazi, S., Wirgoatmodjo, N. and Han, Q. (2001) Towards wise coastal management practice in a tropical megacity Jakarta, *Ocean and Coastal Management*, 44, 335–353.

OECD (2008) *Environmental outlook to 2030*, Paris: OECD.

Oke, T.R. (1973) City size and the urban heat island, *Atmospheric Environment* (1967), 7, 769–779.

Oke, T.R. (1987) *Boundary Layer Climates*, London: Methuen.

Olsen, S.B., Sutinen, J.G., Juda, L., Hennessey, T.M. and Grigalunas, T.A. (2006) *A Handbook on Governance and Socioeconomics of Large Marine Ecosystems*, Narragansett: Coastal Resources Center, University of Rhode Island.

Padmalal, D., Maya, K., Sreebha, S. and Sreeja, R. (2008) Environmental effects of river sand mining: a case from the river catchments of Vembanad lake, Southwest coast of India, *Environmental Geology*, 54, 879–889.

Parker, D. and Mitchell, J.K. (1995) Disaster vulnerability of megacities: an expanding problem that requires rethinking and innovative responses, *GeoJournal*, 37, 3, 295–301.

Pauchard, A., Aguayo, M., Pena, E. and Urrutia, R. (2006) Multiple effects of urbanization on the biodiversity of developing countries: The case of a fast-growing metropolitan area (Concepcion, Chile), *Biological Conservation*, 127, 272–281.

Pauleit, S. and Duhme, F. (2000) Assessing the environmental performance of land cover types for urban planning, *Landscape and Urban Planning*, 52, 1–20.

Pavao-Zuckerman, M.A. (2008) The Nature of Urban Soils and their Role in Ecological Restoration in Cities, *Restoration Ecology*, 16, 4, 642–649.

Pelling, M. (2003) *The vulnerability of cities: Natural disasters and social resilience*, London: Earthscan.

Peng, S.T., Dai, M.X., Hu, Y.D., Bai, Z.P. and Zhou, R. (2009) Long-term (1996–2006) variation of nitrogen and phosphorus and their spatial distributions in Tianjin coastal seawater, *Bulletin of Environmental Contamination and Toxicology*, 83, 416–421.

Pezzuto, P.R., Resgalla, C.J., Abreu, J.G.N. and Menezes, J.T. (2006) Environmental Impacts of the Nourishment of Balneario Camboriu Beach, SC, Brazil, *Journal of Coastal Research (Special Issue)*, 39, 863–868.

Phiri, O., Mumba, P., Moyo, B.H.Z. and Kadewa, W. (2005) Assessment of the impact of industrial effluents on water quality of receiving rivers in urban areas of Malawi, *Environmental Science and Technology*, 2, 3, 237–244.

Pilinis, C., Capaldo, K.P., Nenes, A. and Pandis, S.N. (2000) MADM-A New Multicomponent Aerosol Dynamics Model, *Aerosol Science and Technology*, 32, 482–502.

Pimental, D., Zuniga, R. and Morrison, D. (2005) Update on the environmental and economic costs associated with alien-invasive species in the United States, *Ecological Economics*, 52, 273–288.

Pinho, O.S. and Orgaz, M.D.M. (2000) The urban heat island in a small city in coastal Portugal, *International Journal of Biometeorology*, 44, 198–203.

Population Information Program (2002) *Meeting the Urban Challenge*, Population Reports – Special Topics, Series M, Number 16, Baltimore: Center for Communication Programs, The Johns Hopkins University.

Powledge, F. (2006) The Millennium Assessment, *BioScience*, 56, 11, 880.

Prakash, S., Manhart, A., Amoyaw-Osei, Y. and Opoku Agyekum, O. (2010) *Socio-economic assessment and feasibility study on sustainable e-waste management in Ghana*, Freiburg: Öko-Institut e.V.

Primavera, J.H. (2006) Overcoming the impacts of aquaculture on the coastal zone, *Ocean and Coastal Management*, 49, 531–545.

Priyadarsini, R., Hien, W.N. and Wai, D.C.K. (2008) Microclimatic modelling of the urban thermal environment of Singapore to mitigate urban heat island, *Solar Energy*, 82, 727–745.

Pryor, S.C., Barthelmie, R.J., Schoof, J.T., Binkowski, F.S., Delle Monache, L. and Stull, R. (2008) Modeling the impact of sea-spray on particle concentrations in a coastal city, *Science of The Total Environment*, 391, 132–142.

Rees, W.E. (1996) Revisiting carrying capacity: Area-based indicators of sustainability, *Population and Environment: A Journal of Interdisciplinary Studies*, 17, 3, 1–22.

Rees, W.E. (2001) Ecological Footprint, Concept of, in: Levin, S.A. (ed.) *Encyclopaedia of Biodiversity*, second edition, San Diego: Academic Press, 229–244.

Rizwan, A.M., Dennis, L.Y.C. and Liu, C. (2008) A review on the generation, determination and mitigation of Urban Heat Island, *Journal of Environmental Sciences*, 20, 120–128.

Rodriguez-Pose, A. (2008) The rise of the 'city-region' concept and its development policy implications, *European planning studies*, 16, 8, 1025–1046.

Roth, M. (2007) Review of urban climate research in (sub)tropical regions, *International Journal of Climatology*, 27, 1859–1873.

Roy, A. and Alsayad, N. (ed.) (2004) *Urban Informality. Transnational Perspectives from the Middle East, Latin America and South Asia*, Lanham: Lexington Books.

Ruiz, G.M., Carlton, J.T., Grosholz, E.D. and Hines, A.H. (1997) Global Invasions of Marine and Estuarine Habitats by Non-Indigenous Species: Mechanisms, Extent, and Consequences, *American Zoologist*, 37, 621–632.

Saaroni, H., Ben-Dor, E., Bitan, A. and Potchter, O. (2000) Spatial distribution and microscale characteristics of the urban heat island in Tel-Aviv, Israel, *Landscape and Urban Planning*, 48, 1–18.

Sakakibara, Y. and Owa, K. (2005) Urban–rural temperature differences in coastal cities: influence of rural sites, *International Journal of Climatology*, 25, 811–820.

Sanders, R.A. (1986) Urban vegetation impacts on the hydrology of Dayton, Ohio, *Urban Ecology*, 9, 361–376.

Sarrat, C., Lemonsu, A., Masson, V. and Guedalia, D. (2006) Impact of urban heat island on regional atmospheric pollution, *Atmospheric Environment*, 40, 1743–1758.

Sasaki, K., Mochida, A., Yoshino, H., Watanabe, H. and Yoshida, T. (2008) A new method to select appropriate countermeasures against heat-island effects according to the regional characteristics of heat balance mechanism, *Journal of Wind Engineering and Industrial Aerodynamics*, 96, 1629–1639.

Sassen, S. (2010) *Global Cities: New York, London, Tokyo*, Princeton: University Press.

Sato, A. and Mimura, N. (1997) Environmental Problems and Current Management Issues in Coastal Zones of South and Southeast Asian Developing Countries, *Journal of Global Environmental Engineering*, 3, 163–181.

Scalenghe, R. and Marsan, F.A. (2009) The anthropogenic sealing of soils in urban areas, *Landscape and Urban Planning*, 90, 1–2, 1–10.

Schönwiese, C. (2008) *Klimatologie*, third edition, Stuttgart: UTB.

Schuetze, M., Robleto, G., Leon, C. and Rodriguez, I. (2009) *Sustainable water management in the emerging megacity of Lima – Based on macro-modelling and participation*, Fifth Urban Research Forum of the World Bank: Cities and Climate Change: Responding to an Urgent Agenda, Marseille, 28–30 June.

Scorer, R.S. (1978) *Environmental Aerodynamics*, Chichester: John Wiley and Sons Inc.

Scott, A.J. and Storper, M. (2003) Regions, globalisation, development, *Regional Studies*, 37, 579–593.

Sekovski, I., Newton, A. and Dennison, W.C. (2011) Megacities in the coastal zone: Using a driver-pressure-state-impact-response framework to address complex environmental problems, *Estuarine, Coastal and Shelf Science*, 1–12.

Seto, K.C. (2011) Exploring the dynamics of migration to mega-delta cities in Asia and Africa: Contemporary drivers and future scenarios, *Global Environmental Change*, 21, 94–107.

Seto, K.C., Sánchez-Rodríguez, R. and Fragkias, M. (2010) The New Geography of Contemporary Urbanization and the Environment, *Annual Review of Environment and Resources*, 35, 1, 167–194.

Shair, F.H., Sasaki, E.J., Carlan, D.E., Cass, G.R., Goodin, W.R., Edinger, J.G. and Schacher, G.E. (1982) Transport and dispersion of airborne pollutants associated with the land breeze-sea breeze system, *Atmospheric Environment* (1967), 16, 2043–2053.

Shao, M., Tang, X., Zhang, Y. and Li, W. (2006) City clusters in China: air and surface water pollution, *Frontiers in Ecology and the Environment*, 4, 353–361.

Shaw, B.J. (2011) Agriculture, development and Southeast Asian Megacities, in: Tonts, M. and Siddique, M.A.B. (eds) *Globalisation, Agriculture and Development: Perspectives from the Asia-Pacific*, Cheltenham: Edward Elgar Publishing, 160–174.

Sien, C.L. (2001) *Overview of Impact of Sewage on the Marine Environment of East Asia: Social and Economic Opportunities*, Bangkok: United Nations Environment Programme.

Sintusingha, S. (2011) Bangkok's Urban Evolution: Challenges and Opportunities for Urban Sustainability, in: Sorensen, A. and Okata, J. (eds) *Megacities. Urban Form, Governance, and Sustainability*, cSUR-UT Series: Library for Sustainable Urban Regeneration, Volume 10, New York: Springer, 133–162.

Sorensen, A. (2011) Megacity Sustainability: Urban Form, Development, and Governance, in: Sorensen, A. and Okata, J. (eds) *Megacities. Urban Form, Governance, and Sustainability*, cSUR-UT Series: Library for Sustainable Urban Regeneration, Volume 10. New York: Springer, 397–418.

Sorensen, A. and Okata, J. (2011) Introduction: Megacities, Urban Form, and Sustainability, in: Sorensen, A. and Okata, J. (eds) *Megacities. Urban Form, Governance, and Sustainability*, cSUR-UT Series: Library for Sustainable Urban Regeneration, Volume 10, New York: Springer, 1–12.

Srinivasan, V., Gorelick, S. M. and Goulder, L. (2010) Sustainable urban water supply in south India: Desalination, efficiency improvement, or rainwater harvesting?, *Water Resources Research*, 46, W10504.

Streets, D.G., Carmichael, G.R. and Arndt, R.L. (1997) Sulfur dioxide emissions and sulfur deposition from international shipping in Asian waters, *Atmospheric Environment*, 31, 1573–1582.

Streutker, D.R. (2003) Satellite-measured growth of the urban heat island of Houston, Texas, *Remote Sensing of Environment*, 85, 282–289.

Su, W., Gu, C., Yang, G., Chen, S. and Zhen, F. (2010) Measuring the impact of urban sprawl on natural landscape pattern of the Western Taihu Lake watershed, China, *Landscape and Urban Planning*, 95, 61–67.

Sukopp, H. (1998) Urban ecology – scientific and practical aspects, in: Breuste, J., Feldmann, H. and Uhlmann, O. (eds) *Urban ecology*, Berlin Heidelberg: Springer, 3–16.

Swaney, D.P., Santoro, R.L., Howarth, R.W., Hong, B. and Donaghy, K.P. (2011) Historical changes in the food and water supply systems of the New York City Metropolitan Area, *Regional Environmental Change*.

Syvitski, J.P.M., Vörösmarty, C.J., Kettner, A.J. and Green, P. (2005) Impact of humans on the flux of terrestrial sediment to the global coastal ocean, *Science*, 308, 5720, 376–380.

Syvitski, J.P.M., Kettner, A.J., Overeem, I., Hutton, E.W.H., Hannon, M.T., Brakenridge, G.R., Day, J., Vörösmarty, C., Saito, Y., Giosan, L. and Nicholls, R.J. (2009) Sinking deltas due to human activities, *Nature Geoscience*, 2, 10, 681–686.

Taha, H., Konopacki, S. and Akbari, H. (1998) Impacts of lowered urban air-temperatures on precursor emission and ozone air quality, *Journal of the Air & Waste Management Association*, 48, 860–865.

Tang, B., Tong, L., Kang, S. and Zhang, L. (2011) Impacts of climate variability on reference evapotranspiration over 58 years in the Haihe river basin of north China, *Agricultural Water Management*, 98, 10, 1660–1670.

Taniguchi, M., Burnett, W. and Ness, G. (2009a) Editorial: Human impacts on urban subsurface environments, *Science of the Total Environment (Special Issue)*, 407, 3073–3074.

Taniguchi, M., Burnett, W. and Ness, G. (2009b) Erratum to Integrated research on subsurface environments in Asian urban areas, *Science of the Total Environment (Special Issue)*, 407, 3076–3088.

Taniguchi, M., Shimada, J., Fukuda, Y., Yamano, M., Onodera, S., Kaneko, S. and Yoshikoshi, A. (2009c) Anthropogenic effects on the subsurface thermal and groundwater environments in Osaka, Japan and Bangkok, Thailand, *Science of the Total Environment (Special Issue)*, 407, 3153–3164.

Tibbetts, J. (2002) Coastal cities – Living on the edge, *Environmental Health Perspectives*, 110, 11, a674–a681.

Tirabassi, T., Fortezza, F. and Vandini, W. (1990) Wind circulation and air pollutant concentration in the coastal city of Ravenna, *Energy and Buildings*, 16, 1–2, 699–704.

Tran, H., Uchihama, D., Ochi, S. and Yasuoka, Y. (2006) Assessment with satellite data of the urban heat island effects in Asian mega cities, *International Journal of Applied Earth Observation and Geoinformation*, 8, 34–48.

Turner II, B.L., Kasperson, R.E., Meyer, W.B., Dow, K.M., Golding, D., Kasperson, J.X., Mitchell, R.C. and Ratick, S.J. (1990) Two types of global environmental change:

Definitional and spatial-scale issues in their human dimensions, *Global Environmental Change*, 1, 1, 14–22.

Turner, R.K., van den Bergh, J.C.J.M., Söderqvist, T., Barendregt, A., van der Straten, J., Maltby, E. and van Ierland, E.C. (2000) The values of wetlands: Landscape and institutional ecological-economic analysis of wetlands: scientific integration for management and policy, *Ecological Economics*, 35, 7–23.

Turok, I. (2009) Limits to the Mega-City Region: Conflicting Local and Regional Needs, *Regional Studies*, 43, 6, 845–862.

UN (2010) *World Urbanization Prospects: The 2009 Revision*, New York: United Nations-DESA, Population Division.

UNCHS (2001) *The state of the world's cities 2001*. Nairobi: United Nations Centre for Human Settlements.

UN-DESA (2007) *Indicators of Sustainable Development: Guidelines and Methodologies*, third edition, New York: United Nations, Department of Economic and Social Affairs.

UNESCO (2000) *Reducing megacity impacts on the coastal environment – Alternative livelihoods and waste management in Jakarta and the Seribu Islands*, Coastal Region and Small Island Papers 6, Paris: UNESCO.

UNFPA (2007) *State of world population 2007. Unleashing the potential of urban growth*, New York: United Nations Population Fund.

UN-HABITAT (2010) *State of the World's Cities 2010/2011. Bridging The Urban Divide*, London: Earthscan – United Nations Human Settlements Programme.

Vallejo, B.M., Aloy, A.B. and Ong, P.S. (2009) The distribution, abundance and diversity of birds in Manila's last greenspaces, *Landscape and Urban Planning*, 89, 75–85.

Varis, O., Biswas, A.K., Tortajada, C. and Lundqvist, J. (2006) Megacities and Water Management, *Water Resources Development*, 22, 2, 377–394.

Velazquez-Lozada, A., Gonzalez, J.E. and Winter, A. (2006) Urban heat island effect analysis for San Juan, Puerto Rico, *Atmospheric Environment*, 40, 1731–1741.

Villa, F., Wilson, M.A., Groot, R. de, Farber, S., Costanza, R. and Boumans, R.M.J. (2002) Designing an integrated knowledge base to support ecosystem services valuation, *Ecological Economics*, 41, 445–456.

Vitousek, P.M., Mooney, H.A., Lubchenco, J. and Melillo, J.M. (1997) Human Domination of Earth's Ecosystems, *Science*, 277, 5325, 494–499.

Vörösmarty, C.J., Lévêque, C. and Revenga, C. (2005) Fresh Water, in: Hassan, R., Scholes, R. and Ash, N. (eds) *Ecosystems and Human Well-being: Current State and Trends*, Washington, DC: Island Press, 795–825.

Vörösmarty, C.J., Green, G., Salisbury, J. and Lammers, R.B. (2000) Global Water Resources: Vulnerability from Climate Change and Population Growth, *Science*, 289, 5477, 284–288.

Wackernagel, M., Onisto, L., Bello, P., Callejas Linares, A., López Falfán, I.S., Méndez Garcı_a, J., Suárez Guerrero, A.I. and Guadalupe Suárez Guerrero, M. (1999) National natural capital accounting with the ecological footprint concept, *Ecological Economics*, 29, 375–390.

Warren-Rhodes, K. and Koenig, A. (2001) Escalating trends in the urban metabolism of Hong Kong: 1971–1997, *Ambio*, 30, 7, 429–438.

Weischet, W. (2008) *Einführung in die Allgemeine Klimatologie*, seventh edition, Berlin Stuttgart: Borntraeger.

Weng, Q. and Yang, S. (2006) Urban Air Pollution Patterns, Land Use, and Thermal Landscape: An Examination of the Linkage Using GIS, *Environmental Monitoring and Assessment*, 117, 463–489.

Werner, A.D. (2010) A review of seawater intrusion and its management in Australia, *Hydrogeology Journal*, 18, 1, 281–285.

Wheeler, S. (2009) Regions, Megaregions, and Sustainability, *Regional Studies*, 43, 6, 863–876.

Wichmann, H.-E., Spix, C., Tuch, T., Wölke, G., Peters, A., Heinrich, J., Kreyling, W.G. and Heyder, J. (2000) *Daily Mortality and Fine and Ultrafine Particles in Erfurt, Germany*, Part I: Role of Particle Number and Particle Mass, Research Report/ Health Effects Institute, 98, 5–86.

Wilson, M.A., Costanza, R., Boumans, R. and Liu, S. (2003) Integrating Assessment and valuation of ecosystem goods and services provided by coastal systems, in: Wilson, J.G. (ed.), *The Intertidal Ecosystem: The Value of Ireland's Shores*, Dublin: Royal Irish Academy, 1–24.

Wisner, B. (2003) Disaster Risk Reduction in Megacities: Making the Most of Human and Social Capital, in Kreimer, A., Arnold, M. and Carlin, A. (ed.) *Building Safer Cities: The Future of Disaster Risk*, Disaster Risk Management Series No. 3, Washington, DC: World Bank Publications, 181–196.

Wisner, B., Cannon, T., David, I., Blaikie, P. (2004) *At Risk: natural hazards, people's vulnerability and disasters*. 2nd ed. London: Routledge.

Wolanski, E. (2006) *Increasing Trade and Urbanisation of the Asia Pacific Coast. The Environment in Asia Pacific Harbour*, Dordrecht: Springer, 1–13.

Wong, M.S., Nichol, J.E., To, P.H. and Wang, J. (2010) A simple method for designation of urban ventilation corridors and its application to urban heat island analysis, *Building and Environment*, 45, 1880–1889.

World Health Organization, WHO (United Nations Environment Programme) (2002) *The World Health Report 2002. Reducing Risks, Promoting Healthy Life*, WHO Geneva: World Health Organization.

World Health Organization, WHO (United Nations Environment Programme) (2008) *Our cities, our health, our future. Acting on social determinants for health equity in urban settings*, Kobe: WHO Centre for Health Development, 199.

Wu, J.G., Jenerette, G.D., Buyantuyev, A. and Redman, C.L. (2011) Quantifying spatiotemporal patterns of urbanization: The case of the two fastest growing metropolitan regions in the United States, *Ecological Complexity*, 8, 1–8.

Xu, L.B., Gong, T., Jiang, Chen, D.L. and Singh, V.P. (2006) Analysis of spatial distribution and temporal trend of reference evapotranspiration and pan evaporation in Changjiang (Yangtze River) catchment, *Journal of Hydrology*, 327, 81–93.

Xu, X., Peng, H., Xu, Q., Xiao, H. and Benoit, G. (2009) Land Changes and Conflicts Coordination in Coastal Urbanization: A Case Study of the Shandong Peninsula in China, *Coastal Management*, 37, 1, 54–69.

Yamaguchi, T. (2005) Influence of urbanization on ant distribution in parks of Tokyo and Chiba City, Japan II. Analysis of species, *Entomological Science*, 8, 17–25.

Yang, W., Li, F., Wang, R. and Hu, D. (2011) Ecological benefits assessment and spatial modelling of urban ecosystem for controlling urban sprawl in Eastern Beijing, China, *Ecological Complexity*, 8, 153–160.

Yimin, M. and Lyons, T.J. (2003) Recirculation of coastal urban air pollution under a synoptic scale thermal trough in Perth, Western Australia, *Atmospheric Environment*, 37, 443–454.

Yin, Z.-Y., Walcott, S., Kaplan, B., Cao, J., Lin, W., Chen, M., Liu, D. and Ning, Y. (2005) An analysis of the relationship between spatial patterns of water quality and urban development in Shanghai, China, *Computers, Environment and Urban Systems*, 29, 2, 197–221.

Yoshikado, H. and Tsuchida, M. (1996) High Levels of Winter Air Pollution under the Influence of the Urban Heat Island along the Shore of Tokyo Bay, *Journal of Applied Meteorology*, 35, 1804–1813.

Yoshikoshi, A. (2011) 'Urban development and water environment changes in Asian megacities', in Taniguchi, M. (ed.) *Groundwater and subsurface environments – Human impacts in Asian coastal cities*, Tokyo: Springer.

Yoshikoshi, A., Adachi, I., Taniguchi, T., Kagawa, Y., Kato, M., Yamashita, A., Todokoro, T. and Taniguchi, M. (2009) Hydro-environmental changes and their influence on the subsurface environment in the context of urban development, *Science of the Total Environment (Special Issue)*, 407, 3105–3111.

Zhang, H. and Huang, G.H. (2011) Assessment of non-point source pollution using a spatial multicriteria analysis approach, *Ecological Modelling*, 222, 2, 313–321.

Zhu, K., Blum, P., Ferguson, G., Balke, K.D. and Bayer, P. (2010) The geothermal potential of urban heat islands, *Environmental Research Letters*, 5, 1–6.

Chapter 3

# Coastal megacities, environmental hazards and global environmental change

Lead author: *Andrea Ferraz Young (UNICAMP)*

Contributing authors: *Antje Bruns (Humboldt Universität zu Berlin, Germany), Maida Zahid (Pakistan Meteorological Department), Ralf Weisse HZG, Institute for Coastal Research, Germany), Ivonne Radjawane (Bandung Institute of Technology, Indonesia), Ramesh Ramachandran (Anna University, India) Poh Poh Wong (National University, Singapore), Gordon McGranahan (IIED, UK), Chia-Yang Tsai, Wen-Yan Chiau (National Taiwan Ocean University, Taiwan)*

## 1 Introduction

Coastal zones are exposed to a wide range of hazards including cyclones, flooding, tsunamis, heat waves and sea-level rise (Klein *et al.*, 2003). Recent global environmental change not only exacerbates existing sources of risk but introduces new ones as well and the extent and scale of these alterations leads to uneven outcomes across ecosystems with unequal consequences for populations (MEA, 2005; O'Brien, 2006; Stern *et al.*, 1992; Redclift, 1992).

Over the last 30 years UNISDR calculates the proportion of world population living in flood-prone river basins has increased by 114 percent, while those living on cyclone-exposed coastlines have grown by 192 percent (UNISDR, 2011). Many large cities are located near rivers and in low altitudes within a 100-kilometer range off the coastline and population density along the coast is likely to double by 2050 (Copertino *et al.*, 2010a, 2010b). Observations at the global scale indicate that anthropogenic warming over the last three decades has had a discernible, if uneven, influence on the biophysical environment (Solomon *et al.*, 2007), affecting ecosystem services and core functions of urban systems (O'Brien and Leichenko, 2006). A better understanding of the range of hazards, their construction and impact is necessary for the formulation of policies aimed at risk reduction.

In addition to biophysical hazards, coastal megacities face several urban processes outlined in detail in Chapters 1 and 2, which exert pressures on coastal ecosystems and geomorphology, with both short- and long-term implications for sustainability. Particularly the growing concentration of population and assets along the coastal strip intensifies hazard exposure. Inadequate access to critical infrastructure (clean water provision, waste treatment, quality housing) in many low- and middle-income countries increases vulnerability. For example, deficient drainage systems may exacerbate urban flooding and cause serious public health crises (World Bank, 2012). Further, the resilience of populations able to recover from disaster events may be compromised

(Adger, 2000). Thus, coastlines under stress from human activities may become even more susceptible to global environmental change impacts in a vicious cycle of risk, maldevelopment and loss. Against this background, the present chapter describes these impacts on megacities, which face a dual challenge: urban systems act both as drivers for global change and need to adapt to its many impacts.

### *1.1 Global environmental change*

Global environmental change (GEC) is an umbrella term used to encompass a range of environmental transitions. Its key facets are the uneven risks associated with climate change and cumulative environmental changes such as natural resource depletion, loss of biodiversity and environmental degradation (Kasperson and Kasperson, 2001; MEA, 2005).

Though widespread, these changes differ considerably across regions and interact with local environmental variability, and regional climate systems such as the El Niño-Southern Oscillation (ENSO). Further, they are difficult to project and model and thus increase uncertainty for planners. Human activities including population growth, resource extraction, energy consumption, urbanisation, technological change and changes in consumer demands are key drivers behind GEC (O'Brien and Leichenko, 2006). As a result, crucial ecosystem functions and services are altered, as outlined in more detail in Chapter 2.

Coastal regions are considered to be particularly exposed to GEC impacts felt through coastal erosion, sea-level rise and increase in hazard extremes (Nicholls *et al.*, 2007). Continuous population growth along the coastlines adds additional pressures due to land-use patterns and cumulative environmental degradation, while the adaptive capacity of most urban centers remains limited (McGranahan *et al.*, 2007).

### *1.2 The nature of urban risks*

The ongoing urbanisation rates of coastal regions calls for a detailed understanding of the vulnerabilities to extreme events as well as the drivers of creeping environmental stresses characteristic to most large cities in the context of urban poverty (Parnell *et al.*, 2007; Wisner *et al.*, 2004; Parker and Mitchell, 1995). Research and public debate on the impact of GEC for urban areas is only recently emerging and largely concentrated on climate change adaptation and on the development and disaster risk literature (for example, Grimm *et al.*, 2008; Betsill and Bulkeley, 2007; Simon and Leck, 2010; Kreimer *et al.*, 2003; Bai, 2008; Satterthwaite *et al.*, 2007; Pelling and Wisner, 2009).

The environmental and economic risks projected for the twenty-first century are considerable. Coastal industries particularly affected by extreme events include the oil and gas industry, shipping, fishing, sea ports as well as wider urban infrastructures. Many large and megacities around the world have already experienced some of the extreme events outlined in this chapter. Recent examples include São Paulo and Buenos Aires (floods in February 2010), Rio de Janeiro (floods and landslides in April 2010), London (heat wave in April 2010 and July 2013) or New York (storms and floods in August 2011 and October 2012).

In discussions about natural hazard risks to coastal megacities it is useful to distinguish between those originating in the atmosphere, hydrosphere and geosphere – categorized here as atmospheric hazards, coastal hazards, and geohazards. This chapter addresses each of these in turn, in addition to environmental health hazards, which are particularly significant for urban populations living in low- and middle-income countries. For each hazard type, the hazard context, key outcomes, and current trends in distribution, magnitude, and frequency will be outlined. Further, we explore their main impact on urban systems, such as settlements, infrastructure and economies and reflect on social–ecological systems.

Each hazard is the product of a complex intersection between different spheres, including feedback loops. For example, melting of permafrost in arctic regions causes ground subsidence, which can increase flood risk, and also releases additional methane (a strong greenhouse gas), which may accentuate global warming (Solomon *et al.*, 2007; Zimov *et al.*, 2006). However, since existing knowledge of these complex interactions is limited, modeling the extent and direction of global climate change – and anticipating shifting patterns in hazard exposure – remains a challenge (Kennedy *et al.*, 2002). Furthermore, environmental shifts are often non-linear (i.e. the outcome is not directly proportional to the input), such that outcomes are not readily predicted from historical observations. Changes may be extreme, with rapid transition from one stable state to another (ibid.).

## 2 Atmospheric hazards

### *2.1 Tropical cyclones*

Tropical cyclones are non-frontal synoptic-scale low-pressure systems over tropical or subtropical waters where sea surface temperatures exceed 26°C. They are characterized by organized convection (thunderstorms) and cyclonic surface wind circulation (Shepherd and Knutson, 2007; Webster *et al.*, 2005). Being highly intense storm systems they result in strong wind, extreme precipitation, increased wave height, as well as storm surges. Once mature, wind speeds reach at least 33 m/s [meters per second] and the outer circulation can extend more than 1,000 kilometers (Montgomery and Farrell, 1993). As a result, these phenomena may cause significant wind and flood damage affecting entire city regions.

There has been much debate about the possible impact of GEC on tropical cyclone behavior, with suggested shifts in frequency, intensity, duration or preferred storm tracks (Easterling, 2000). Until recently, research suggested an increase in tropical cyclone activity as a result of elevated global temperatures (for example Webster *et al.*, 2005). However, new evidence suggests that the number of tropical cyclones reported has in fact been remarkably steady since the advance of geostationary satellites in the mid-1970s. The IPCC Special Report for Managing the Risks of Extreme Events and Disasters to Advance Climate Change Adaptation (SREX) indicates a high probability that the frequency of tropical cyclones will either decrease or remain stable throughout the coming century (IPCC, 2012).

Whilst frequency may be relatively stable, high-resolution models and observations suggest an increase in tropical cyclone *intensity* and *duration* over certain ocean basins since the 1970s, despite significant inter-year variations (IPCC, 2012; Knutson *et al.*,

2010), culminating in increased storm destructiveness (Emanuel, 2005). SREX (2012) observes a clear increase in the number of Category 4 and 5 storms over this period, accompanied by increases in the Power Dissipation Index partly due to increased storm duration. Thus far, the largest increase in high-intensity storm events has been observed in the North Pacific, Indian and Southwest Pacific Oceans (Webster *et al.*, 2005). There is also evidence to suggest a poleward migration of storm paths (Neu, 2009).

The growing proportion of high-intensity tropical cyclone events poses a significant threat to vulnerable coastal megacities and urban regions. For example, in July 2005 Mumbai was struck by a cyclone leading to extensive flooding and Shanghai monsoon-related storms causing storm surges and increased surface runoff (De Sherbinin *et al.*, 2007). An increase in frequency and intensity of cyclone events may lead to more flood and wind damages, particularly in areas with limited adaptive capacities (EPA, 2010).

## 2.2 Extra-tropical cyclones

Extra-tropical cyclones are large-scale low-pressure systems developing along a boundary between warm and cold air masses. These types of cyclonic disturbances are the dominant weather phenomenon at mid- and high latitudes, but pressure gradients and near-surface wind speeds are typically lower compared to tropical cyclones. They tend to be more pronounced during the cold season, and are often associated with severe weather such as high wind speeds or heavy precipitation. Extra-tropical cyclones generally propagate eastwards in both hemispheres and are characterized by sharp transition zones between different fronts (air masses) (IPCC, 2012). These transition zones are characterized by strong temperature and pressure gradients and are often associated with severe weather. Onshore, the impacts of extra-tropical cyclones range from wind damage, flooding caused by heavy precipitation to rapid temperature decrease in combination with severe snow conditions and heavy icing (blizzards), while offshore hazards are mainly wind related, leading to extreme wave conditions or storm surges (Weisse and von Storch, 2009).

Depending on location, coastal megacities may experience shifts in extra-tropical storm activity. Similar to tropical cyclones, most models project a poleward shift of the mid-latitude storm tracks in both hemispheres, with little or no change expected in extra-tropical storm frequency and intensity (IPCC, 2012; Solomon *et al.*, 2007).

## 2.3 Macro-climate systems

### 2.3.1 Monsoon climate

Monsoons are regional-scale wind systems associated with large-scale seasonal temperature and pressure differences and are characterized by strong rainfall. The monsoons are strongest in South and East Asia but also occur along the coasts of a few other tropical regions (Glickman, 2000). Because of their persistence, monsoons may lead to frequent floods caused by heavy precipitation, droughts, or heat waves.

Depending on location, coastal megacities may experience impacts from changes in the monsoon climate in the future. SREX (2012) presents some evidence that an increase in heavy precipitation occurred in all monsoon regimes worldwide for about the last 50 to 60 years. With regard to the Indian monsoon, such an increase is reported over the past 100 years with the increase becoming more pronounced in recent decades (IPCC, 2012). For Africa, a northward migration of rainfall is reported, however both SREX (2012) and the IPCC Fourth Assessment Report (2007) note that there is a considerable lack of understanding with regard to long-term changes. Towards the end of the century increased precipitation due to enhanced moisture convergence is expected, despite a tendency towards weakening of the monsoonal flows (Solomon *et al.*, 2007). However, the IPCC also highlights that many aspects of tropical climatic responses remain uncertain.

### *2.3.2 Circulation patterns and ENSO*

A prominent aspect of weather and climate is its variability, stemming from the complexity of weather patterns and the fact that circulation systems are dynamic over multiple timescales. Circulation patterns are a symptom of the Earth's atmospheric system and many of them are on a global scale, affecting entire ocean basins and continents (NOAA, 2011a, 2011b, 2011c, 2011d).

Some patterns, particularly those over the North Pacific region, may be driven by changes in tropical sea-surface temperatures and tropical convection associated with the ENSO cycle (Wang, 2001; NOAA, 2011a). La Niña represents the cool phase of ENSO. Occurring periodically every two to seven years, El Niño and La Niña bring changes in Pacific Ocean temperatures and affect the patterns of tropical rainfall from Indonesia to the west coast of South America (every three to five years), however, during ENSO-neutral periods, neither El Niño nor La Niña is present (NOAA, 2011a). Krishnamurthy and Kirtman (2003) study the Indian Ocean variability and its seasonal relation with monsoon and ENSO. The North Atlantic Oscillation, one of the most important circulation patterns, combines parts of the East Atlantic and West Atlantic (IPCC, 2012). Projections for the likelihood of more extreme climate episodes such as ENSO-related variations in regional climate patterns are crucial for mitigation actions as they affect a wide range of economic sectors and livelihoods.

## **2.4 Heat wave**

Though various indices of heat waves are available there is now solid evidence for global increases in their frequency, intensity and duration linked to global climate change (IPCC, 2012). Recent regional climate simulations support this analysis, such as for Europe (Fischer and Schär, 2010), North America (Kunkel *et al.*, 2010), and Australia (Wang *et al.*, 2011). Since the 1950s the number of heat extremes has become more common with climate models projecting more frequent hot days throughout the twenty-first century (IPCC, 2012).

Intense episodes of heat waves have direct and indirect impact on urban systems. The heat poses serious health risks for vulnerable populations (Semenza *et al.*, 1996). Its health impacts can be significant, particularly in countries not usually exposed to sustained higher temperatures, as experienced during the European heat wave in 2003 or in Russia in 2010 (Campbell-Lendrum and Corvalán, 2007; Kovats and Hajat,

2008). Health risks for vulnerable populations (elderly, chronically ill, infants) include cardiovascular and respiratory diseases, chronic illness and death (Rosenzweig *et al.*, 2011).

Heat waves are exacerbated by the urban heat island (UHI) effect, introduced in detail in Chapter 2. Further, higher consumption needs during sustained hot days may impact water quality and its very provision, already a major problem for many urban areas in developing countries. Heat waves linked to drought episodes increase the likelihood of wood fires, threatening urban areas and diminishing air quality (Solomon *et al.*, 2007; Kreimer *et al.*, 2003). Other adverse impacts relate to the productivity of agriculture and fisheries (Parry *et al.*, 2007), thus possibly leading to changes in food systems, i.e. production and consumption patterns.

### 2.5 Extreme precipitation and flooding

Changes in temperature patterns affect rainfall distributions, causing more intense and frequent precipitation events (Solomon *et al.*, 2007). Positives have been observed particularly in Europe and Central and South Asia (Klein *et al.*, 2003; Brunetti *et al.*, 2010; Rahimzadeh *et al.*, 2009; Manton *et al.*, 2001; Klein Tank *et al.*, 2006; Zahid and Rasul, 2011). This shift became more pronounced in the second half of the twentieth century, and is anticipated to increase the frequency of floods in the future (Frich *et al.*, 2002).

Heavy precipitation events increase the hazard risks in urban areas, including flash floods, subsidence and landslides. Coastal cities are further exposed to serious flooding when heavy rains combine with high tides or storm surges (Jha *et al.*, 2011; Parry *et al.*, 2007). For example, Pakistan experienced severe flooding in 2010 as a result of abnormally high magnitude precipitation events over high-altitude areas, which were anomalous monsoonal 'pulses' (Webster *et al.*, 2011; Houze *et al.*, 2011). Vulnerability factors here included preceding drought and extensive deforestation (Webster *et al.*, 2011). Analysis of extreme precipitation events in Karachi between 1965 and 2009 reveal a statistically significant and sharp increasing trend for all criteria (50mm, ≥100mm, and ≥150mm), thus indicating a high vulnerability to future hazard events (Zahid and Rasul, 2012).

Urban flooding has multiple, often interacting causes, mainly storm surges, river overflows or intense rainfall. Coastal regions are particularly vulnerable to the impact of urban flooding, due to a combination of natural and man-made factors (Cashman, 2011; Nicholls, 2004; Wisner *et al.*, 2004). Drains are often blocked by solid waste and debris, drainage systems are inadequate, and urban land cover is generally impermeable and compact, thus limiting space for water storage (Sherbini *et al.*, 2007). Floods can also increase pollutant levels in water supplies (e.g. reservoirs, rivers), and disproportionately affect those settlements located on floodplains (Rosenzweig *et al.*, 2011).

It should be noted that floods, storm surges, and other weather-related hazards are normally directly modified by human activities in and outside the urban area:

- excessive groundwater withdrawal may cause subsidence, a major issue already in low-lying coastal cities in Asia (Klein *et al.*, 2003);

- land use of intertidal areas due to urban expansion, lack of law enforcement on new structures owing to economic or political factors, or capacity, which increases flood risk as natural water flow is obstructed;
- erosion and deposition problems due to hydrological modifications, land use, and changes in subsurface environments.

Particularly groundwater exploitation is a major problem in many rapidly growing urban centers, resulting in subsidence with the related damage to key infrastructures and buildings. Chapter 2, Section 4, outlines in detail the vulnerabilities stemming from human interference in subsurface and soil environments.

Particularly the high growth rate of Asian megacities correlates with coastal subsidence: in Manila between five and nine centimeters a year, in Osaka around 8.2 centimeters a year, in Shanghai six centimeters a year, and in Tokyo 2.7 centimeters a year (Rodolfo and Siringan, 2006). Frequently, the rates of subsidence exceed the maximum projected rates of sea level rise, which increases the intensity of negative impacts on coastal megacities, especially in low-lying areas. For example in Bangkok it is predicted that land subsidence will result in a trebling of the flood damage increases due to other causes (World Bank, 2010). Large deltas are sinking at rates of at up to six centimeters per year due to land compaction or extraction of groundwater. The Pearl River and Mekong River deltas are particularly vulnerable (Fuchs, 2010).

Yet, while subsidence rates can be rapid, reaching up to one meter per decade, better groundwater management can reverse this trend. Osaka, Tokyo and Shanghai have successfully managed to address this problem, while recent efforts by Jakarta and Bangkok authorities to regulate inner-city groundwater use highlight the role of socioeconomic factors in triggering subsidence (Klein *et al.*, 2003; Rodolfo and Siringan, 2006; Fuchs, 2010; World Bank, 2010).

## 3 Coastal hazards

### 3.1 Relative sea level rise

Over the twentieth century, global mean sea level (also termed relative sea level) rise is estimated at an average of one to two millimeters per year (Church *et al.*, 2001; Bindoff *et al.*, 2007; Jevrejeva *et al.*, 2008). Since the start of the 'satellite era' in monitoring from 1993 onwards, this has increased to approximately three millimeters per year, with consistency between both satellite and tide-gauge data (Jevrejeva *et al.*, 2008; Church *et al.*, 2008). There is considerable regional variability in the observed rates of sea-level change, as well as temporal shifts for example due to ENSO described above.

For the end of the twenty-first century a further increase in global mean sea level is anticipated, primarily as a consequence of ocean thermal expansion related to rising temperatures, and freshwater input as a result of melting ice sheets, glaciers, and ice caps. Contributions from melting ice sheets are thought to have been minor during the twentieth century but are anticipated to be a major contributor in the future (Church *et al.*, 2008). The IPCC Fourth Assessment Report projects an increase in global mean sea level of 18–59 centimeters (up to 79 centimeters with the inclusion of Antarctic and Greenland ice sheets) by the end of the twenty-first century (Meehl *et al.*, 2007).

These figures echo the values reported in the IPCC Third Assessment Report; in particular for the upper end of the projected range. Regionally, substantial deviations from these estimates may occur, and so far models of the relevant physical processes and their implementation have had limited success in making accurate projections (Church *et al.*, 2010). As a result one finds a wide range of estimated rises in mean sea level in the literature, reflecting the degree of scientific uncertainty.

In some regions including North America and Europe, mean sea level rise has been offset by an isostatic lift of the land mass in response to pressure released from ice melt, which has occurred since the end of the last ice age. However, under scenarios of accelerated sea-level rise, such coastlines are anticipated to be subject to increased sedimentary erosion, which will reduce this 'offset' effect in the future (de Groot, 1992).

With regard to the consequences of mean sea-level rise, the most obvious is the submergence and the heightened risks of coastal flooding from storm surges during cyclonic events, as well as increased wave intensity and height. Further, sea-level rise accelerates coastal erosion, which over time can destabilize coastal structures and is a major risk to coastal wetlands. Strong wave incidence may result in disruption of natural barriers, mangroves, post-adjustment of shorelines (e.g. dunes, beaches, and deltas) (Muehe, 2010). The erosion can affect lagoons, deltas, and estuaries (transport of sediments and siltation).

Many of the world's largest cities are located on or near deltas. Megacities located within deltas (for example Dhaka or Lagos) are particularly threatened by relative sea-level rise due to their low elevation combined with dual exposure to river and coastal flooding, thus increasing the risk of inundation. Further, in combination with land subsidence relative sea-level rise is a consequence of land submergence, thus further exacerbating hazard exposure. Syvitski *et al.*, (2009) examine the linkages between human activities and their vulnerability to flooding as a result of sediment compaction, trapping of sediments upstream, and water infrastructure engineering. In Europe, river floods, which are the most common natural disasters, result in high economic losses up to €15 billion (Ciscar *et al.*, 2011). Nicholls *et al.*, (2008) carried out a global study assessing the vulnerability of the world's largest port cities to coastal flooding.

However, there are major uncertainties in how global mean sea-level rise will manifest itself on a regional and local scale. Confidence in projections of regional sea-level changes is low, although it is clear that they are important with respect to impacts of sea-level rise on coastal zones (Klein *et al.*, 2003).

### 3.2 Storm surge

Many megacities on the coast are at risk from ocean flooding, particularly when storm surges combine with high tides and/or high-volume river flows (McGranahan *et al.*, 2007). Storm surge is defined as an abnormal rise of water generated by a storm, over and above the predicted astronomical tides. It generally occurs as a result of water being pushed onshore by cyclical wind movement around a central storm system (NOAA, 2011b). De Gaetano (2008) argues that the adverse effects of some storms surges can exceed those caused by hurricanes alone, given their frequency, long duration, and wave intensity. Storm surge should not be confused with storm tide, the latter of which is defined as the water level rise due to the combination of storm surge and the astronomical tide (NOAA, 2011b). According to De Gaetano (2008), the rise in water level associated with wind pressure can cause extreme flooding on the coast, particularly when storm surges coincide with the arrival of storm systems themselves. Often, coastal flooding is the product of the intense precipitation, wave intensity, and low-altitude topography of the coastal plains, which also results in difficulty of water discharge due to the blocking of channels outlets by waves and tides.

Table 3.1 below summarizes projected changes in extreme atmospheric and coastal hazards and consequences for coastal megacities. It is difficult to draw conclusions

*Table 3.1* Projected changes in extreme atmospheric and coastal hazards and consequences for coastal megacities

| *Projected change in extreme climate phenomena and likelihood* | *Consequences in coastal megacities* |
|---|---|
| Warmer, with more frequent hot days and nights<br><br>(certain) | • Urban heat island effect<br>• Declining air quality<br>• Increased demand for cooling (energy use)<br>• Changes in transport needs |
| Heat waves, frequency increases over most land areas<br><br>(very likely) | • Increased water demand<br>• Water and sanitation quality problems<br>• Increased mortality risk for vulnerable groups (elderly, very young, socially isolated)<br>• Changing housing needs |
| Heavy precipitation events, frequency increases over most land areas | • Impact on water supply<br>• Adverse effects on surface and groundwater quality<br>• Increase in water-borne diseases, risk of infectious, skin diseases<br>• Large displacement of people |
| (very likely) | • Disruption of urban infrastructures, commerce, transport, settlements due to flooding<br>• Loss of assets and property |
| | • Reduction in water stress (short-term) |
| Intense tropical cyclone activity<br><br>(uncertain) | • Disruption of water supply, power<br>• Distress migration<br>• Increased risk of deaths, water and food-borne diseases, post-traumatic stress<br>• Disruption of urban infrastructures by flood and high winds<br>• Loss of assets and property<br>• Withdrawal of risk coverage in vulnerable areas by private insurers |
| Increased incidence of extreme high sea level (excluding tsunamis)<br><br>(likely) | • Decreased freshwater availability due to saltwater intrusion<br>• Increased risk of deaths and injuries by drowning and migration-related health effects<br>• Loss of assets, property, and livelihood<br>• Permanent erosion and land submersion<br>• Costs of coastal protection versus costs of land-use relocation<br>• Potential for movement of people and infrastructure, commerce |

from this global review for individual coastal megacities. However, it becomes clear that research efforts should be focused on better understanding and planning for hazards and risk associated with increased temperatures, precipitation, and sea-level rise, as well as subsidence (see Chapter 2) and geohazards as described below and the interaction between these hazards and their underlying linkages to the development processes.

## 4 Geohazards

### 4.1 Landslides

Landslides have several triggers, the most significant being earthquakes, large storms with intense precipitation, rapid snowmelt, and volcanic activity (Malamud *et al.*, 2004). The slide occurs as a result of soil instability, whereby shear stress (under gravity) exceeds the forces maintaining the material's aggregate properties. Chau *et al.*, (2004) and Guzzetti (2000) observe that landslides can manifest in many different forms, including rockfalls, rockslides, rock avalanches, debris flows, soils slips, and mud flows, all of which have the potential for catastrophic outcomes.

Literature on landslide hazards, risk analysis, and management is wide (see for example Antonini *et al.*, 2002; Chau *et al.*, 2004; Hervás, 2007; Ardizzone *et al.*, 2008; Guzzetti *et al.*, 2009; Reichenback and Gunther, 2010; Brunetti *et al.*, 2010; Martelloni *et al.*, 2011), reflecting the challenge of making accurate risk assessments. The high degree of uncertainty and causal complexity in landslide mechanisms are common themes.

Risk assessment typically involves determining location, frequency, magnitude, and timing of future landslide events and the analysis of local geology, hydrological conditions, and landscape typology and environmental context. The local stress conditions such as erosion, deposition, and groundwater fluctuation are also important, i.e. in urban contexts. Guzzetti (2000) provides an estimative of landslide risk by analyzing historical events that have caused loss of life, missing persons, injuries, and homelessness.

Rainfall intensity is a strong predictor of falls, flows, and earth-mass mechanisms, as is heavy or irregular land use of mountains and hills at risk. The projected intensification of precipitation patterns may therefore intensify landslide risks. This is of particular concern in densely populated areas, especially those containing low-quality housing located on steep slopes, and in coastal areas, which experience high levels of relief rainfall. These conditions coincide, for example, in Rio de Janeiro. Here, high landslide vulnerability is the product of the high number of densely populated informal *favela* settlements located on steep slopes, housing vulnerable households. Being economically and socially marginalized from the city proper, such households have a weak recovery capacity, lack asset insurance and have low access to emergency relief in case of a disaster.

### 4.2 Earthquakes

Earthquake hazards mainly refer to the effect of ground shaking, ground displacement along a fault, flooding, and fires. Plate 5 by Jackson (2006) shows (a) cities with populations more than five million (small squares) or ten million (large squares) while (b) shows earthquakes of the last 1,000 years known to have killed more than 10,000 (red dots) or 100,000 (yellow dots) people; 113 earthquakes are listed in this figure, many of their symbols overlying each other. A total of 34 of these earthquakes occurred in the last 100 years alone. It is clear that the odds of being hit by an earthquake are not evenly distributed (Dilley, 2005): 85 percent of all earthquake fatalities occurred along the Alpine/Himalayan fault belt between Western Europe and Eastern Asia, while 12 percent are found along the eastern edges of the Pacific (Bilham, 2009). That is, major earthquakes are concentrated in less than 12 percent of the Earth's surface area, making a few countries particularly vulnerable to such event. China, Japan, Italy, Iran, Indonesia, and Turkey are the least fortunate of these. This figure clearly highlights the vulnerability of cities along the earthquake belt where tectonic plates meet. The histogram (c) shows the number of earthquakes killing more than 10,000 (grey) or 50,000 (red) people per century and illustrates the steep increase in fatalities following the global population expansion after 1600. Coastal areas are not more at risk from earthquakes than others (Hewitt, 1983) but the concentration of people, assets, and commerce in megacities implies a different, magnified fragility of urban systems and infrastructure that stretches far beyond the locality itself.

The effect of earthquakes is largely about its fatality rate and property destruction. As building collapse causes most fatalities, knowledge about the micro-geography of settlements and appropriate planning and enforcement of building regulations are thus regarded as the single most important seismic risk-reduction measure (Bilham, 2009; Wisner *et al.*, 2004). Earthquake hazard risks can greatly be reduced, as demonstrated by California or Chile where magnitude 6–7 earthquakes are now mainly about economic loss, whereas they routinely kill thousands in areas elsewhere. Tucker (cited by Huppert and Sparks, 2006) points out 'the costs of the 1989 and 1994 Californian earthquakes were 0.2 and 1 percent of the regional (not national) GDP, while the cost of the 1972 Nicaragua and 1986 El Salvador earthquakes were 40 and 30 percent of those countries' *entire* GDP.' Thus, vulnerability greatly varies depending on location, temporal characteristics, building structures, and the type of protective measures undertaken. Wisner *et al.* (2004: 274) stress that while earthquakes do occur irrespective of human activities and modification of the environment, 'human action and inaction can nevertheless impact upon the *outcomes* of earthquakes and volcanic eruptions.'

## 4.3 Earthquake-induced tsunami

Tsunamis are composed of seismic sea waves of extremely long wave length. They are generated by disturbances that cause deformation of the sea floor and/or mass displacement of seawater and are associated primarily with submarine earthquakes, but can also be triggered by landslide and volcanic activity (NOAA 2011c, 2011d; State of California Seismic Safety Commission, 2005).

Tsunami waves are a wall of turbulent water mass and are extremely destructive, particularly since earthquake-generated tsunamis allow for very little warning ahead. They occur globally, although the Pacific Ocean is particularly susceptible because it is surrounded by a series of mountain chains, deep ocean trenches and island arcs that are associated with large earthquakes. This region is popularly referred to as the 'Pacific ring of fire'. Tsunamis may cause damage long distances away from the source event and thus can take hours until they hit coastlines (State of California Seismic Safety Commission, 2005).

Tsunamis tend to have a high mortality rate and cause devastating economic losses, as a result of the short warning time, limits of hazard mitigation, and the magnitude of impact. The difficulty of tsunami risk reduction is illustrated by the failure to reduce the number of tsunami-related deaths over the last 40 years. For example, on 27 July 1976, a 7.5 magnitude earthquake in Tangshan, China caused a registered 255,000 deaths. More recently, on 26 December 2004, a 9.1 magnitude earthquake off the coast of Sumatra generated a tsunami with a height of up to 25 meters, causing destruction in 11 countries in the Indian Ocean with a recorded 227,898 fatalities (Bird *et al.*, 2007).

The catastrophic impact and cascades of secondary hazards that have occurred in many regions around the world as a result of earthquake-induced tsunamis in the past decade highlights the vulnerability of megacities to tsunami-induced disasters.

## 5 Environmental health risks

Climate change has already contributed to an increase in disease and premature death and altered the spread of infections and vector-borne diseases as well as heat-wave related death rates (Parry *et al.*, 2007). The increase in extreme weather events outlined above also worsens existing health challenges faced by the urban poor, the elderly, and children. This is a highly inequitable process where those 'at greatest risk are also those who have contributed the least to greenhouse gas emissions' (Dodman, 2009: 13).

The effects of environmental change are manifold and often difficult to directly attribute. Increased heat exposure is likely to be particularly severe among low-income populations in tropical cities. Hajat *et al.*, (2005) find that heat-wave-related mortality in Delhi persisted several times longer than in London, and unlike there it was not soon followed by a comparable decline in mortality. This sort of difference could be explained by the higher prevalence of infectious diseases and child mortality in low-income settings.

Storms and flooding, which are a particular hazard in coastal cities, also have appreciable health impacts, to some degree through injuries, but more importantly through the increased transmission of infectious diseases due to contamination of water supplies (Kovats and Akhtar, 2008). As outlined earlier, a high share of impermeable land cover and poor drainage increase the risk of flooding. Inadequate water and sanitation facilities increase the likelihood of water and sanitation related diseases, already a leading risk factor especially among vulnerable populations (Cairncross and Alvarinho, 2006; Morgan *et al.*, 2005).

Many of the plausible routes through which these and other global environmental changes are likely to affect human health involve health hazards currently associated with urban poverty and informality, such as inadequate water and sanitation, poor drainage, overcrowding, the settlement of flood- or landslide-prone lands. UN-HABITAT (2009) estimates that in 2005 one in three people in cities in the developing world lived in slum conditions, characterized by a lack of durable housing, insufficient living space, adequate access to water and sanitation, and/or secure tenure (Sverdlik, 2011).

Lopez *et al.* (2006) estimate that exposure to air pollution in capital cities and urban centers with populations over 100,000 results in the loss of over half a million lives a year, and accounts for over half a percent of the total burden of disease. There are several routes through which environmental change can be expected to exacerbate urban air pollution, including ground-level ozone and particulates (Spickett *et al.*, 2011). On the other hand, measures to mitigate GHG emissions can also reduce hazardous urban air pollution (Haines *et al.*, 2009).

Densely populated areas may provide breeding grounds for mosquitoes and increased temperatures and result in higher humidity levels conducive to species expansion such as those carrying malaria, dengue, or filariasis (Haines *et al.*, 2006).

Global environmental change will also change rural conditions, thus indirectly affecting urban health through changes in, for example, food security (Friel *et al.*, 2011). It may create new health risks, and lead to the emergence of new diseases (Slenning, 2010). It remains difficult, however, to even conjecture about the health consequences of these indirect and uncertain risks for human health in coastal cities. What is clear, however, is that improvements in public health services and environmental management reduce the vulnerability to climate-related health risks in the future (Grönwall *et al.*, 2010). It is not the rapid and dense urbanisation process itself that determines vulnerability but rather the location of settlements in coastal risk zones, hillsides, and flood plains.

## 6 Risk impacts in coastal megacities

Urban systems consist of both bio-geophysical and human structures and processes and can be understood as socio-ecological systems as outlined in Chapter 2. In this chapter we have so far focused on bio-geophysical systems, outlining the hazards to which coastal megacities are exposed, and the implications of GEC for the changing frequency, behavior, and distribution of these. Coastal megacities have been shown to be particularly vulnerable due to their location and environmental conditions. In addition, coastal cities themselves act as stressors for the environment, which may lead to hazardous feedbacks.

This section focuses on the impact of GEC on social and economic components. These are the urban territory itself (e.g., territory loss as a result of coastal erosion), housing and infrastructure, the various economic sectors, and the social–political system (governance and human well-being in general). The distinct vulnerabilities of individual coastal communities are caused by a mix of various drivers, such as urbanisation, globalization processes, and the aforementioned dimensions of environmental change.

### *6.1 Impacts on the stability of human systems*

The following subsections focus on the functioning of social systems and outline how GEC impacts on aspects of human security, namely migration flows and food systems as these illustrate the complexity of push and pull factors across geographical scales.

### *6.1.1 Shifting migration patterns*

The potential scale of migration flows triggered by environmental change is an emerging issue, often framed as a problem or threat, based on the mapping of likely climate change impacts in a given area – with little reflection on the feasibility of the projected displacement process itself or possible destination routes (ADB, 2011; Stern, 2007). Migration trends are highly context specific and the relationship to environmental triggers is complex. Accordingly, there is a consensus in the social science literature that environmental change impacts sit alongside a diverse set of social, economic, demographic, and political push and pull factors (Warner *et al.*, 2009; Findlay, 2011). Both the recent Foresight Review (2011) and Gemene (2011) find little empirical evidence for popular policy scenarios predicting a surge in 'environmental refugees,' highlighting that a migration decision is shaped by multiple drivers.

Climate change may affect migration often indirectly, by changing livelihoods or political drivers such as conflict over scarce resources. Several recent studies develop frameworks seeking to capture the difficult relationships between single environmental drivers and human movement (McLeman and Smith, 2006; Perch-Nielsen *et al.*, 2008, Piguet, 2008, Warner *et al.*, 2009). For example, Lille Lilleør and van den Broeck (2011) assess the evidence for climate change effects with regard to temperature variability and rainfall for some migration drivers such as income variability in origin area, and highlight that evidence for isolated drivers is poor, echoing several other studies (e.g. Hugo, 2011 on demographics; Raleigh, 2011 on conflict). The degree of livelihood dependency on natural resources is an important driver for those groups possessing limited human and social capital to diversify income at home, but these movements tend to stay within rural areas, and are often seasonal or circular (Tacoli, 2009). Both Warner *et al.* (2009) and Black *et al.* (2011) focus on existing migration systems to analyze environmental drivers and find that voluntary migration as an adaptation strategy follows familiar destinations and within social networks.

The cumulative impact of hazard events may change this but it is likely that mobility patterns will continue to be shaped by interaction with non-environmental drivers (Black *et al.*, 2011). Thus while the figures for internally displaced populations following disaster events show that natural hazards displace more people than violent conflict (UNHCR, 2012), many movements will be temporary and local (Brown, 2007, cited by Black *et al.*, 2008: 20).

It is clear that a changing climate exposes new populations to new types of risks from natural hazards. Planned migration may be a response strategy for single communities to slow-onset disasters, but is generally considered to be unlikely as a dominant scenario for large populations and urban centers, as the costs (emotional as well as economic) may make migration harder, particularly when this involves crossing international borders (Foresight, 2011; ADB, 2012). In fact, people may also move to areas of high risk, particularly coastal cities and delta regions for livelihood opportunities (Seto, 2011). The review of the drivers of migration for Asian and African megacities highlights how economic opportunities clearly outweigh climate-related hazard risks, offering important insights for adaptation policies. The Foresight Review (2011) places coastal and deltaic cities in a 'double jeopardy' situation: the need to absorb rural–urban migration and to cope with the additional pressures on infrastructure and services while being themselves particularly at risk from coastal hazards.

Indeed, as rural migrants tend to settle on flood-prone lands and other high-risk locations such as riverbanks in the cities of arrival, they may not be safer themselves and constitute a distinct, additional challenge to disaster mitigation efforts. This is a recognized problem for many coastal cities in Asia where the number of people living on flood-plains is estimated to rise from 18 million in 2000 up to threefold (Foresight, 2011). As a large part of urban growth is driven by informal settlements, where already in 2005 almost one-sixth of the world population was living (UNPF, 2007), the increase in rural-urban migration flows to vulnerable settlements 'may simply mean the translocation (or even escalation) of climate vulnerability from one place to another, especially for the poor' (Black *et al.*, 2008: 23).

### *6.1.2 Food security*

The global food crises of 2007–9 revealed how a combination of factors may threaten food security following reduced access to food: by mid-2008 real food prices were 64 percent above their 2002 levels, leading virtually all developing countries to introduce policies seeking to absorb the effects (FAO, 2010). The crisis highlighted the particular vulnerability of the urban poor, as they tend to be net food buyers. Those living under US$1 a day spend up to 70 percent of their income on food, and are thus hit hardest by an increase of food staples such as maize, rice, or wheat (Banerjee and Duflo, 2007; ADB, 2012). Ill-absorbed urbanisation – a key feature of the rapidly growing megacity – is thus tightly linked to sudden shocks in agricultural markets (Matuschke, 2009).

Attaining food security is largely a problem of *access* to food, as world production is more than enough to feed everyone; yet, close to one billion people are undernourished, the vast majority living in Asia (FAO, 2010). While people have adjusted to changing environments well over the past millennia, the scope and scale of interactions in the context of urbanisation and globalization mean that lines of production, distribution, governance, and consumption are changing profoundly. Beddington *et al.* (2012: 10) observe that 'total global food demand, GHG emissions and land competition will increase as more of the world's consumers switch to diets that are richer in meat, dairy products and processed foods.'

Not surprisingly, concerns about the additional challenges that GEC may entail for food security have entered public policy and science agendas in the past decade, such as the Global Environmental Change and Food Systems (GECAFS) international research project (Ingram *et al.*, 2010). Research on climate change impacts is concerned with questions at a range of different scales, moving from crop production and yields itself to distribution and regulatory questions (Ingram *et al.*, 2010; Matuschke, 2009; Bayo, 2006), and a host of recent assessment reports argues for reforming the current food system (for example: ADB, 2012; Foresight, 2011; MEA, 2005).

Ingram *et al.* (2010) and Ericksen (2008) conceptualize the strength of a respective food system vis-à-vis global environmental change, i.e. climate change, as the outcome of the interplay of dynamic components: the socioeconomic conditions (e.g. populations, regulatory governance, market structure, technology), climate (e.g. trends, variability), and other bio-geophysical conditions (e.g. water- and land-use structure). The respective weight and character of each of these varies considerably across localities, as there are so many interactions that influence food system outcomes (Ingram and Brklacich, 2002). For example, distinct types of ecosystem services determine food availability from local sources such as fishing. Similarly, while rice farming in the Philippines is likely to stay a viable sector despite strongly felt climate change impacts thanks to innovations in agri-science, this may not halt the decline in rural employment as rural livelihoods are less resilient to the increase in extreme weather events leading to emigration. Foreign rice imports from China may further weaken domestic markets in the long term and increase vulnerability to regional market volatilities. Research on vulnerabilities to food security in the context of GEC increasingly acknowledges the reality of multiple stressors (Adger *et al.*, 2005; O'Brien and Leichenko, 2006) and shifts the focus away from GEC as a primary trigger to other factors such as institutional arrangements or policy failures that inhibit adaptive capacity and act across multiple scales.

Seeking to understand why people and ecosystems are vulnerable to shocks and stresses such as storm surges or climate change impacts calls for a combination of actor-oriented approaches with ecological approaches that focus on systems. In their review of international assessments on the linkages between GEC and food, Wood *et al.*, (2010) observe grave uncertainties in knowledge about the many linkages among environmental drivers of change and food system activities. Further, how such changes influence livelihoods – such as the impact of urban floods on food distribution systems – remains unclear (ibid., 2010). Given the many trade-offs between attaining food security, i.e. its reliance on technology and economic progress, and ecosystem services degradation – as demonstrated by coastal China over the past three decades – one risks limiting one type of vulnerability by creating a new one (Ericksen *et al.*, 2011). For example, urban gardening carries considerable health risks; similarly, ever more remote food production sites imply more complex transport systems that impact on food quality and regulatory oversight, and, lastly, infrastructure investments may carry considerable environmental risks. In this sense, the issue of urban food security in the context of GEC is a prime example of a coupled socio-ecological system as outlined in Chapter 2 (Sundkvist *et al.*, 2005).

We may experience a change in the geography of food production due to more frequent weather extremes, which, for example, cause the decline in a given local livestock industry or alter the productivity of fishery systems. Its consequence, however, will not be felt equally within the same locality. Spatial or common food-chain analyses are thus not particularly helpful; instead, a system approach may 'connect the activities of food producers, processors, distributors, retailers and consumers involved in food systems to the food security and environmental outcomes' (Ingram *et al.*, 2010: 83). For example, while urban areas are usually well connected to their rural, food-producing hinterlands, the nature of global and domestic markets inhibits price transmissions along local, domestic, and global markets. Agricultural output itself is just one factor in determining food security as food procession may take place across different localities. Particularly in urban areas it is private sector activities (i.e., retail) that increasingly dominate food markets and heavily influence dietary choices (Beddington *et al.*, 2012). In addition, markets in developing countries often lack the capacity to absorb shocks such as extreme flooding, leading to high domestic price volatilities even if global conditions are calm.

The particular vulnerability of coastal urban populations themselves depends on the respective nature and the manifold dynamics of a given food system coupled with its exposure to the GEC risks described in the previous sections. Lipper *et al.* (2010) estimate that 70 percent of urban households in developing countries participate in some agricultural activities. As food-insecure populations increasingly migrate to cities, their lack of social entitlements often leads to peri-urban and urban agricultures (ADB, 2012). Informal agriculture may indeed offer part solutions for GEC – reducing travel costs and potentially the embodied carbon of intensive industrial agriculture. But without regulation and support they can also lead to the entry of polluted foodstuffs into the human food chain. Coastal cities, and megacities in particular, have a specific health risk from the consumption of contaminated fish and shellfish impacted by urban coastal pollution. There is very little research on the scale and pathways of consumption related to urban fisheries, formal or informal.

Thus, analysis in GEC and food security in the coastal megacity context spans the global scale (in the form of trade regimes such as the WTO or GHG emission trading schemes), regional arrangements (such as Mercosur or the EU seeking to reduce trade barriers), national governance (such as for environmental regulations or sector strategy formulation) and down to the local. It calls for cross-disciplinary research that analyses food security beyond the dominant focus on agriculture; further, distinct scenarios and models are currently very limited. The degree to which coastal megacities, or any larger urban area, may possess attributes that enhance or inhibit their resilience to food shortages remains uncertain; the fact that they have different bargaining powers due to their economic size certainly influences future distributional conflicts. Ingram *et al.* (2010: 39) comment on the lack of detailed studies on 'how sea-level rise would affect food systems, especially the interaction between sea-level rise, more frequent storms, the vulnerability of large cities (many located on coasts) and food security.'

## 6.2 Impacts on socioecological systems

### 6.2.1 Ecosystem health and ecosystem services

Healthy ecosystems provide a wide range of services that play a vital role in supporting the lives and economic activities of urban populations (Klein *et al.*, 2003). Decker *et al.* (2000: 719) regard ecosystem services as including 'air filtering, climate regulation, and sewage treatment,' which despite being vital for the survival and well-being of urban populations, tend to have no market value 'until they must be replaced by energy-intensive anthropogenic substitutes.' However, as outlined in Chapter 2, substantial evidence suggests that anthropogenic pressures and global environmental change are causing extensive loss of marine and terrestrial biodiversity impairing the ocean's ability to provide food, maintain water quality, and recover from pollution (Worm *et al.*, 2006).

For example, many sea grass habitats have been lost due to poor water quality surrounding coastal megacities, primarily caused by high nutrient runoff and sediment loadings. Other drivers of the degradation and loss of marine ecosystems include the expansion of coastal aquaculture, overfishing and destructive fishing methods, and, most recently, climate change (Crooks *et al.*, 2011). The loss and deterioration of coastal ecosystems represents the removal of carbon sinks and thus contributes to climate change, and denies future generations access to valuable ecosystem services. These include the rich services offered by coastal wetlands as well as others derived from access to clean fresh water supplies, clean air, and recreational space in the city proper. A rich literature on urban ecology provides important insights into how urban centers could be managed in such a way as to promote biodiversity and ecosystem health (Francis and Hoggart, 2008).

## 6.3 Impacts on economic systems

Environmental risks, particularly the new and uncertain pressures associated with climate change, pose a substantial threat to infrastructure, global trade, and security, with serious implications for human development and prosperity (UN-DESA, 2009). It has implications for both public and private spheres and at all scales, from individuals, households, and communities through to megacities, countries, and regions (Rosenzweig *et al.*, 2011). As a result of relative concentrations of high-value assets, high-income countries are likely to face greater absolute economic loss. However, these countries are also the ones whose citizens and businesses are more likely to have household insurance and access to capital investment allowing post-disaster reconstruction. The economic losses in developing countries, whilst in absolute terms may be lower, frequently have much longer-term impacts (Schipper and Pelling, 2006).

Climate change is inextricably linked to socioeconomic development trajectories due to its impact upon industries, human security, and health (Parry, 2007; IPCC, 2012). From this perspective the resilience of megacities will be the product of their ability to adapt to additional, non climate-driven pressures, the focus of Chapter 5. Resilience here refers to the ability of urban systems to withstand and recover from a broad range of stresses as a means to reduce baseline vulnerability to the additional climate stressors introduced in the previous sections. Measures could include seeking to improve water and sanitation supply and facilities, investing in environmental sustainability and energy efficiency. Such an approach demands a systemic framework of adaptation, based upon an understanding of the complex interactions and synergies between system components, particularly the influence of feedbacks and non-linear responses.

An exemplary selection of other sector-specific impacts is outlined below.

### 6.3.1 Energy

Most economic activity is dependent on energy and many of the hazards outlined in the above sections may disrupt urban energy supply and consumption infrastructures whose impacts may be felt across all geographical scales. An increase in storm surges, for example, could disrupt local power plants that provide energy for an entire region. Drought as much as extreme precipitation, and underlying long-term trends of change, can destabilize water catchment management for megacities that increasingly rely on surface water collection. As the 2011 tsunami in Japan has shown, urban centers can both generate risk out of the city and themselves become exposed to hazards when energy production is located on the coast even when this is geographically quite distant. There is no simple solution to these risks, but it does seem that conventional decision-support tools have not adequately captured the emerging regime of hazard and vulnerability to which they and surrounding areas are exposed.

Similarly, sustained higher temperatures can increase energy demand for cooling effects placing unplanned-for pressure on supply systems (Rosenzweig *et al.*, 2011). Both GEC and climate change impacts increase the vulnerability for energy supply systems such as thermoelectric power generation and distribution, natural gas distribution infrastructure as well as renewable power generation.

Forecasts predict increases in urban energy demand, largely stemming from non-OECD countries (IEA, 2012) and making up to 80 percent of all energy use by 2040.

Both projected increases in energy demand and the realization that placing energy production outside of urban areas may not diminish risk requires megacities to consider additional or alternative energy sources in the future. This transition could be facilitated by opportunities for renewable energy generation afforded by coastal locations. Options that may even be enhanced as a result of global environmental change – particularly wave-generated energy. The exploitation of these new opportunities is exemplified by the offshore wind farm development in the North Sea, which is driven by mitigation targets and could have a positive effect on the green economy in coastal cities including London around the North Sea and beyond (Bruns and Ohlhorst, 2011).

### *6.3.2 International maritime industry: ports*

Ports are key nodes in the global supply chain and often form a central part of coastal megacities' integration into world trade. Maritime transportation (shipping and freight) carries over 80 percent of the volume of world trade (UNCTAD, 2009). Similar to other sectors it is both a driver of global warming – shipping contributes around 3 percent of global CO2 emissions – and is affected by its impact. Yet, very little research has been done on ports (Esteban *et al.*, 2012). Nicholls *et al.*, (2008) assess coastal flood risks to 136 world port cities in terms of population and assets; particularly those cities in low-lying areas in the coastal zone, such as Shanghai and Rio de Janeiro, are vulnerable to inundations (Department of Climate Change and Energy Efficiency, 2011; Young, 2011).

Global seaports are exposed, to varying degrees, to the extreme weather events outlined above. The projected increase in frequency and intensity of tropical cyclones is particularly dangerous for breakwaters. Illustrating this challenge, Esteban *et al.* (2012) describes how wave-breaking patterns and depth is a key factor in breakwater design and the uncertainty on sea-level-rise projections coupled with local environmental change impact carries substantially higher costs. Investigating the effects of an increase in various weather extremes for Asian ports, i.e. port downtime, Esteban *et al.* (2012) concludes that future ports will need to be bigger to manage increased climate uncertainties. In their study on adaptation strategies for the 100 biggest seaports, Esteban *et al.* (2012) calculate the costs for port elevation or dyking, which raises important questions for public–private partnerships in key urban infrastructure investments. Further, maritime infrastructures and assets have long lifetimes and are based on past climate and weather patterns, highlighting the challenge to respond to future uncertainty and non-linearity.

Yet, port facilities are only part of a bigger picture and thus illustrate the protracted challenge for any mitigation and adaptation efforts. Inland infrastructure and supply chains not able to adapt to climate impacts, such as rail networks along the coast or inland water transport, may constitute additional vulnerabilities (Crist, 2011; Stenek, 2011). Further, the supply cargo itself may be affected by extreme events, coastal erosion, and the opening of new sea routes, and faces competitive risks due to its dependency on fossil fuels (though still only a third of the energy used by land-based transport).

### *6.3.3 Tourism and recreation*

The recreation and tourism industry constitutes a diverse group of businesses that includes the airline industry, travel agents, tour operators, rental car companies, convention organizers, and resorts/hotels. It is amongst the world's biggest industries and for many regions is the most important source of income (De Freitas, 2003: 45–54).

Exposure to sources of environmental risk has a direct impact on the tourist industry, in both the long and short term. The immediate costs as a result of damage to tourist infrastructure and facilities, in Thailand following the Asian tsunami in 2004, and in Cancun, Mexico, as a result of Hurricane Wilma in 2005, were enormous. In developing countries in particular, a major challenge is governing the process of reconstruction following a disaster. Reconstruction is in itself extremely expensive (Ehmer and Heymann 2008), and often becomes co-opted by private developers exploiting the space where businesses have been destroyed and the previous owners have insufficient capital to rebuild. Naomi Klein (2007) is highly critical of this practice – termed disaster capitalism – arguing that it represents a form of gentrifying land-grab, whereby land is transferred by default from local landowners to international corporations, with negative implications for local and national economies. She has since been criticized for simplistic treatment of the private sector, however, her central observation of the complexity and challenges associated with reconstruction, combined with the need for strong governance to ensure the equity of the rebuilding process, is astute.

The impacts of environmental risk on the tourism industry can be long lasting because of the persistence of risk perceptions (Birkland, *et al.*, 2006). The natural environment is often very important in determining the demand for tourism, especially for nature-based tourism. The impact of climate change on biodiversity and natural landscapes may have a negative influence on visitor numbers (Scott, 2008). Similarly De Freitas (2003: 45–54) observes that whilst the characteristics of weather and climate are not necessarily determinants of tourism, they nevertheless constitute an important factor in both financial terms for tourism operators and the personal experiences of tourists.

Longer heat waves and dry seasons associated with climate change could cause (or aggravate) natural disasters such as fires or water scarcity, which are negative factors for attractiveness. Already, longer dry periods are causing difficulties for water supplies and increasing costs in some tourist regions (e.g. southern Spain, North Africa) particularly for tourist facilities such as swimming pools or golf courses (Ehmer and Heymann, 2008). This is in addition to competition for water from agriculture, industry, and services, which is likely to promote the privatization of water supplies and/or drive up the cost of water.

## 7 Conclusion

Environmental risk is concentrated in coastal megacities as a result of the concentration of population and assets, combined with the multiplicity of hazards culminating at the interface between land and ocean. As demonstrated in this chapter, the range of hazards affecting coastal megacities is wide, and many regions are anticipated to be affected by the changing behavior and frequency of hazards associated with global warming.

Many coastal regions are already experiencing the effects of GEC. A major challenge, however, is to understand and manage the complexity of connections among different meteorological, oceanographic, and geophysical phenomena and the anthropogenic processes affecting the shoreline. As shown by the Australian Government, Department of Climate Change and Energy Efficiency (2011), changes to ocean temperature, currents, winds and extreme weather conditions are expected to impact on fisheries and aquaculture industries. Compounding this are pollutants emitted from megacities, which further damage fish populations as well as other coastal ecosystems such as wetlands, which are of important market and non-market value to urban populations. For this reason, the damage of ecosystems, by both urban and environmental pressures, constitutes a key feedback mechanism in the construction of city–coast risks.

However, vulnerability to environmental risk in megacities is dependent upon factors in addition to those increasing direct exposure, including urban poverty and structural social inequalities. Deficiencies in core urban infrastructure and public services poses serious challenges with regard to the adaptive capacity of megacities. The implementation of appropriate risk management interventions is difficult due to the diversity of interest groups, lack of coordination of planning decisions and jurisdictions (both across space and scale), and finally because of great uncertainty surrounding the timing and magnitude of GEC, particularly at city and local levels.

Long-term adaptation strategies, explored further in Chapter 5, are required to deal with the unavoidable impacts of climate change. In terms of research priorities, current literature on GEC impacts on coastal cities is scattered and fails to offer solid projections in terms of geographical patterns and variation in the timescale of (coastal) city regions specifically. It is therefore problematic to draw general conclusions about the fate of coastal megacities in high-risk contexts. These uncertainties call for research on those vulnerable coastal areas where climate change puts livelihoods at risk. In addition, greater emphasis should be put on broader linkages between social–ecological systems and the particular function of coastal cities.

## References


ADB (2011) *Climate change and food security in the Pacific. Rethinking the options*, Mandaluyong City: Asian Development Bank.

ADB (2012) *Addressing Climate Change and Migration in Asia and the Pacific*, Mandaluyong City: Asian Development Bank.

Adger, N.W. (2000) Social and ecological resilience: are they related? *Progress in Human Geography*, 24 (3), 347–364.

Adger W.N., Arnell N.W., Tompkins E.L., (2005) Successful adaptation to climate change across scales, *Global Environmental Change*, 15 (2), 77–86.

Antonini, G., Ardizzone, F., Cardinali, M., Galli, M., Guzzetti, F., and Reichenbach, P. (2002) Surface Deposits and Landslide Inventory Map at the Area Affected by the 1997 Umbria-Marche Earthquakes, *Boll Sociology Geology*, Special Volume (1), 843–853.

Ardizzone, F., Cardinali, M., Guzzetti, F., and Reichenbach, P. (2008) Landslide Hazard Assessment: Estimation and Risk Evaluation at the Basin Scale, *Proceedings of The First World Landslide Forum*, Tokyo. United Nations University, November, ISDR, 70–74.

Bai, X. (2008) Integrating Global Environmental Concerns into Urban Management: The Scale and Readiness Arguments, *Journal of Industrial Ecology*, 11 (2), 15–29.

Banerjee, A.V. and Duflo, E. (2007) The Economic Lives of the Poor, *Journal of Economic Perspectives*, 21 (1), 141–167.

Bayo, A.O. (2006) City planning, city growth and food security: The inevitable trinity in the Nigerian food equation, *Agricultural Journal*, 1 (3), 113–118.

Becker, A., Inoue, S., Fischer, M., and Schwegler, B. (2011) Climate change impacts on international seaports: knowledge, perceptions, and planning efforts among port administrators, *Climatic Change* DOI 10.1007/s10584-011-0043-7

Beddington J., Asaduzzaman, M., Clark, M., Fernández, A., Guillou, M., Jahn, M., Erda, L., Mamo, T., Van Bo, N., Nobre, C.A., Scholes, R., Sharma, R., and Wakhungu, J. (2012) Achieving food security in the face of climate change: Final report from the Commission on Sustainable Agriculture and Climate Change. CGIAR Research Program on Climate Change, Agriculture and Food Security (CCAFS), Copenhagen, www.ccafs.cgiar.org/commission (last accessed 27 March 2012).

Betsill, M. and Bulkeley, H. (2007) Looking Back and Thinking Ahead: A Decade of Cities and Climate Change Research, *Local Environment*, 12 (5), 447–456.

Bilham, R. (2009) The seismic future of cities, *Bulletin of Earthquake Engineering*, 7 (4), 839–887.

Bindoff, N.L., Willebrand, J., Artale, V., Cazenave, A., Gregory, J., Gulev, S., Hanawa, K., Le Quéré, C., Levitus, S., Nojiri, Y., Shum, C.K., Talley, L.D., and Unnikrishnan, A. (2007) Observations: Oceanic Climate Change and Sea Level In: *Climate Change 2007: The Physical Science Basis, Contribution of Working Group I to the Fourth Assessment Report of the Intergovernmental Panel on Climate*, Cambridge: Cambridge University Press, Cambridge, United Kingdom and New York, NY, USA. S. Solomon, D. Qin, M. Manning, Z. Chen, M. Marquis, K.B. Avery, M. Tignor, and H.L. Miller (eds).

Bird, M., Cowie, S., Hawkes, A., Horton, B., MacGregor, C., Ong, J.E., Hwai, A.T.S., Sa, T.T., and Yasin, Z. (2007) Indian Ocean tsunamis: environmental and socio-economic impacts in Langkawi, Malaysia, *The Geographical Journal*, 173 (2), 103–117.

Birkland, T., Herabat, P., Little, R., and Wallace, W. (2006) The impact of the December 2004 Indian Ocean Tsunami on tourism in Thailand, *Earthquake Spectra*, 22 (S3), S889–S900.

Black, R., Adger, W.N., Arnell, N.W., Dercon, S., Geddes, A., and Thomas, D. (2011) The effect of environmental change on human migration, *Global Environmental Change*, 21 (S1), S3–S11.

Black, R., Kniveton D., Schmidt-Verkerk, K., (2011) Migration and climate change: towards an integrated assessment of sensitivity. *Environment and Planning* A, volume 43(2) 431–450.

Brunetti, M.T., Peruccacci, S., Rossi, M., Luciani, S., Valigi, D., and Guzzetti F. (2010) Rainfall thresholds for the possible occurrence of landslides in Italy, *Natural Hazards Earth System Science*, 10, 447–458.

Bruns, E. and Ohlhorst, D. (2011) Wind Power Generation in Germany: a transdisciplinary view on the innovation biography, *The Journal of Transdisciplinary Environmental Studies*, 10, 45–67.

Cairncross, S. and Alvarinho, M. (2006) The Mozambique floods of 2000: Health impact and response, in: R. Few and F. Matthies (eds) *Flood hazards and health: responding to present and future risks*, London: Earthscan, 111–127.

Campbell-Lendrum, D. and Corvalán, C. (2007) Climate Change and Developing-Country Cities: Implications For Environmental Health and Equity, *Journal of Urban Health*, 84 (S1), 109–117.

Cashman, A.C. (2011) Case study of institutional and social responses to flooding: reforming for resilience? *Journal of Flood Risk Management*, 4 (1), 33–41.

Chau, K.T., Sze, Y.L., Fung, M.K., Wong, W.Y., Fong, E.L., and Chan, L.C.P. (2004) Landslide hazard analysis for Hong Kong using landslide inventory and GIS, *Computers and Geosciences*, 30, 429–443.

Church, J.A., Gregory, J.M., Huybrechts, P., Kuhn, M., Lambeck, C., Nhuan, M.T., Qin, D., Woodworth, P.L., (2001): Changes in sea level, in: Houghton, J.T., Ding, Y., Griggs, D.J., Noguer, M., Van der Linden, P.J., Dai, X., Maskell, K., and Johnson, C.A. (eds): *Climate Change 2001: The Scientific Basis: Contribution of Working Group I to the Third Assessment Report of the Intergovernmental Panel on Climate Change*, Cambridge University Press (Cambridge, New York), 639–694.

Church J.A., Gregory, J.M., Huybrechts, P., Kuhn, M., Lambeck, K., Nhuan, M.T., Qin, D., and Woodworth, P.L. (2001) Changes in Sea Level, in: G. Houghton and Ding (eds) *Climate Change 2001: The Scientific Basis*, Cambridge: Cambridge University Press.

Church, J.A., White, N.J., Hunter, J.R., and Lambeck, K. (2008) *A post-IPCC AR4 update on sea level rise*, Hobart: Antarctic Climate and Ecosystems Cooperative Research Centre.

Church, J.A., Woodworth, P.L., Aarup, T., and Wilson, W.S. (2010) *Understanding sea-level rise and variability*, London: Wiley-Blackwell.

Ciscara, J.-C., Iglesiasb, A., Feyenc, L., Szabóa, L., Van Regemortera, D., Amelunge, B., Nicholls, R.J., Watkissh, P., Christenseni, O.B., Christenseni, R.B., Dankersc, R., Garrotek, L., Goodess, C.M., Huntm, A., Morenoe, A., Richards, J., and Soriaa, A. (2011) Physical and economic consequences of climate change in Europe, *PNAS*, www.pnas.org/cgi/doi/10.1073/pnas.1011612108 (last accessed 27 March 2013).

Copertino, M.S., Garcia, A.M., Muelbert, J.H., and Garcia, C.A.E. (2010) Introduction to Special Issue on Climate Change and Brazilian Coastal Zone, *Pan American Journal of Aquatic Science*, 5 (2), I–VIII.

Crist, P. (2011) Transport Infrastructure and Network Adaptation to Climate Change: Issues and Strategies for Ports, in: *Ad Hoc Expert Meeting on Climate Change Impacts and Adaptation: A Challenge for Global Ports*, 29–30 September, Geneva: UNCTAD.

Crooks, S., Herr, D., Tamelander, J., Laffoley, D., and Vandever, J. (2011) Mitigating Climate Change through Restoration and Management of Coastal Wetlands and Near-shore Marine Ecosystems Challenges and Opportunities, *Marine Ecosystem Series*, 60578, Paper number 121, 01–59. World Bank. www.worldbank.org/environment/publications.

Decker, E., Elliott, S., Smith, F., Blake, D., and Sherwood Rowland, F. (2000) Energy and material flow through the urban ecosystem, *Annual Review of Environment and Resources*, 25, 685–740.

De Freitas, C.R. (2003) Tourism climatology: evaluating environmental information for decision making and business planning in the recreation and tourism sector, *International Journal of Biometeorology*, 48, 45–54.

De Gaetano, A.T. (2008) Predictability of seasonal east coast winter storm surge impacts with application to New York's Long Island, *Meteorological Applications*, 15, 231–242.

De Groot, R.S. (1992) *Functions of Nature: evaluation of nature in environmental planning, management and decision making*, Gronnigen: Wolters Noordhoff.

De Sherbinin, A., Schiller, A., and Pulsipher, A. (2007) The vulnerability of global cities to climate hazards, *Environment and Urbanization*, 19, 39–64.

Dilley, M. (2005) *Natural disaster hotspots: a global risk analysis*, Washington, DC: World Bank.

Dodman, D. (2009) Urban density and climate change. Analytical review of Interaction between urban growth trends and environmental changes, *UNFPA*, www.unfpa.org/webdav/site/global/users/schensul/public/CCPD/papers/Dodman%20Paper.pdf (last accessed 29 May 2013).

Easterling, D.R. (2000) Climate Extremes: Observations, Modeling, and Impacts, *Science*, 289 (5487), 2068–2074.

Ehmer, P. and Heymann, E. (2008) Climate Change and Tourism: where will the journey lead? *Energy and Climate Change Current Issues* (series), published 11 April 2008. Deutche Bank Research: Frankfurt (Germany). Available from: www.dbresearch.com (last accessed 29 May 2013).

Emanuel, K. (2005) Increasing Destructiveness of Tropical Cyclones over the past 30 years, *Nature*, 436, 686–688.

EPA (2010) *Sea level change*, http://epa.gov/climatechange/science/indicators/ (last accessed 27 March 2013).

Ericksen, P.J. (2008) What is the vulnerability of a food system to global environmental change? *Ecology and Society*, 13 (2), 14, www.ecologyandsociety.org/vol13/iss2/art14/ (last accessed 27 March 2013).

Ericksen, P., Thornton, P., Notenbaert, A., Cramer, L., Jones, P., and Herrero, M. (2011) Mapping hotspots of climate change and food insecurity in the global tropics, CGIAR Research Program on Climate Change, Agriculture and Food Security (*CCAFS*): Copenhagen (Denmark), Report number 5. Available online at: www.ccaf.cgiar.org.

Esteban, M., Nguyen, D.T., Takagi, H., and Shibayama, T. (2012) "Increase in Port Downtime and Damage in Vietnam Due to a Potential Increase in Tropical Cyclone Intensity in Filho, W. L. (eds), *Climate Change and the Sustainable Use of Water Resources*, Series: Climate Change Management, Part 2, Springer-Verlag Berlin Heidelberg, 101–125.

FAO (2010) *The State of Food Insecurity in the World 2010*, Rome: Food and Agriculture Organization.

Findlay, A. (2011) Migrants destinations in an era of environmental change, *Global Environmental Change*, 21 (1), S50–S58.

Fischer, E.M. and Schär, C. (2010) Consistent geographical patterns of changes in high-impact European heatwaves, *Nature Geoscience*, 3 (6), 398–403.

Foresight (2011) *Migration and Global Environmental Change*, Final Project Report, London: The Government Office for Science.

Francis, R. and Hoggart, S. (2008) Waste Not, Want Not: The Need to Utilize Existing Artificial Structures for Habitat Improvement along Urban Rivers, *Restoration Ecology*, 16 (3), 373–381.

Frich, P., Alexander, L.V., Della-Marta, P., Gleason, B., Haylock, M., Tank, A.M.G.K., and Peterson, T. (2002) Observed coherent changes in climatic extremes during the second half of the twentieth century, *Climate Research*, 19 (3), 193–212.

Friel, S., Hancock, T., Kjellstrom, T., McGranahan, G., Monge, P., and Roy, J. (2011) Urban Health Inequities and the Added Pressure of Climate Change: An Action-Oriented Research Agenda, *Journal of Urban Health*, 88 (5), 886–895.

Fuchs, R. (2010) Cities at risk: Asia's coastal cities in an age of climate change, *Asia Pacific Issues*, 96, 1–2.

Gemene, F. (2011) Why the numbers don't add up. A review of estimates and predictions of people displaced by environmental change, *Global Environmental Change*, 21, S41–S49.

Glickman, T.S., (2000) *Glossary of Meteorology*, published by American Meteorological Society. The University of Chicago Press: Chicago (USA).

Grimm, N.B., Faeth, S.H., Golubiewski, N.E., Redman, C.L., Wu, J., Bai, X., Briggs, J.M. (2008) Global Change and the Ecology of Cities, *Science*, 319 (5864), 756–760.

Grönwall, J.T., Mulenga, M., and McGranahan, G. (2010) *Groundwater, self-supply and poor urban dwellers: A review with case studies of Bangalore and Lusaka*, Human Settlements Working Paper Series Water and Sanitation #26, London: IIED.

Guzzetti, F. (2000) Landslide fatalities and evaluation of landslide risk in Italy, *Engineering Geology*, 58, 89–107.

Guzzetti, F., Esposito, E., Balducci, V., Porfido, S., Cardinali, M., Crescenzo, V., Fiorucci, F., Sacchi, M., Ardizzone, F., Mordini, A., Reichenbach, P., and Rossi, M. (2009) Central Italy Seismic Sequences-Induced Landsliding: 1997–1998 Umbria-Marche and 2008–2009 L'Aquila Cases. The Next Generation of Research on Earthquake-induced Landslides: An International Conference in Commemoration of 10th Anniversary of the Chi-Chi Earthquake, *Proceedings Conference*, Number 300, Taiwan: the National Central University.

Haines, A., Kovats, R.S., Campbell-Lendrum, D., and Corvalan, C. (2006) Climate change and human health: impacts, vulnerability, and mitigation, *The Lancet*, 367 (9528), 2101–2109.

Haines, A., McMichael, A.J., Smith, K.R., Roberts, I., Woodcock, J., Markandya, A., and Wilkinson, P. (2009) Public health benefits of strategies to reduce greenhouse-gas emissions: overview and implications for policy makers, *The Lancet*, 374 (9707), 2104–2114.

Hajat, S., Armstrong, B.G., Gouveia, N., and Wilkinson, P. (2005) Mortality displacement of heat-related deaths: A comparison of Delhi, Sao Paulo, and London, *Epidemiology*, 16, 613–620.

Hervás, J. (2007) Guidelines for Mapping Areas at Risk of Landslides in Europe, *Proceedings of the Experts Meeting*, October, Ispra: Institute for Environment and Sustainability Joint Research Centre (JRC).

Hewitt, K. (ed.) (1983) *Interpretations of Calamity: from the viewpoint of human ecology*, Allen Unwin: London.

Houze, R., Rasmussen, K., Medina, S., Brodzik, S., and Romatschike, U. (2011) Anomalous atmospheric events leading to the summer 2010 floods in Pakistan, *Bulletin of the American Meteorological Society*, 92(3), 291–298.

Hugo, G. (2011) Future demographic change and its interactions with migration and climate change, *Global Environmental Change*, Supplement 1, S21–S33.

Huppert, H.E. and Sparks, R.S.J. (2006) Extreme Natural Hazards: population growth, globalization and environmental change, *Philosophical Transactions of the Royal Society*, 364, 1875–1888.

IEA (2012) *World Energy Outlook 2012*, www.worldenergyoutlook.org/ (last accessed 27 March 2013).

Ingram, J. and Brklacich, M. (2002) Global Environmental Change and Food Systems – GECAFS: a new interdisciplinary research project, *die Erde*, 113 (4), 427–435.

Ingram, J., Ericksen, P., and Liverman, D. (eds) (2010) *Food Security and Global Environmental Change*, London: Earthscan.

IPCC (2012) Managing the Risks of Extreme Events and Disasters to Advance Climate Change Adaptation, in: C.B. Fields, V. Barros, T.F. Stocker, D. Qin, D.J. Dokken, K.L. Ebi, M.D.

Mastrandrea, K.J. Mach, G.-K. Plattner, S.K. Allen, M. Tignor, and P.M. Midgley (eds) *A Special Report of Working Groups I and II of the Intergovernmental Panel on Climate Change*, Cambridge: Cambridge University Press.

Jackson, J. (2006) Fatal attraction: living with earthquakes, the growth of villages into megacities, and earthquake vulnerability in the modern world, *Philosophical Transactions of the Royal Society A*, 364, 1911–1925.

Jevrejeva, S., Moore, J.C., Grinsted, A., Woodworth, P.L. (2008) Recent global sea level acceleration started over 200 years ago? *Geophysical Research Letters*, 35, L08715.

Jha, A., Lamond, J., Bloch, R., Bhattacharya, N., Lopez, A., Papachristodoulou, N., Bird, A., Proverbs, D., Davies, J., and Barker, R. (2011) Five Feet High and Rising: Cities and Flooding in the 21st Century, *Policy Research Working Paper 5648*, published by The World Bank DOI: 10.1596/1813-9450-5648. The World Bank East Asia and Pacific Region Transport, Energy and Urban Sustainable Development Unit.

Kasperson, R.E. and Kasperson, J. X. (2001) *Climate change, vulnerability and social justice*, Stockholm: Stockholm Environment Institute.

Kennedy, V.S., Twilley, R.R., Kleypas, J.A., Cowan, J.H. Jr., James, H., and Hare, S.R. (2002) "Coastal and marine ecosystems & Global climate change. Potential Effects on U.S. Resources", Pew Center on Global Climate Change: Arlington, Virginia (USA), p. 64.

Klein, N. (2007) *The Shock Doctrine: the rise of disaster capitalism*, London: Penguin.

Klein, R.J.T., Nicholls, R.J., and Thomalla, F. (2003) The Resilience of Coastal Megacities to Weather-Related Hazards, in: M. Arnold and A. Carling (eds) *Building Safer Cities: The Future of Disaster Risk*, Disaster Risk Management Series No. 3, Washington, DC: The World Bank, Disaster Management Facility, 101–120.

Klein Tank, A.M.G., Peterson, T.C., Quadir, D.A., Dorji, S., Zou, X., Tang, H., Santhosh, K., Joshi, U.R., Jaswal, A.K., Kolli, R.K., Sikder, A.B., Deshpande, N.R., Revadekar, J.V., Yeleuova, K., Vandasheva, S., Faleyeva, M., Gomboluudev, P., Budhathoki, K.P., Hussain, A., Afzaal, M., Chandrapala, L., Anvar, H., Amanmurad, D., Asanova, V.S., Jones, P.D., New, M.G., and Spektorman T. (2006) Changes in Daily Temperature and Precipitation Extremes in Central and South Asia, *Journal of Geophysical Research*, Volume 111 Issue D16. 105, p. 1–8.

Knutson, T.R., John, L., McBride, J.L., Chan, J., Emanuel, K., Holland, G., Landsea, C., Held, I., James, P., Kossin, J.P., Srivastava, A.K., and Masato Sugi, M. (2010) Tropical cyclones and climate change, *Nature Geoscience*, 3, 157–163.

Kovats, R.S. and Akhtar, R. (2008) Climate, climate change and human health in Asian cities, *Environment and Urbanization*, 20 (1), 165–175.

Kovats, R.S. and Hajat, S. (2008) Heat stress and public health: A critical review, *Annual Review of Public Health*, 29, 41–55.

Kreimer, A., Arnold, M., and Carlin, A. (2003) *Building Safer Cities: The Future of Disaster Risk*, World Bank: Washington, D.C. (USA). World Bank, Disaster Risk Management Series, vol. 3, 1–299.

Krishnamurthy, V. and Kirtman, B. (2003) Variability of the Indian Ocean: Relation to monsoon and ENSO, *Quarterly Journal of the Royal Meteorological Society*, 129, December, 1623–1646.

Kunkel, K.E., Liang, X.-Z., and Zhu, J. (2010) Regional Climate Model Projections and Uncertainties of U.S. Summer Heat Waves, *Journal of Climate*, 23 (16), 4447–4458.

Lilleør, H.B. and Van den Broeck, K. (2011) Economic drivers of migration and climate change in LDCs, *Global Environmental Change*, Supplement 1, S70–S81.

Lipper, L., Mann, W., Meybeck, A., and Sessa, R. (2010) *'Climate-smart' agriculture: policies, practices and financing for food security, adaptation and mitigation*, Rome: Food and Agriculture Organization, United Nations.

Lopez, A.D., Mathers, C.D., Ezzati, M., Jamison, D.T., and Murray, C.J.L. (2006) Global and regional burden of disease and risk factors: systematic analysis of population health data, *The Lancet*, 367 (9524), 1747–1757.

Malamud, B.D., Turcotte, D.L., Guzzetti, F., and Reichnbach, P. (2004) Landslide, earthquakes, and erosion, *Earth and Planetary Science Letter*, 229, 45–59.

Manton, M.J., Della-Marta, P.M., Haylock, M.R., Hennessy, K.J., Nicholls, N., Chambers, L.E., Collins, D.A., Daw, G., Finet, A., Gunawan, D., Inape, K., Isobe, H., Kestin, T.S., Lefale, P., Leyu, C.H., Win, T.L., Maitrepierre, L., Ouprasitwong, N., Page, C.M., Pahalad, J., Plummer, N., Salinger, M.J., Suppiah, R., Tran, V.L., Trewin, B., Tibig, I., and Yee, D. (2001) Trends in extreme daily rainfall and temperature in Southeast Asia and the South Pacific: 1961–1998, *International Journal of Climatology*, 21 (3), 269–284.

Martelloni, G., Segoni, S., Fanti, R., and Catani, F. (2011) Rainfall thresholds for the forecasting of landslide occurrence at regional scale, *Landslides*, http://rd.springer.com/article/10.1007/s10346-011-0308-2 (last accessed 27 March 2013).

Matuschke, I. (2009) *Rapid urbanization and food security: Using food density maps to identify future food security hotspots*, Rome: Food and Agriculture Organization of the United Nations.

McGranahan, G., Balk, D., and Anderson, B. (2007) The rising tide: assessing the risks of climate change and human settlements in low elevation coastal zones, *Environment and Urbanization*, 19 (1), 17–37.

McLeman, R. and Smit, B. (2006) Migration as an adaptation to climate change, *Climatic Change*, 76, 31–53.

MEA (2005) *Millennium Ecosystem Assessment*, http://millenniumassessment.org/en/index.aspx (last accessed 27 March 2013).

Meehl, G., Stocker, T., Collins, W., Friedlingstein, P., Gaye, A., Gregory, J., Kitoh, A., Knutti, R., Murphy, J., Noda, A., Raper, S., Watterson, I., Weaver, A., and Zhao, Z. (2007) *Global Climate Projections. Climate Change 2007: The Physical Science Basis. Contribution of Working Group I to the Fourth Assessment Report of the Intergovernmental Panel on Climate Change*, Cambridge: Cambridge University Press.

Montgomery, M. and Farrell, B. (1993) Tropical Cyclone Formation, *Journal of the Atmospheric Sciences,* 50 (2), 285–310.

Morgan, O., Ahern, M., and Cairncross, S. (2005) Revisiting the Tsunami: Health Consequences of Flooding, *PLoS Medicine, online journal*, 2(6), e184.

Muehe, D. (2010) Brazilian coastal vulnerability to climate change, *Pan American Journal of Aquatic Science,* 5 (2), 173–183.

Neu, U. (2009) Influence of Global Warming on Extratropical Storms, *Swiss Re*, www.scribd.com/doc/54511974/Influence-of-Global-Warming-on-Extra-Tropical-Storms (last accessed 27 March 2013).

Nicholls, R.J. (2004) Coastal flooding and wetland loss in the 21st century: changes under the SRES climate and socio-economic scenarios, *Global Environmental Change*, 14 (1), 69–86.

Nicholls, R.J., Hanson, S., Herweijer, C., Patmore, N., Hallegatte, S., Corfee-Morlot, J., Chateau, J., and Muir-Wood, R. (2008) *Ranking Port Cities with High Exposure and Vulnerability to Climate Extremes: Exposure Estimates*, OECD Environmental Working paper 1. www.oecd-ilibrary.org/environment/ranking-port-cities-with-high-exposure-and-vulnerability-to-climate-extremes_011766488208

Nicholls, R.J., Wong, P.P., Burkett, V.R., Codignotto, J.O., Hay, J.E., McLean, R.F., Ragoonaden, S., Woodroffe, C.D. (2007) Coastal systems and low-lying areas, in: M.L. Parry, O.F. Canziani, J.P. Palutikof, P.J. van der Linden and C.E. Hanson (eds) *Climate Change 2007: Impacts, Adaptation and Vulnerability. Contribution of Working Group II to the*

*Fourth Assessment Report of the Intergovernmental Panel on Climate Change*, Cambridge: Cambridge University Press, 315–356.

NOAA, National Oceanic and Atmospheric Administration (2011a) *La niña*, www.elnino.noaa.gov/lanina_new_faq.html (last accessed 27 March 2013).

NOAA, National Oceanic and Atmospheric Administration (2011b) *Storm Surge*, www.nhc.noaa.gov/surge/ (last accessed 27 March 2013).

NOAA, National Oceanic and Atmospheric Administration (2011c) *Tsunami*, www.tsunami.noaa.gov/ (last accessed 27 March 2013).

NOAA, National Oceanic and Atmospheric Administration (2011d) *Tsunami model*, www.tsunami.noaa.gov/research_modeling.html (last accessed 27 March 2013).

O'Brien, K. (2006) Are we missing the point? Global environmental change as an issue of human security, *Global Environmental Change*, 16 (1), 1–3.

O'Brien, K. and Leichenko, R. (2006) Climate Change, Equity and Human Security, *Die Erde*, 137 (3), 223–240.

Parker, D. and Mitchell, J.K. (1995) Disaster vulnerability of megacities: an expanding problem that requires re-thinking and innovative responses, *GeoJournal*, 37, 295–301.

Parnell, S., Simon, D., and Vogel, C. (2007) Global environmental change: conceptualising the growing challenge for cities in poor countries, *Area*, 39 (2), 357–369.

Parry, M.L., Canziani, O.F., Palutikof, J.P., van der Linden, P.J., and Hanson C.E. (2007) Impacts, Adaptation and Vulnerability, *Contribution of Working Group II to the Fourth Assessment Report of the Intergovernmental Panel on Climate Change 2007*, Cambridge: Cambridge University Press.

Pelling, M. and Wisner, B. (eds) (2009) *Disaster Risk Reduction: cases from urban Africa*, London: Earthscan.

Perch-Nielsen, S., Bättig, M.B., and Imboden, D. (2008) Exploring the link between climate change and migration, *Climatic Change*, 91 (3–4), 375–393.

Piguet, E. (2008) *Climate Change and Forced Migration*. New Issues in Refugee Research, Research Paper # 153, Geneva: United Nations High Commissioner for Refugees, www.unhcr.org/47a316182.html (last accessed 27 March 2013).

Rahimzadeh, F., Asgari, A., and Fattahi, E. (2009) Variability of extreme temperature and precipitation in Iran during recent decades, *International Journal of Climatology*, 29 (3), 329–343.

Raleigh, C. (2011) The search for safety: the effects of conflict, poverty and ecological influences on migration in the developing world, *Global Environmental Change*, advance online publication.

Redclift, M. (1992) Sustainable development and global environmental change, *Global Environmental Change*, 2 (1), 32–42.

Reichenbach, P. and Gunther, A. (2010) Methods and strategies to evaluate landslide hazard and risk, *Natural Hazards and Earth System Sciences*, 10, 2197–2198.

Rodolfo, K.S. and Siringan, F.P. (2006) Global sea-level rise is recognised, but flooding from anthropogenic land subsidence is ignored around northern Manila Bay, Philippines, *Disasters*, 30, 118–139.

Rosenzweig C., Solecki, W., and De Gaetano, A. (2011) Responding to Climate Change in New York State: The ClimAID Integrated Assessment for Effective Climate Change Adaptation in New York State, *Final Report*. NYSERDA-Report 11-18. New York State Energy Research & Development Authority (NYSERDA): New York (USA).

Satterthwaite, D., Huq, S., Pelling, M., Reid, H., and Lankao, R. (2007) *Adapting to Climate Change in Urban Areas: The Possibilities and Constraints in Low- and Middle-Income Countries*, Human Settlements Climate Change and Cities Discussion Series 1, London: IIED.

Schipper, L. and Pelling, M. (2006) Disaster risk, climate change and international development: scope for, and challenges to, integration, *Disasters*, 30 (1), 19–38.

Scott, D., Amelung, B., Becken, S., Ceron, J. P, Gössling, S., Peeters, P., (2008) *Climate Change and Tourism – Responding to Global Challenges*, Madrid: World Tourism Organization and United Nations Environment Programme.

Semenza, J.C., Rubin, C.H., Falter, K.H., Selanikio, J.D., Flanders, D., Howe, H.L., and Wilhelm, J.L. (1996) Heat-related deaths during the July 1995 heat wave in Chicago, *New England Journal of Medicine*, 335: 84–90.

Seto, K. (2011) Exploring the dynamic of migration to mega-delta cities in Asia and Africa: contemporary drivers and future scenarios, *Global Environmental Change*, 21 (1), S94–S107.

Shepherd, J.M. and Knutson, T. (2007) The current debate on the linkage between global warming and hurricanes, *Geography Compass*, 1 (1), 1–24.

Simon, D. and Leck, H. (2010) Urbanizing the global environmental change and human security agendas, *Climate and Development*, 2 (3), 263–275.

Slenning, B.D. (2010) Global Climate Change and Implications for Disease Emergence, *Veterinary Pathology Online*, 47 (1), 28–33.

Solomon, S., Qin, D., Manning, M., Chen, Z., Marquis, M., Averyt, K.B., Tignor, M., and Miller, H.L. (2007) The Physical Science Basis, *Contribution of Working Group I to the Fourth Assessment Report of the Intergovernmental Panel on Climate Change – 2007*, Cambridge: Cambridge University Press.

Spickett, J.T., Brown, H.L., and Rumchev, K. (2011) Climate Change and Air Quality: The Potential Impact on Health, *Asia-Pacific Journal of Public Health*, 23 (2 suppl), 37S–45S.

State of California Seismic Safety Commission (2005) The Tsunami Threat to California, *Finding and Recommendations on Tsunami Hazards and Risk*, State of California Seismic Safety Commission. Sacramento, California (USA).

Stenek, V. (2011) *Climate Risk and Ports*, UNCTAD, http://unctad.org/sections/wcmu/docs/AHM2011_2_08_Stenek_en.pdf (last accessed 27 March 2013).

Stern, N. (2007) *The economics of climate change: the Stern review*, Cambridge: Cambridge University Press.

Stern, P.C., Young, O.R., and Druckman, D. (eds) (1992) *Global environmental change: understanding the human dimensions*, Washington, DC: National Academy Press.

Sundkvist, A., Milestad, R., and Jansson, A. (2005) On the importance of tightening feedback loops for sustainable development of food systems, *Food Policy*, 30, 224–239.

Sverdlik, A. (2011) Ill-health and poverty: a literature review on health in informal settlements, *Environment and Urbanization*, 23 (1), 123–155.

Syvitski, J., Kettner, A.J., Overeem, I., Hutton, E.W.H., Hannon, M.T., Brakenridge, G.W., Day, J., Vörösmarty, C., Saito, Y., Giosan, L., and Nicholls, R.J. (2009) Sinking deltas due to human activities, *Nature: Geoscience*, 2, 681–686.

Tacoli, C. (2009) Crisis or adaptation? Migration and climate change in a context of high mobility, *Environment and Urbanization*, 21 (2), 513–525.

UN-DESA (2009) *World Urbanization Prospects: The 2009 Revision*, UN DESA Population Division, New York: United Nations, http://esa.un.org/unpd/wup/index.htm (last accessed 27 March 2013).

UN-HABITAT (2009) *Slums: levels and trends, 1990–2005*, Nairobi: UN-HABITAT.

UNCTAD (2009) Multi-Year Expert Meeting on Transport and Trade Facilitation: Maritime Transport and the Climate Change Challenge, *Summary of Proceedings*, Geneva: United Nations Conference on Trade and Development.

UNISDR (2011) *Global Assessment Report 2011: Revealing Risk, Redefining Development*, Geneva: UNISDR.

UNHCR (2012) *The State of the World's Refugees*, Geneva: UNHCR.

UNPF (2007) *The State of the World Population: Unleashing the Potential of Urban Growth*, New York: United Nations Population Fund.

Wang, C. (2001) Atlantic Climate Variability and Its Associated Atmospheric Circulation Cells, *Journal of Climate*, 15, 1516–1536.
Wang, X.Y., Barnett, A.G., Yu, W., FitzGerald, G., Tippett, V., Aitken, P., Neville, G., McRae, D., Verrall, K., and Tong, S. (2011) The impact of heatwaves on mortality and emergency hospital admissions from non-external causes in Brisbane, Australia, *Occupational and Environmental Medicine*, 69 (3), 163–169.
Warner, K., Ehrhart, C., de Sherbinin, A., Adamo, S., and Chai-Onn, T. (2009) *In Search of Shelter: Mapping the Effects of Climate Change on Human Migration and Displacement*, www.careclimatechange.org (last accessed 27 March 2013).
Webster, P., Holland, G., Curry, J., Chang, H. (2005) Changes in Tropical Cyclone Number, Duration and Intensity in a Warming Environment, *Science*, 309, 1844–1846.
Webster, P., Toma, V., and Kim, H-M. (2011) Were the 2010 Pakistan floods predictable? *Geophysical Research Letters*, 38, 1–5.
Weisse, R. and von Storch, H. 2009: Marine Climate and Climate Change. Storm Surges. Springer Praxis Books: Chichester (UK). 200pp.
Wisner, B., Cannon, T., Blaikie, P., and David, I. (2004) *At Risk*, London: Routledge.
Wood, S., Ericksen, P., Stewart, B., Thornton, P., and Anderson, M. (2010) Lessons Learned from International Assessments, in: J. Ingram, P. Ericksen, and D. Liverman (eds) *Food Security and Global Environmental Change*, London: Earthscan, 46–62.
World Bank (2010) *Climate Risks and Adaptation in Asian Coastal Megacities: A synthesis report*, Washington, DC: World Bank.
World Bank (2012) *Integrated urban flood risk management for the 21st century: a practitioner's handbook*, Washington, DC: World Bank.
Worm, B., Barbier, E., Beaumont, N., Duffy, J., Folke, C., Halpern, B., Jackson, J., Lotze, H., Micheli, F., Palumbi, S., Sala, E., Selkoe, K., Stachowicz, J., and Watson, R. (2006) Impacts of biodiversity loss on ocean ecosystem services, *Science*, 314, 787–790.
Young, A.F. (2011) Sea Level Rise and Flood Risk in Rio de Janeiro City: Challenges of Global Climate Change. Opportunities and Challenges for Sustainability in an Urbanizing World: Selections from the UGEC. 2010. Conference. *UGEC Viewpoints*, 5, April, 33–36.
Zahid, M. and Rasul, G. (2012) Changing trends of thermal extremes in Pakistan, *Climatic Change*, 113 (3–4), 883–896.
Zimov, S., Schuur, E., and Chapin, S. (2006) Permafrost and the Global Carbon Budget, *Science*, 312, 1612–1613.

Chapter 4

# Contributions of coastal megacities to environmental changes at regional and global scales

Lead author: *C.S.B. Grimmond (King's College London, UK)*

Contributing authors: *Alexander Baklanov (Danish Meteorological Institute, Denmark), Martin Best (Met Office, UK), Richard Betts (Met Office, UK), Johannes Feddema (University of Kansas, USA), Pete Falloon (Met Office, UK), Gerd Folberth (Met Office, UK), Rob Francis (King's College London), Teddy Holt (Naval Research Laboratory, USA), Trisha Jackson (South Dakota State University, USA), Elizabeth Larson (King's College London, UK), Mark McCarthy (Met Office, UK), Talmor Meir (Stevens Institute of Technology, USA), Naho Mirumachi (King's College London, UK), Keith Oleson (NCAR, USA), Philip Orton (Stevens Institute of Technology, USA), Julie Pullen (Stevens Institute of Technology, USA)*

## 1 Introduction

In Chapter 1, 23 megacities with populations in excess of 10 million were identified, there are a further 30 with populations greater than seven million (UN, 2011). More than two-thirds of these megacities are coastal (Plate 6) and 80 per cent have a coastal influence. By 2025 the population in megacities in the coastal zone is expected to increase from 220.7 million to 301.7 million people (UN, 2010). These cities are experiencing growth at a rapid rate, above that of other urban areas (Seto *et al.*, 2011).

The direct and indirect impacts of cities on the environment at the local, regional and global scales are well established; examples include the well-documented urban heat island (UHI), poorer air quality within the city and downwind, enhanced emissions of greenhouse gases (GHG) as cities account for a large (but debated) fraction of global primary energy demand (IEA, 2008; Satterthwaite, 2008), depletion and pollution of ground and surface water, and demands for goods and resources that drive regional land-use change.

The impact of coastal megacities can be considered in terms of the significant resources diverted into them to sustain them. These 'inputs' – of water, energy, food etc. – alter the regions from which they are derived – at multiple scales. Second, 'outputs' from these cities impact regions 'downwind' or 'downstream'. These impacts and their footprints can be both observed and modelled. In this chapter we consider some examples of the impact of megacities regionally and globally. The approach is not exhaustive but provides examples of impact on weather, climate, air quality, water and ecology.

## 2 The effect on the atmosphere

The impacts of megacities on regional and global climate and weather depend on their locations, the climate they are exposed to, building design and urban planning, building-construction materials, and level of socioeconomic development. While these directly influence GHG emissions by determining traffic patterns, material flows and transport-related energy consumption (Mills, 2007), equally they influence the urban climate, water consumption, energy usage and citizen well-being. As outlined in Chapter 2, urban geography varies widely due to regional climatic conditions, topography, socioeconomic factors, available resources and cultural and aesthetic values.

The following subsections address the distinct physical processes characteristic of coastal megacities, and the resulting weather, climate and air-quality issues that impact regional- and global-scale processes. As computing capacity is only beginning to capture urban processes at global scales the number of completed modelling studies is relatively small.

### *2.1 Urban–coastal interactions and transport processes*

Micro- and meso-scale atmospheric circulations and boundary layer dynamics as well as atmospheric pollution processes for megacities in coastal zones are complex. Coastal meteorology is subject to sharp changes in heat, moisture and momentum transfer through contrasts in heating, surface roughness and moisture supply between the ocean and different land cover types. Many coastal cities are located near or across different elevation zones, often bounded inland by mountains (e.g., Los Angeles or Hong Kong). There are also changes in surface radiation related to coastal clouds, haze and fog; and meteorological phenomena such as sea/land breezes, related thunderstorms, coastal fronts, orographically trapped winds and low-level jets. Thus, thermal atmospheric circulations around coastal megacities involve at least three interacting nonlinear mechanisms within the planetary boundary layer: (1) urban effects and the related UHI, (2) coastal phenomena, e.g. sea/land breezes, and (3) (often) katabatic and anabatic slope winds and mountain circulations.

As outlined in detail in Chapter 2, a wide range of urban and coastal features influence atmospheric flow, turbulence and micro-climate. In response, the transport, dispersion and deposition of atmospheric pollutants are modified both within the urban area and downwind (e.g. through acid rains). The release of pollutants (including aerosols) impact radiative transfer, convective processes and clouds and precipitation formation; warm air rising from a city may interact with moist air from the nearby sea, releasing precipitation downwind of the city. The warm air and precipitation can affect winds for hundreds of kilometres.

Given the dense nature of many coastal megacities, there are also pronounced urban influences on anthropogenic heat emissions and surface roughness. Anthropogenic heat fluxes for megacities can be very high (50–500 W $m^{-2}$ e.g. Flanner, 2009; Allen *et al.*, 2011 globally; Iamarino *et al.*, 2011 for London; Quah and Roth, 2012 for Singapore; or locally reaching 1500 W $m^{-2}$ Ichinose *et al.*, 1999 in a small area of Tokyo). Under clear skies and light wind conditions, megacities can be more than 10°C warmer than surrounding rural environments, called the UHI intensity (Oke, 1982).

Although the interaction of sea breezes, urban drag and the UHI has been studied (e.g. Ado, 1992; Yoshikado, 1994; Melas *et al.*, 1995; Yoshikado and Tsuchida, 1996, Khan and Simpson, 2001; Ohashi and Kida, 2002; Gedzelman *et al.*, 2003; Freitas *et al.*, 2007; Thompson *et al.*, 2007; Dandou *et al.*, 2009, Calmet and Mestayer, 2009) the role of the UHI on the speed of the sea-breeze front remains uncertain (Freitas *et al.*, 2007). The enhanced drag generated by the city reduces the speed of the sea-breeze front (Loose and Bornstein, 1977; Bornstein and Thompson, 1981; Martilli, 2003; Thompson *et al.*, 2007) and the UHI circulation can cause the sea breeze to stop, resulting in flow convergence. The UHI can cause sea breezes to occur hours earlier (Khan and Simpson, 2001), whereas the sea breeze can change the spatial characteristics of the UHI (Gedzelman *et al.*, 2003; Dandou *et al.*, 2009).

Thus there is a delicate balance between the UHI and sea-breeze circulations (Yoshikado, 1992, 1994; Mestayer *et al.*, 2005). Given coastal cities are often located in topographically complex terrain (e.g. multiple coastal boundaries), multiple sea-breeze boundaries (e.g. New York City, Colle *et al.*, 2003), slope winds and mountain circulations (e.g. Athens, Melas *et al.*, 1995) may interact with the UHI and sea-breeze circulations.

### 2.2 Impact on regional weather

The fine-scale atmospheric flow field within the coastal urban boundary layer (UBL), and its subsequent interaction and effect on transport and dispersion of air pollutants and aerosols are not well understood and are highly dependent on local terrain/orography. Accurate assessments of the 'spatial footprint' of a megacity are difficult due to the complexity of interactions between the UBL with its surrounding ("rural") environment. These atmospheric flow changes can make important contributions to regional climate change through their impact on: cloud microphysical processes through modified low-level atmospheric convergence; large aerosol radiative forcing of the atmosphere and surface (solar radiation dimming); and surface energy balances via modifications to coastal vertical circulations (e.g., land–sea breezes).

High-resolution numerical weather prediction simulations for two coastal mega-cities, New York City (NYC) and Tokyo, illustrate the potential impact large urban areas could have on regional weather through modifications to the fine-scale atmospheric flow within the UBL (Holt and Pullen, 2007; Thompson *et al.*, 2007; Holt *et al.*, 2009). Numerical simulations for the densely urbanised area of Manhattan (NYC) indicate that the building height and areal plan fraction (or urban footprint) of the urban landscape are equally important in modifying the UBL wind speed. The evolution of a 2011 high-heat event in the NYC region and its interaction with the sea-breeze circulation has been examined using a ~1 km model (Plate 7). The urbanised version of the *Coupled Ocean-Atmosphere Mesoscale Prediction System (COAMPS)* accounts for city heating and morphology effects on the atmosphere through the implementation of single and multilayer urban canopy parameterization options (Holt and Pullen, 2007). The *National Weather Service's 12 km resolution North American Model (not urbanized)* does not show the fine-scale urban footprint nor create the sharp wind and air temperature gradients of COAMPS that are confirmed by observations in the coastal zone (Plate 7). Furthermore, sensitivity studies utilizing this model configuration demonstrate the positive impact of incorporating ocean information. In particular, Pullen *et al.* (2007b) link COAMPS to a local ultra-high resolution ocean circulation model of the wider New York region. This ocean model produces realistic patterns of upwelling colder water in the harbour area, along the New Jersey coast and Long Island Sound with near-surface air temperatures cooled by 1–2 °C and reduced near-surface wind speeds (15–20 per cent). In contrast the control simulation (without the high-resolution ocean model input) less closely followed the observations. More research into the links between urban structure and offshore marine boundary layer modification are needed.

Simulations, in which passive tracers are released within the UBL to emulate the effect of aerosol and pollutants, indicate that the extent of the tracer footprint varies tremendously depending upon the extent and characterization of the megacity. The land–sea breeze circulation is often a dominant wind-flow pattern for coastal regions, significantly influencing the regional climate. In simulations with the megacity removed (termed 'anilhilation') the land–sea breeze circulations are generally accelerated horizontally downwind, and decelerated vertically (as it is more stable with reduced mechanical mixing), resulting in generally increased tracer concentrations near the surface. Thus, the primary impact of the megacity is to spatially diffuse the tracer, both horizontally and vertically – reducing the near-surface concentration at any one point within the UBL. The location of the horizontal front of the sea-breeze circulation is very sensitive to megacity structure and the development and strength of the UHI. From ensemble numerical weather prediction simulations for Tokyo (Holt *et al.*, 2009) the location covered the ~30-kilometre width of the urban centre (Plate 8). In the Tokyo region the prevalence of sea-breeze conditions modifies the UHI and can mitigate temperature extremes. The reduced and inland displaced summertime UHI, relative to winter, is associated with stronger summertime sea breezes (Ichinose *et al.*, 1999). Numerical simulations of offshore and urban tracer releases (Plate 9) vary considerably from small-scale features such as enhanced anthropogenic heating, variations in building heights and coastally upwelled water. With the inclusion of realistic sea surface temperature variations (i.e., warm Tokyo Bay or local upwelling) the simulation shows subtle variations in sea breeze with dramatic impact on tracer distribution (Plate 9).

These results can have tremendous implications for regional weather because of the interaction and interdependence of atmospheric dynamic and thermodynamic parameters. These interactions are driven simply through UBL wind-flow changes.

### 2.3 Influence on regional climate

Similarly, models allow the climate impacts of the urban land surface on the wider environment both within and beyond the city to be estimated (e.g., Trusilova *et al.*, 2008; Oleson *et al.*, 2011). To account for urban processes, the United Kingdom's Met Office Unified Model (UM) currently includes a simple slab representation (Best *et al.*, 2006) as well as a more sophisticated scheme separating roofs and street canyons to parameterize urban thermal properties from simple geometric form (Porson *et al.*, 2010). The coupled atmosphere–urban-soil scheme provides two-way interactions for the exchange of momentum, heat, water and carbon between urban areas and the wider environment. Such two-way coupling is essential to describe the urban environmental impact and how the dynamical atmosphere may transport urban influences outside the confines of the city.

From two 20-year simulations that use the Hadley Centre Regional Climate Model Version 3 (HadRM3) on a rotated pole grid with resolution 0.22° (approximately 25 kilometres) over a European domain, the coastal cities of the UK and Northern Europe are apparent, as are other important urban agglomerations such as Paris, Berlin and the Rhine-Ruhr valley region (Plate 10) (McCarthy *et al.*, 2011; McCarthy and Sanderson, 2011). The urban micro-climate in this model is largely a result of increased surface sensible heating, reduced evaporation and increased surface drag (Best *et al.*, 2006). This results in higher temperatures and reduced humidity (not shown). Further, the additional warming marginally increases the amount of convective precipitation over the cities, although additional non-urban precipitation variability has been captured by the experiments presented in Plate 10. Wind speeds are reduced and there is an increase in plant productivity per unit area of vegetation for this part of Northern Europe in the urbanised run compared to non-urban, although the replacement of vegetation with urban surface outweighs these increases to reduce the overall land carbon sink (not shown).

Plate 10 illustrates aspects of regional urban climate feedbacks as they interact with the energy, water and carbon cycles, all of which have been previously documented as urban climatic features within the wider literature (e.g., Shepherd, 2002; Arnfield, 2003; Trusilova and Churkina, 2008), but have received limited attention in long-integration climate modelling. The continued refinement and evaluation of the urban land surface scheme is an essential requirement in order to ensure that such models provide the right answer for the right physical reason (Grimmond *et al.*, 2011). Such schemes provide a vital climate–land surface connection for inclusion in climate research.

### *2.4 Climate change – global perspective*

To date very few Global Climate Model (GCM) studies have accounted for urban areas (McCarthy *et al.*, 2010; Oleson *et al.*, 2011). The Oleson *et al.* (2008a, 2008b) urban canyon model in the National Center for Atmospheric Research (NCAR) Community Climate System Model (CCSM) can be used with weather information either from the atmospheric model within CCSM or from reanalysis products (Oleson *et al.*, 2008a). Simulations use urban morphology and material properties for 33 regions (Jackson *et al.*, 2010) and the Landscan 2004 population dataset (Landscan 2004; Dobson *et al.*, 2000) is used to regionally define three urban classes (tall-building district: over ten stories; high density: three to ten stories; and medium density: one to three stories). A fourth low-density class (one storey dispersed) was defined but the urban canyon representation was not considered appropriate because in many places tree cover masks many buildings. As modelling capacity allows for only one urban type per model grid cell (as a tiled land unit; see Oleson *et al.*, 2008a, 2011) the dominant density class is simulated, which most frequently is the medium density type.

To test the importance of building materials and internal temperature settings on local climates, Jackson *et al.* (2010) compare simulations globally using settings representing a mid-latitude industrial area in Vancouver (modified from Voogt and Grimmond, 2000) and the spatially varying global parameter settings. They found that the magnitude of the UHI decreased systematically towards the tropics, and increased in higher latitudes when realistic building materials are specified. The changes in UHI simulations relate to multiple factors, including radiative and thermal properties of building materials, as well as human behaviour patterns related to heat and air-conditioning use and thermostat settings in buildings. Hence it is important to include these factors when simulating coastal megacity climates.

In general, urban areas are known to exhibit reduced diurnal temperature ranges (DTR) compared to rural areas as illustrated by the Vancouver simulation where urban properties are held constant in space (Plate 11a). In this simulation the urban DTR decreases more relative to rural DTRs as climates become colder, resulting in a general increasing trend from equator to pole (Plate 11a). Some tropical areas actually experience higher urban DTRs in comparison to rural areas because of well-insulated building properties and changes in albedo. However, the pattern changes significantly if more realistic spatially varying urban parameters are used in the simulation (Plate 11b). Changes in thermal mass of buildings and albedo all contribute to reduced urban DTR values in tropical/sub-tropical urban areas compared to rural areas. There are a few exceptions to this general trend, for example in eastern Brazil, the urban DTRs are larger than rural DTRs. These changes are in large part related to climate differences across the country (see Oleson *et al.*, 2011 for these relations). In the Middle East low urban albedos are in part responsible for larger urban DTRs (see Jackson *et al.*, 2010: Figure 9). Because building codes and types in Europe and North America are similar to Vancouver, results between the simulations are similar in these regions (Plate 11c). These results (Jackson *et al.*, 2010, Oleson *et al.*, 2011) confirm that building material properties are influential in creating different urban climates, which have the potential to alter energy demand for climate control systems in buildings. These results suggest that climate impacts of megacities are likely to change should they be redesigned and new and perhaps more energy-efficient building materials be introduced in the future.

The Hadley Centre Global Environmental Model Version 2 (HadGEM2) with additional Earth System (-ES) components (Jones *et al.*, 2011) is included in the Fifth Coupled Model Inter-comparison Project (CMIP5) and contributing to the fifth Assessment Report (AR5) of the Intergovernmental Panel on Climate Change (IPCC). As part of this, four 'representative concentration pathways' (RCPs) of future emissions of GHG have been generated for CMIP5, which are labelled based on their approximate radiative forcing at year 2100. The associated land-use harmonization project provides historical and future projections of land use consistent with these RCPs (Hurtt *et al.*, 2011). RCP8.5 represents a high population and urban growth scenario (A2r, Riahi *et al.*, 2007) and equivalent anthropogenic global radiative forcing of approximately 8.5 W $m^{-2}$ in the year 2100.

McCarthy *et al.* (2010) and Oleson *et al.* (2011) predict that the warming rates of urban areas differ from rural areas in response to radiatively forced climate change. This implies that even without changes in urban extent, the environmental impact of urban areas is not static over time in a non-stationary climate. Plate 12 shows the change in mean UHI from HadGEM2-ES for 2081–2100 compared to 1981–2000 in the RCP8.5 scenario. Overall there is a reduction in the global average $UHI_r$ of 4.8 per cent, with the largest reductions through sub-Saharan, Central and Southern

Africa, South Asia and South America but increases through Central Europe and the Mediterranean, the East Coast of the US, parts of Central Asia and the Southeast Coast of Australia. However, the city interaction with mesoscale meteorological features is not well resolved by a GCM, such as the sea-breeze effect (e.g. Lebassi *et al.*, 2010) (Sections 2.1 and 2.2).

The environmental impact of major coastal cities will change in the twenty-first century as a result of changes to the background climate due to radiatively forced climate change, and as a result of changes in the local climate from further urbanisation. Changes in urban planning and innovations in building materials as well as any feedbacks with mesoscale features resulting from the impact of local urban climate forcings at the interface of land–sea–atmosphere in regions of complex orography are further contributing factors (See sections 2.1 and 2.2). The interactions of these components across a range of space and time scales pose a particular challenge for defining future urban environmental impacts. For instance, from an aggregate of 16 GCM models Horton *et al.* (2011) show a stronger ~2 °C coastal land/sea temperature gradient along the US in the 2080s relative to the 1971–2000 period (Plate 13). And, Jacobson and Ten Hoeve (2011), using a model with a simple ocean mixed layer formulation, recently estimated the global UHI effect to be 2–4 per cent of gross global warming. These factors may superimpose in nonlinear ways to contribute to climate change on multiple scales.

## 2.5 Present-day and future radiative forcing from megacity emissions on the global scale

In many coastal megacities the proximity of ports and heavy industrial activity provide additional sources of air pollutants. Nitrogen oxides ($NO_x$), volatile organic compounds (VOC) and particulate matter (PM) have the potential to impact both climate and health. $NO_x$ and VOC contribute to tropospheric ozone ($O_3$) formation and affect the lifetime of long-lived GHG such as, for instance, methane. The sign of the impact is determined by the $NO_x$/VOC ratio. This ratio is highly variable, both temporally and spatially. PM has natural and anthropogenic sources. PM emissions comprise sulphate (predominately emitted as $SO_2$ and transformed chemically to sulphate particles), black carbon (soot) and organic particulate matter. The backscattering of shortwave radiation is highly efficient by sulphate aerosols, causing cooling, whereas organic particles are less efficient. The efficient absorption of longwave radiation by black carbon aerosols has a warming effect.

There are direct, indirect and semi-direct effects from aerosols (sea salt, dust, primary and secondary particles of anthropogenic and natural origin) and chemical species. GHG and some aerosol particle components (black carbon, iron, aluminium, polycyclic and nitrated aromatic compounds) absorb radiation (short, long wave), leading to warmer air. Some aerosols (water, sulphate, nitrate, most organic compounds) backscatter incident short-wave radiation into space, which cools the air. Korsholm (2009) and Baklanov *et al.* (2008) demonstrate that urban aerosol indirectly affects modulate dispersion by affecting atmospheric stability and precipitation (difference in deposition fields in specific cases are up to 7 per cent). The 100–200-metre effect on the nocturnal boundary layer may be of similar order of magnitude as the effects of the UHI.

The climate impact of megacity pollutants has been studied with the HadGEM2 Earth System Model with contributions from the Max Plank Institute for Chemistry's MATCH-MPIC chemistry-transport model (Collins, 2009; Folberth *et al.*, 2010, 2011). The climate impact in this context is interpreted as the global annual mean of the direct top of the atmosphere (TOA), radiative forcing due to megacity emissions of long-lived greenhouse gases (LLGHG) and pollutants that indirectly affect LLGHG (such as $NO_x$ or VOC). The TOA direct radiative forcing of aerosols that are emitted from megacities has also been taken into account. Analysis comprises simulations of present-day and future conditions with a control and an 'annihilation' scenario in each case. The annihilation scenario has been constructed by removing anthropogenic emissions of LLGHG and short-lived species (excluding methane) from megacities using the megacity mask of Butler *et al.* (2008). Anthropogenic and wildfire emission scenarios are based on the historical (up to the year 2000; Lamarque *et al.*, 2010) and RCP8.5 (Riahi *et al.*, 2007) CMIP5 emission scenarios for LLGHG and pollutants. The model interactively calculates emissions of biogenic VOC at each time step.

The radiative forcing effects are calculated from megacity emissions, for: methane with the change in chemical lifetime and Ramaswamy *et al.*'s (2001) simplified expressions; and ozone with the change in burden assuming Gauss *et al.*'s (2006) scaling factors. To diagnose the direct radiative forcing (DRF) from megacity aerosols HadGEM2 is run twice (Bellouin *et al.*, 2007). Plate 14 shows the short-wave forcing from megacity aerosols that is mostly driven by changes in sulphate. These show a cooling effect (a reduction in the TOA radiative forcing through backscatter of short-wave radiation) around the megacity locations. Plate 14 shows a diminishing cooling effect for the future scenario. This is mainly caused by a reduction in the emission of sulphate aerosol precursors ($SO_2$) due to air-quality directives. Consequently, less sulphate aerosol is formed in the atmosphere and the amount of shortwave radiation backscattered is also reduced yielding a net warming effect from present-day to future conditions.

Under RCP8.5 it is expected that megacities will play a prime role in reducing global pollutant emissions. The species considered (Table 4.1) show a significant reduction. The contribution from megacities is persistently below the global average for every anthropogenic pollutant emission source. While pollutant emissions seem to have a negligible impact on climate, GHG emissions from megacities represent an important contribution. At this level megacities will contribute to the global temperature rise by 2100 amounting to approximately one-quarter of a Kelvin in terms of $GWP_{100}$. Emissions of long-lived GHG are expected to double by 2050 under RCP8.5 and have an even higher impact by the end of the century. $GWP_{100}$ – global-warming potential – the amount of heat trapped relative to $CO_2$. GWP =1 is the same as $CO_2$. The subscript 100 refers to the number of years the calculation is based on.

Table 4.1 summarizes the DRF by megacity emissions including both long-lived species and pollutants. The combined total annual mean radiative forcing from megacity pollutants in 2000 is $-1.0\pm0.32$ mW $m^{-2}$. From Table 4.1 it can be seen that in the case of the climate impact of pollutant emissions individual terms tend to cancel each other out. For example, the slight cooling of $-1.9\pm0.04$ mW $m^{-2}$ from a reduced methane lifetime when megacity pollutant emissions are included is nearly compensated by the slight warming effect of $+1.5\pm0.01$ mW $m^{-2}$ due to black carbon emitted from megacities. Overall, the effect of LLGHG emissions from megacities dominates

*Table 4.1* Present and future direct radiative forcing due to pollutant emissions from megacities

| *Species* | *2005 DRF ($mW\ m^{-2}$)* | *2050 DRF ($mW\ m^{-2}$)* |
|---|---|---|
| Megacity Direct Radiative Forcing due to *Long-lived* GHG | | |
| $CO_2$ (tot TOA) | +120.0 | +254.0 |
| $CH_4$ (tot TOA) | +28.4 | +59.8 |
| $N_2O$ (tot TOA) | +3.3 | +8.8 |
| *Total forcing (TOA)* | *+151.7* | +322.6 |
| Megacity Direct Radiative Forcing due to *Short-lived* GHG and Pollutants | | |
| $CH_4$ (tot TOA) | −1.9 ± 0.04 | −0.7 ± 0.02 |
| $O_3$ (tot TOA) | +5.7 ± 0.02 | +2.8 ± 0.02 |
| [1]SWas (tot TOA) | −6.1 ± 0.21 | −2.2 ± 0.10 |
| [2]LWcs (tot TOA) | +1.5 ± 0.01 | +0.6 ± 0.01 |
| *Total forcing (TOA)* | *−0.8 ± 0.24* | *+0.5 ± 0.09* |
| Combined Direct Radiative Forcing | | |
| *Total forcing (TOA)* | *+150.9 ± 0.24* | *+323.1± 0.09* |

[1]shortwave all-sky
[2]longwave clear-sky

Notes: Current best estimate of present-day and future annual global mean direct radiative forcing ($mW\ m^{-2}$) due to pollutant emissions from megacities computed with HadGEM2 (Folberth *et al.*, 2011; personal communication).

the climate effect at present-day condition. Under 2050 conditions the climate impact of LLGHG emissions seems to increase whereas the climate impact of pollutants has diminished in good agreement with the reduced pollutant emissions that follow air-quality directives.

## 2.6 Effects on precipitation processes and pollutant removal processes

Megacity–coastal interactions may also impact the hydrological cycle and pollutant removal processes through the development of fog, clouds and precipitation in and around urban and coastal areas. The UHI can significantly influence mesoscale circulations and resulting convection, e.g. warm seasonal rainfall increases of 7 per cent–25 per cent within and 50–75 kilometre downwind of megacities (Landsberg, 1970; Huff, 1986; Changnon *et al.*, 1991; Ohashi and Kida, 2002; Shepherd and Burian, 2003).

There are three possible causes of urban-induced convective phenomena (Hidalgo *et al.*, 2008): (1) changes in circulation that are induced by the UHI and sea breeze, (2) an increase in urban surface roughness, and (3) an increase in the concentration of condensation nuclei (CCN; e.g., particles which absorb water to form cloud droplets) from megacity air pollution. The mechanisms of urban effects on rainfall (especially in coastal areas) are complex, sometimes contradicting and their understanding is far from complete (Shepherd and Burian, 2003). Cloud microphysics may *reduce* rainfall via increased urban aerosols (Rosenfield, 1999; Ramanathan *et al.*, 2001) *or* may *enhance* urban rainfall through local dynamics and thermodynamics related to the UHI-induced convergence zone and destabilized boundary layers (Changnon and Westcott, 2002; Ohashi and Kida, 2002; Shepherd *et al.*, 2002).

The UHI can create a *downwind* updraft cell, which is consistent with surface meteorological data analysis that shows a convergence zone that favours convective thunderstorm development (Bornstein and Lin, 2000). Warm urban air may collide with moist sea air, releasing precipitation downwind of the megacity. The increased roughness induces convergence on the upwind side of urban areas. A large quantity of aerosols (small liquid or solid particles) from megacities can physically have the same effect as GHG via a direct feedback mechanism that is caused by a change in radiation properties (Section 2.5). Their presence absorbs or reflects the solar radiation, limiting the amount of sunlight received at the surface, sometimes by more than 10 per cent. However, urban- and marine-originated aerosols also actively influence processes of cloud formation (forming CCN; indirect feedback mechanisms). The effect of CCN could be different for rain than for ice precipitation. Aerosols can suppress rain and snow by increasing the concentration of CCN, which causes nucleation of many small cloud droplets that coalesce inefficiently into raindrops (Rosenfeld, 2000); or reduced cloud droplet size can delay the onset of precipitation, which could allow more intense updrafts and consequently more intense ice precipitation, hail and lightning.

There is increasing evidence that coastal megacities like Tokyo (and other large coastal urban areas) can influence weather (rainfall, lightning, etc.) through complex urban land-use–weather–climate feedbacks (Bouvette *et al.* 1982; Ohashi and Kida, 2002). Orville *et al.* (2001) find the highest annual and summer flash densities were over and downwind of Houston. The lightning distribution was hypothesized to be either UHI-induced convergence and/or increased urban aerosols causing enhanced lightning efficiency.

## 2.7 Effects of climate change on coastal megacities regions

Similar climate-change-related heat impacts can be expected in a number of regions. Cities in the Mediterranean and Europe as a whole, and the Southern and Western US can expect more frequent and more intense heat-wave conditions, which are likely to be exacerbated by additional UHI impacts (Meehl and Tibaldi, 2004). The coastal location of megacities increases their exposure and potential vulnerability to a wide array of meteorological hazards. Socioeconomic development and future urbanisation trends can either exacerbate or reduce this vulnerability, while climate change may influence the recurrence of specific hazards, resulting in changing risk for coastal megacities through the twenty-first century.

Sea-level rise as a consequence of anthropogenic climate change (Church *et al.*, 2006) would be expected to increase extreme coastal high water levels such as during storm surge events (e.g. Lowe and Gregory, 2005). Those cities located in low-lying deltas are at particular risk (Nicholls *et al.*, 2008). It is likely that the proportion of total rainfall from intense or heavy rain events will increase in a changing climate (e.g. Emori and Brown, 2005; Meehl *et al.*, 2005). The large impervious surface cover in urban areas means this would contribute to increased pluvial flood risk and stress on existing urban drainage and sewage systems, as described in Chapter 2. Projected changes in fluvial flood risk have low confidence (IPCC, 2012).

The response of tropical and extra-tropical storms to climate change are uncertain, particularly for individual ocean basins. However, future projections do consistently suggest globally a shift towards stronger tropical storms with higher peak wind speeds and precipitation rates, but a global reduction in total storm numbers (Knutson *et al.*, 2010).

The IPCC special report on managing risks of extreme events (SREX, 2012) concludes 'that it is *virtually certain* that increases in the frequency and magnitude of warm daily temperature extremes and decreases in cold extremes will occur in the 21st century at the global scale' (see also Meehl and Tebaldi, 2004; Clark *et al.*, 2006; Kharin *et al.* 2007). Recent results from global climate models also show that globally increases in extreme hot nights and heat stress are larger for an urban surface than rural surfaces under doubled $CO_2$ climate change (McCarthy *et al.*, 2010; Fischer *et al.*, 2012). In contrast, analysis of observations from two urban coastal California air basins observations for the period 1948–2005 showed cooling summer daytime temperatures in coastal megacities (Lebassi *et al.*, 2009). The mechanism proposed was that increased GHG-induced inland temperatures lead to increased pressure and temperature gradients (coast to inland) and correspondingly increased sea-breeze frequency, intensity and penetration. This resulted in coastal areas with *reduced* summer daytime maximum temperatures and more foggy days. These local and regional climate feedbacks require continued detailed research to understand the interplay between regional and local mechanisms that will contribute to future change in the world's largest coastal megacities.

Climate scenarios for NYC predict that with increasing temperatures heat waves will increase (frequency, intensity, duration); sea-level rise will increase coastal flooding frequency, even if storm intensity is unchanged; flooding from rainfall could threaten drinking water quality and increase coastal pollution; and increased warm season drought may cause water shortages.

Numerous coastal megacities are located in areas that are prone to hurricanes/typhoons (e.g. Seoul, Miami). Most of the damage associated with these storms stems from the storm surge (e.g. NYC). After coupling four general circulation models to two hydrodynamic models, Lin *et al.* (2012) evaluated the frequency of storm surges. After combining the changes in storm climatology with a one-metre sea-level rise, they conclude that the current '100-year surge flooding' would have a three to 20 year return period and the current 500-year event would be a 25–240 year event by 2100 (Lin *et al.*, 2012).

## 2.8 Air pollution in and from megacities

The scale at which pollutants have an effect extends from local (human health, building materials, ecosystem) to regional (acidification, euthrophication) to global (atmospheric composition, climate change). The impacts of megacities vary in space and time as a function of the prevailing winds. The mean transport distance for most megacities (from source location to pollution plume centre) for black carbon and other primary fine PM is of the order of 100–200 kilometres (MegaPoli, 2011). Whereas secondary PM species can be transported much further with sulphate and secondary organic aerosol often over 400 kilometres on average. Maximum transport distances are significantly higher, with secondary particulate matter impacts reaching as far as 2,000 kilometres away. As one example the Boston–New York–Washington region has a relatively weak influence on the European atmosphere. To date no significant above-average concentration event has been observed associated with pollution coming from there (Eckhardt *et al.*, 2011). The range of megacities' effect on regional air quality may be limited by orography; e.g. Po Valley, one of the most polluted areas in Europe, rarely impacts air quality outside the confined north Italy region. The location and surrounding topography of the coastal megacities relative to predominant synoptic conditions impact the region that is influenced. For example, Los Angeles with mountains to the west has a large amount of recirculation back onto the urban area itself.

The impacts on $O_3$ levels are usually much smaller than those for PM and primary pollutants like NOx. Megacities actually tend to cause a decrease of $O_3$ mixing ratios directly in the cities (due to its reaction with NO), while increasing $O_3$ downwind of the cities (by up to 10 ppb). Globally the impacts of megacities on the total burdens of the directly emitted gases are comparable to or slightly smaller than their relative contribution to the global total emissions (several per cent), while the impact on global tropospheric $O_3$ is much smaller (< 1 per cent). Interestingly, this impact tends to become even smaller in future scenario simulations, and the ozone chemistry of megacities tends to become more strongly influenced by upwind pollutant emissions relative to the emissions in the megacities themselves (MegaPoli, 2011).

While the effects of megacities on regional pollutant concentrations levels are limited in spatial extent, most of the pollutants are still exported over long distances. For example, 17–34 per cent of the mass of pollutants with a lifetime of ten days is exported beyond 1,000 kilometres and remains in the lowest one kilometre, for middle- and high-latitude Eurasian megacities (Lawrence *et al.*, 2007). Vertical mixing plays a large role in determining the pollution build-up near the surface, both immediately in and around the megacity, as well as thousands of kilometres downwind. In the tropics, strong vertical transport and scavenging of soluble aerosols lead to a smaller range of the impacts by the megacities. About 50 per cent of the emitted mass of $PM_{2.5}$ is transported prior to deposition more than 1,000 kilometres from the source over large areas making it more dilute. The greatest exposure to high aerosol concentrations tends to still be within the bounds of the cities themselves. However, for only a few existing megacities is the total population exposure to megacity aerosol pollution as large, or larger, than the population exposure within the cities themselves.

The largest cities in Europe are the most significant contributors to deposition of aerosols (especially absorbing aerosols like soot) in the Arctic, regardless of whether considering the annual or only the wintertime deposition (Cassiani *et al.*, 2011). Among these cities the largest contributors are Saint Petersburg, Moscow and the Ruhr Valley (Germany). Roughly, one gram of soot emission in Saint Petersburg results in a deposition in the Arctic, which is equivalent to 30 grams emitted in Beijing, 100 grams from Seoul and 200 grams from Shanghai. This suggests that in order to most effectively reduce soot deposition in the Arctic, the focus needs to be on northern Eurasian sources (MegaPoli, 2011).

## 3 Water resources and coastal megacities in developing countries

Coastal megacities are intrinsically linked to the water resources that surround them. Freshwater demands to sustain the population raise concern about water scarcity, while the concentrated socioeconomic activities of the megacity impact water resources and aquatic environment. Climate change will also impact both the environmental 'input' and 'output' of these coastal megacities in the context of water resources. Megacities also create indirect effects on water resources, dams used to store water for urban use affect river flows, ground water recharge rates and sediment flows, which affect both water quality and quantities in the areas surrounding the cities. Indirectly coastal cities will affect sediment and nutrient flows into water bodies that in turn affect coastal erosion rates and water quality in near-shore ocean environments.

For example, the World Wildlife Fund (2011) found Dhaka to be the coastal megacity that is most vulnerable to the effects of climate change including sea level rise in the Ganges-Brahmaputra mega delta, storm surge and flooding in low-lying areas. Dhaka sits only metres above sea level and is sinking due to subsidence. Subsidence combined with sea level rise has led to salt-water intrusion, which in turn affects rice production, a major source of food for the region (World Wildlife Fund, 2011). Dhaka is an extreme example of the kinds of effects that water resources have on the vitality of megacities, and how climate change can exacerbate those effects. Although not limited to megacities, especially in developing countries, consideration of adequate socioeconomic responses are needed to manage these effects.

### *3.1 Freshwater demands*

In coastal megacities the demands for freshwater include for household use, food production both in the form of urban gardens and food imported from outside the city, and energy production (Varis, 2006). 'Small water' provision, or water for drinking and sanitation and industrial purposes (Allan, 2011), requires both sufficient quantity and quality. In 1997, Jakarta introduced water privatization to accelerate provision of water for this growing city. However, Lanti (2006) reports in 2002 coverage targets were not met from the original 40 per cent, with eastern Jakarta being just above 60 per cent and western Jakarta about 42 per cent. 'Big water' provision, or water for food production (Allan, 2011), is another major source of increased demand. Cities can either rely on domestic agricultural production or import food, or virtual water, to feed its population. While not unique to coastal megacities, expansion of market-led agricultural production is required in megacities and surrounding regions (Varis, 2006). However, urban expansion or sprawl may limit domestic agricultural production capacities or induce a shift to high-value products, not necessarily required for subsistence (Strzepek and Boehlert, 2010).

Surface water and groundwater are used to supply coastal megacities. For example, Bangkok and its metropolitan surroundings have extracted water from the Chao Phraya river, thus competing with the irrigation sector (Molle 2004, 2007), and from groundwater.

Many megacities rely on groundwater, which has resulted in excessive groundwater extraction (see Chapter 2) and growing extraction rates; for example, in Dhaka, Bangkok, Manila and Tokyo (Taniguchi *et al.*, 2009; Pandey *et al.*, 2011; Sekovski *et al.*, 2011). Regulation of groundwater extraction may have strong opposition from the industrial sector (Molle, 2007). While water policy reforms have taken place to meet the increasing and competing water demands between the urban and rural sector, consideration for *equity* and *rights* to water are undermined (Molle, 2004).

The Chao Phraya River is becoming a 'closed' river with very little water available for allocation to new uses. Consequently, to sustain freshwater demand, there have been proposals to transfer water from other parts of the country and from the transboundary Mekong River basin, which would require international negotiation between Thailand and those sharing the river (Molle and Floch, 2008).

### 3.2 Adverse impacts to water resources

Pollution from megacities has made a major impact on the quality of surface and groundwater resources, sometimes with international and political–economic effects. The level of dissolved inorganic nitrogen (DIN) in Changjaing estuary near Shanghai, for example, tripled from 1980–2000 (Sekovski *et al.*, 2012). A three-fold increase in inorganic fertilizer use, to meet growing demand for food, municipal and industrial waste and changes in land use coupled with a lack of water-quality regulation caused the heightened level of DIN that in turn contributed to the decline of fisheries in Shanghai (Zhang *et al.*, 2006).

Many coastal megacities are also major ports. The presence of international cargo ships in the surrounding water resources and heavy marine traffic can lead to increased turbidity, a change in sediment loads, and the introduction of invasive species to the environment through the release of ballast (Tamburri *et al.*, 2002). Marine traffic has taken a toll on water resources such as the Changjaing estuary near Shanghai, where the pollution of metals and dredging activities have caused benthic fauna to decrease from 150–160 species in the 1970s to 10–20 species in 2002 (Zhang *et al.*, 2006).

Declining water quality and lack of water quality regulation or enforcement in coastal megacities have a serious impact on human health. For example, untreated wastewater discharges from Mumbai have led to coliform bacteria contents of 600 and 1,900 count $l^{-1}$ in Mahim Bay (Murthy *et al.*, 2001). Waterborne diseases in coastal megacities are mainly related to inadequate water supply due to a lack of potable water infrastructure, the contamination of drinking water and solid-waste generation. Nearly 30 per cent of all diseases in Mumbai are waterborne, such as diarrhoea, malaria, typhoid and intestinal worms (Sekovski *et al.*, 2012). Similarly, high levels of iron, arsenic and chromium were detected in Kolkata's drinking water reservoir from untreated household runoff from solid waste or chemicals and industrial runoff (Sekovski *et al.*, 2012). Urban soils store heavy metals deposited from transportation, industry, etc. The soils provide a buffer for urban contamination, but when the buffering capacity is reached, the soils themselves become a source of contamination to air and water. The lack of environmental regulation can have implications for industry and wider economic activity; changes in water chemistry can lead to large-scale algal blooms, which can affect the productivity of industries that rely on the water resources.

Storm surges and monsoon flooding are already major concerns for coastal megacities. Increasing urbanisation and related land-use changes may make delta regions more vulnerable to flooding caused by the effects of climate change (Nicholls *et al.*, 2007). In addition, the expansion of coastal megacities can be at the cost of the riparian environment (see Section 4). The removal of mangroves and other aquatic habitats can have effects on erosion, saltwater intrusion and juvenile fish stocks (Sekovski *et al.*, 2012). In addition, the removal of any type of coastal vegetation decreases potential carbon sinks, thus contributing to global climate change (Zhao *et al.*, 2006).

## 4 Ecology of coastal megacities: contributions to changes at regional and global scales

Megacities can have substantial impacts on regional and global ecology, mainly via linked effects on biodiversity and ecosystem services (Folke *et al.*, 1997; Grimm *et al.*, 2008). Although some trends are common to all megacities, coastal megacities in particular have a suite of characteristics that influence the nature and level of impacts on regional and global ecology. Many of these remain hypothetical and/or unquantified, and what observations do exist are based on case studies (e.g. Table 4.2) with few comparative studies.

Megacities are positioned in ecotonal/ecoclinal areas (an ecotone is the area of transition between biomes; an ecocline is the variation of a physiochemical characteristic e.g., thermocline), such as estuaries or directly on the coast, which can be particularly biodiverse, productive or contain rare species (MEA, 2005). The importance of such areas is reflected in the distribution of global biodiversity hotspots, many of which are coastal and coincide with areas of mega urbanisation, including the California Floristic Region (San Francisco, Los Angeles), and a concentration of diversity and urbanisation in Southeast Asia (Myers *et al.*, 2000). The MEA (2005) notes that coastal ecosystems provide services disproportional to their area. However, estimates of loss in recent decades include 35 per cent of mangrove areas and 20 per cent of coral reefs (MEA, 2005). Though not solely due to urbanisation, much exploitation, habitat destruction, introductions of invasive species and pollution associated with ecosystem impacts may result from the ecological footprints of urban areas. Coastal megacities are likely to be amongst the most significant of these.

Wetland and estuarine areas are amongst the first to be lost during urbanisation, especially where they are seen as unproductive (Pauchard *et al.*, 2006). Forman (2008) suggests that any city region with a population greater than eight million will contain no major wetlands. Attempts to restore, rehabilitate or recreate wetland systems in coastal urban areas generally have very limited success, with substitute systems often being of reduced ecological quality compared to pre-urban wetlands, and efforts tending to focus on the creation of riparian woodland at the expense of other forms of freshwater wetland (e.g. Allen and Feddema, 1996; Sudol and Ambrose, 2002). Consequently, a coastal megacity region may have a greatly reduced abundance or absence of estuarine or wetland habitat, such as saltmarsh, mudflats, dunes and other

*Table 4.2* Examples of ecological case studies that are relevant for coastal megacities

| *Reference* | *Broad topic* | *Location* |
|---|---|---|
| *General* | | |
| Nixon 1995 | Urban regions and coastal eutrophication | |
| Smith 2003 | Urban regions and coastal eutrophication | |
| Derraik 2002 | Review of marine plastics | |
| Dennison 2008 | Review of impacts on coastal systems, some reference to urban | |
| Jenerette *et al.* 2006a | Valuation of water footprints for 121 US cities | |
| Jenerette *et al.* 2006b | Comparison of water footprints in Chinese/ US cities | |
| Folke *et al.* 1998 | Marine ecological footprints | |
| Folke *et al.*, 1997 | Footprints of 29 European cities | |
| Kaye *et al.* 2006 | Altered biogeochemical cycles within cities | |
| Duncan *et al.* 2011 | Plant extinction in 11 global cities | |
| Martinez *et al.* 1997 | Coastal ecosystems review | |
| Alberti 2005 | Urban ecosystem structure and function | |
| *Coastal city* | | |
| Faerge *et al.* 2001 | Much nitrogen and phosphorus lost via waterways | Bangkok |
| Goudey & Moran 2005 | Aquaculture in harbour | Boston |
| Ahn *et al.* 2005 | Storm-water pollutants extend 100 km$^2$ from river mouth | California |
| Zhang *et al.* 2011 | Ecosystem services wetlands | Guangzhou |
| Li and Lee 1997 | Mangroves | Hong Kong |
| Lam 2009 | Seawalls | Hong Kong |
| Liu and Kueh 2005 | Coastal pollution | Hong Kong |
| Cleary *et al.* 2006 | Impact on coral reefs | Jakarta |
| Hu *et al.* 2008a | Ecological footprints | Jiangyin |
| Hu *et al.* 2008b | Water footprint | Jiangyin |
| Moore *et al.* 2011 | Plastics in urban rivers | Los Angeles |
| Carnahan *et al.* 2008 | Coastal pollution | Miami |
| Matlack 1997 | Forest clearance aided by port access over 350 years | New York |
| Ascione *et al.* 2011 | Water footprint | Rome |
| Greer and Stowe 2003 | Community change in urban lagoon | San Diego |
| Xie 2011 | Urban development and coastal wetlands | Tianjin |
| Peng 2009 | Coastal pollution | Tianjin |
| Nakayama *et al.* 2007 | Changes to coastal hydrology | Tokyo |
| Ribic *et al.* 2010 | Plastic deposition on beaches | US East Coast |
| Brown *et al.* 1998 | Urban marine pollution | US West Coast |

inter-tidal zones. However, actual measurements are rare, which is particularly significant given the value of these systems for ecosystem services. Costanza *et al.* (1997) quantify coastal services (estuaries, sea grass/algae beds, coral reefs and coastal shelf) as almost 38 per cent of the global ecosystem services value of the biosphere, despite only covering 6 per cent of the global surface. This is particularly of concern as coastal cities in biodiversity hotspots are among those most projected to grow in the coming decades (UN, 2005). Such cities therefore play a major role in the decline of regional and local abundance and quality of regional and global coastal ecosystems – their footprint impacts particularly hard on these areas (Folke *et al.*, 1998).

The geographical position of maritime megacities restricts the expansion–coalescence phases of urban development (Dietzel *et al.*, 2005) from broadly concentric expansion possible for more continental cities. Expansion occurs inland but also particularly along the coast, further degrading these coastal habitats (Costanza, 1997; MEA, 2005). Further habitat degradation occurs due to the construction of coastal or estuarine flood defence walls, groynes and docks, which destroy habitat and can increase physical stresses on biota (Francis *et al.*, 2008; Chapman and Underwood, 2011), and the appropriation of coastal areas for aquaculture and fisheries (Folke, 1998).

Ecological 'footprint' analyses are designed to demonstrate the larger area urbanisation impacts through calculation of the amount of resources consumed including ecosystem services that typically go well beyond the city boundaries (Folke *et al.*, 1998; McGranahan *et al.*, 2005; Grimm *et al.*, 2008; Rees and Wackernagel, 2008).

All cities have a regional footprint of land-use change, resource extraction and ecosystem fragmentation that compromises local species populations and leads to a decline in both populations and biodiversity, with fragmentation in particular due to both land conversion and the construction of transportation corridors leading to reduced patch size and reduced permeability to species movements (Forman, 2008; Murray-Smith *et al.*, 2009). Demand for ecosystem resources such as water, fossil fuels, biomass (for fuel or food) and so on causes ecosystem degradation across the region and throughout the world, though impacts vary geographically (for example in Africa most urban fuel [59 per cent] comes from biomass [wood], which is regionally sourced; Clancy, 2008). Footprints have been calculated in detail for some cities (e.g., Jiangyin, Beijing (China) Hu *et al.* 2008, Hubacek *et al.*, 2009; water footprints in the US, Jenerette *et al.*, 2006a; several cities in Europe, Folke *et al.*, 1997), but less so for coastal cities.

Habitat loss reduces the regional species pool (the general 'biodiversity source') that species can colonize extant, recreate or novel habitat from, and influence global diversity. For example, migratory bird diversity in southern latitudes has declined in response to urbanisation in northern latitudes (Valiela and Martinetto, 2007). It is yet to be determined if similar impacts may be found for coastal/marine species. For long-established cities, port access may have accelerated the land-cover clearance (e.g. of forest; Matlack, 1997) and associated ecosystem degradation.

Coastal megacities also extend their influence into the oceans, with urbanised rivers delivering pollutants from throughout the region directly into estuarine and coastal ecosystems. Impacts from various pollutants can extend up to 100 kilometres squared from the coast (Ahn *et al.*, 2005) and can significantly lower water quality, for example causing eutrophication and algal blooms that reduce oxygen and lower diversity (e.g., Santi *et al.* [2006] report reduced species richness for Rio de Janiero in areas of the inner Bay), particularly of e.g. coral beds (e.g., Cleary *et al.*'s [2006] study of reduced cover and different community composition in zones closest to the city in Jakarta, though this is not solely due to urbanisation) (Nixon, 1995; Smith, 2003; MEA, 2005;

Dennison, 2008). MEA (2005) reports that nitrogen input to the oceans increased by 80 per cent between 1860–1990 – this is a global value, the proportion that could be attributed to activity and/or consumption footprints in coastal megacities is not clear.

Algae blooms in coastal waters are often driven by upwelling-favourable winds that typically occur in the summertime and deliver nutrient-rich water to the surface. However, they can be fuelled by sewage outfalls and pollution from urban centres upstream. The unprecedented August 2011 New Jersey coast algal bloom event (Plate 15) is believed to have been partially associated with a massive accidental untreated sewage spill (760,000 metres cubed) into the Hudson River in late July when the Harlem treatment plant caught on fire. Numerical simulations of the fate of the sewage outfall contributed to the safeguarding of human health. The regional impact of a bloom can be quite severe. In Manila Bay, four days of algal blooms led to a 75 per cent drop in fish price and a loss of USD $2 million in the local shellfish industry (Jaing *et al.*, 2001). Such events are projected to increase with urbanisation and constitute significant ocean impact engendered by megacities.

Megacities may also represent a major source of marine debris (e.g. plastics) that can reduce habitat quality (Derraik, 2002; Ribic *et al.*, 2010; Moore *et al.*, 2011). Their regional and global ecological impact from aquatic pollution is therefore significant. Although there are some estimates of system use and flows of e.g. water/nutrients/pollutants in megacities (Folke *et al.*, 1997; Grimm *et al.*, 2008; Ascione *et al.*, 2011; Faerge *et al.*, 2011), there is limited quantification of flows of species into and out of cities, with perhaps the exception of species delivered in ballast (see below). Coastal megacity rivers and estuaries may also contain physical or environmental barriers that can reduce movement of species and thereby reduce connectivity throughout river catchments – particularly, for example, diadromous fish. Although 'ecological footprints' provide a measure of resource use; the impacts to species or functioning at large scales is not easily determined. Loss of coastal ecosystems reduces flood mitigation services and lowers regional ecological resilience, but there is limited information to quantify this.

Cities are hubs for species introductions. Coastal cities are key sources of accidental aquatic introductions from shipping traffic, such as hull fouling, ballast water etc. (Mack, 2003). It is estimated that over three to 12 billion tonnes of ballast water is transported every year, with up to 7,000 species in cargo vessels moved every 24 hours (Carlton, 1999; Leppäkoski *et al.*, 2002). Many of these ships will dock in coastal cities (many coastal megacities have a history of sea trade) and as such it is probably common for aquatic invasions to originate from such cities. For example, the Chinese mitten crab has spread to urban estuaries around the world via ballast (Bentley, 2011).

This process contributes to both regional and global homogenization of biota, in particular of aquatic species, and megacities are probably key nodes of exchange between other global cities. These invasions could be likened to the arrival of a virus to a system that is already weakened (i.e. low aquatic diversity, or indeed low biodiversity in general), and causes further harm. Such invasions may also be linked to risk of disease outbreak, e.g. from mosquitoes transported via ships (Nelson *et al.*, 1976). Elton (1958) records the most significant example of this in the twentieth century in Brazil, though not in a specifically urban context. There is also potential for global effects as disease spreads via transport networks (e.g. airport malaria etc.). In Buenos

Aires, there is a large body of evidence linking transport of urban water and mosquitoes (mainly *Aedes egyptii*, originally native to Africa and invasive in North and South America), which is clearly a significant problem.

Cities are contributing to the microevolution of new genotypes/species (see review in Francis and Chadwick, 2012), which emerge due to responses to the urban environment that may sometimes be driven by processes of natural selection. Francis and Chadwick (2012) define synurbic species as those with 'populations with greater urban than rural densities'. Coastal megacities may facilitate this for coastal/estuarine species. Francis and Hoggart (2008) find that artificial structures (e.g. flood defence walls) are more biodiverse than degraded inter-tidal foreshore (a form of coastal/estuarine wetland) through Central London. Most of the plant species found on the walls are broad terrestrial generalist species rather than riparian/wetland species (Francis and Hoggart, 2012).

## 5 Conclusions

Coastal megacities impact the atmosphere, land, oceans, hydrology and ecology of the surrounding regions as well as with effects that also extend globally. These large, dense, rapidly growing cities are often physically constrained by inland topography. Their setting often results in a concentration of industrial activity and emissions with profound implications for air quality and human health. In view of the projected urban growth and further megacity expansion it is crucial to develop a better understanding of the impacts of these city systems on the environment, and to develop methods to evaluate the best mitigation and adaptation strategies to reduce these impacts.

Fundamental to understanding the impact of changes (at both the global or within city scales) is an ability to be able to simulate the processes. To date there are relatively few modelling studies of the impacts of coastal megacities as computer resources are only just becoming available with enough capacity to resolve key details across cities. Atmospheric modelling studies typically treat urban effects (heat island, sea/land-use, roughness, storage heat capacity, soil moisture, precipitation change, etc.), the effects of urban emissions/air pollution (e.g. direct, semi-direct, first and second indirect effects of aerosols) and ecological impacts and land-cover change in the surrounding areas separately. However, in reality these processes continuously interact. Therefore, to understand the mechanisms of interactions and their non-linearity, and correctly model the effects of megacities, online coupled/integrated models with two-way interaction of meteorological and chemical/aerosol processes at different scales should be considered.

Two recent reviews of urban meteorology needs and understanding (Grimmond *et al.*, 2010b; Mills *et al.*, 2010) make key recommendations that are applicable to coastal megacities. From an atmospheric perspective, little work has been done on scaling up the impact of coastal cities to larger areal extents. In contrast, there has been much more work in the other direction; i.e. dynamical downscaling of global models to regional scales. Generally there is better understanding of the processes within an urban area than those impacting the region beyond because of difficulties in observations as well as the lack of computing resources capable to resolve these scales in numerical models. The IPCC assessment reports, for example, so far have not included urban areas within the land surface tiles as their spatial resolution was not sufficient. The next IPCC report will, however, take urban areas into account using some of the runs reported here.

Key findings reported in this Chapter include:

- To correctly model the effects of coastal megacities and to understand the mechanisms of interactions and their non-linearity; online coupled/integrated models with two-way interaction of meteorological and chemical/aerosol processes on different scales are needed. This will enhance our understanding of interacting processes (e.g. coastal winds and air temperature in the vicinity of megacities) through coupled air/sea/urban modelling and impacts (e.g. simulating and predicting algal blooms).
- Models show that megacities will play a prime role in reducing global *atmospheric* pollutant emissions.
- The combined total annual mean radiative forcing from megacity pollutants in 2000 was $-1.0\pm0.32$ mW $m^{-2}$ as individual terms cancel each other out (Table 4.1).
- While atmospheric pollutant emissions seem to have a negligible impact on climate, *GHG* emissions from megacities represent an important contribution and are expected to further contribute to global temperature warming (by 2100 approximately 0.25 K in terms of $GWP_{100}$). Emissions of long-lived GHG are expected to double by 2050 according to scenario RCP8.5, pointing toward an even higher impact by the end of the century.
- The nature of building materials in urban areas influences the strength of temperature differences. Temperature gradients impact the surrounding areas, regionally and beyond.
- The environmental impact of major coastal cities will change in the twenty-first century as a result of changes to the background climate due to radiatively forced climate change, as a result of changes in the local climate from further urbanisation, and any feedbacks with mesoscale features such as sea breezes. The interactions of these components across a range of space and time scales represent a particular challenge for defining future potential urban environmental impacts.
- The continued refinement and evaluation of the urban land surface models is an essential requirement in order to ensure that such models provide the right answer for the right physical reasons.
- Urban systems affect water resources upstream and downstream through such indirect impacts as dam building for water resources, which in turn affect water quality, as well as geomorphic processes related to sediment transport, which can affect ecosystems downstream including shoreline erosion processes and ocean water quality (e.g. algal blooms). Ecological models need to be developed that link all these processes.
- Ecosystem services from coastal areas are important both regionally and globally but coastal megacities are a major driver of their decline and may offer limited potential for system recovery due to intense use and population pressures. Loss of coastal ecosystems reduces flood mitigation services and lowers regional ecological resilience, but there is very limited information to quantify this, particularly at regional or broader scales.

- Coastal urban areas are key locations for non-native species introductions (both aquatic and terrestrial), and those species that become invasive may have dramatic regional impacts on both ecosystems and society. Coastal megacities may represent stepping stones for the global dispersal of species via shipping, and should arguably be a focus for coordinated biosecurity control initiatives.
- Populations of coastal urban species (native or non-native) that are responding to the urban environment may display genetic differentiation and related changes in physiology or behaviour, as exhibited in some large global cities. Megacity populations are probably more likely to exhibit such changes, and synurbic responses of estuarine/marine species will only be seen in coastal cities. These possibilities remain speculative at the present time, but have implications for coastal and urban biodiversity at the regional scale.

## References

Ado, H.Y. (1992) Numerical study of the daytime urban effect and its interaction with the sea-breeze. *Journal of Applied Meteorology* 31: 1146–1164.

Ahn, J.H., Grant, S.B., Surbeck, C.Q., Digiacomo, P.M., Nezlin, N.P. and Jiang, S. (2005) Coastal water quality impact of stormwater runoff from an urban watershed in southern California. *Environmental Science and Technology* 39: 5940–5953.

Alberti, M. (2005) The effects of urban patterns on ecosystem function. *International Regional Science Review* 28(2): 169–192.

Allan, J.A. (2011) *Virtual Water: Tackling the Threat to our Planet's Most Precious Resource*, I.B. Tauris, London.

Allen, A.O. and Feddema, J.J. (1996) Wetland loss and substitution by the Section 404 Permit Program in Southern California, USA. *Environmental Management* 20: 263–274.

Allen, L., Lindberg, F., Grimmond, C.S.B. Global to city scale model for anthropogenic heat flux, International Journal of Climatology DOI: 10.1002/joc.2210

Ascione, M., Bargigli, S., Campanella, L. and Ulgiati, S. (2011) Exploring an urban system's dependence on the environment as a source and a sink: the city of Rome (Italy) across space and time scales. *Chemsuschem* 4: 613–627.

Bentley, M.G. (2011) Eriocheir sinensis H. Milne-Edwards (Chinese mitten crab). In: Francis, R.A. (ed.) *A handbook of global freshwater invasive species*. Routledge, London.

Best, M., Grimmond, C.S.B. and Villani, M. (2006) Evaluation of the urban tile in MOSES using surface energy balance observations. *Boundary-Layer Meteorology* 118: 503–525.

Bornstein, R.D. (1968) Observations of the Urban Heat Island effect in New York City. *J. Appl. Meteor.*, 7, 575–582.

Bornstein, R.D. and Thompson, W.T. (1981) Effects of frictionally retarded sea breeze and synoptic frontal passages on sulfur dioxide concentrations in New York City. *Journal of Applied Meteorology* 20: 843–858.

Brown, D.W., McCain, B.B., Horness, B.H., Sloan, C.A., Tilbury, K.L., Pierce, S.M., Burrows, D.G., Chan, S.L., Landahl, J.T. and Krahn, M.M. (1998) Status, correlations and temporal trends of chemical contaminants in fish and sediment from selected sites on the Pacific coast of the USA. *Marine Pollution Bulletin* 37: 67–85.

Browne M.A., Chapman M.G. (2011) Ecologically informed engineering reduces loss of intertidal biodiversity on artificial shorelines. *Environmental Science and Technology* 45: 8204–8207.

Butler, T.M., Lawrence, M.G., Gurjar, B.R., van Aardenne, J., Schultz, M. and Lelieveld, J. (2008) The representation of emissions from megacities in global emission inventories. *Atmospheric Environments*, 42: 703–719.

Carlton, J.T. (1999) Molluscan invasions in marine and estuarine communities. *Malacologia* 41: 439–454.

Carnahan, E.A., Hoare, A.M., Hallock, P., Lidz, B.H. and Reich, C.D. (2008) Distribution of Heavy Metals and Foraminiferal Assemblages in Sediments of Biscayne Bay, Florida, USA. *Journal of Coastal Research* 24: 159–169.

Cassiani, M., Stohl, S., Eckhardt, S., Sovief, M., Prank, M., Butler, T., Lawrence, M., Collins, W.J., Folberth, G.A., Rumbold, S., Pyle, J.A., Russo, M.R., Stock, Z., Siour, G., Coll, I., D'Allura, A., Finardi, S., Radice, P. and Silibello, C. (2011) Prediction of Megacities Impact on Regional and Global Atmospheric Composition. Deliverable D5.4, *MEGAPOLI Scientific Report 11-11*, MEGAPOLI-37-REP-2011-06

Changnon, S.A., Jr. (1968) The La Porte anomaly – fact or fiction? *Bulletin of the American Meteorological Society*, 49, 4–11.

Chapman, M.G. and Underwood, A.J. (2011) Evaluation of ecological engineering of "armoured" shorelines to improve their value as habitat. *Journal of Experimental Marine Biology and Ecology* 400: 302–313.

Church J.A., White, N.J. (2006) A 20th century acceleration in global sea-level rise, *Geophysics Research Letters*, 33, L01602, doi:10.1029/2005GL024826.

Clancy, J.S. (2008) Urban ecological footprints in Africa. *African Journal of Ecology* 46: 463–470.

Clark, R.T., Brown, S.J. and Murphy, J.M. (2006) Modeling Northern Hemisphere Summer Heat Extreme Changes and Their Uncertainties Using a Physics Ensemble of Climate Sensitivity Experiments. *Journal of Climate* 19: 4418–4435.

Cleary, D.F.R., Suharsono, and Hoeksema, B.W. (2006) Coral diversity across a disturbance gradient in the Pulau Seribu reef complex off Jakarta, Indonesia. *Biodiversity and Conservation* 15: 3653–3674.

Colle, B.A., Olson, J.B. and Tongue, J.S. (2003) Multi-season verification of the MM5: Part II, Evaluation of high resolution precipitation forecasts over the Northeast U.S. *Weather Forecasting* 18: 458–480.

Collins, W.J. (2009) Global radiative forcing from megacity emissions of long-lived greenhouse gases. Deliverable 6.1, *MEGAPOLI Scientific Report 09-01*, MEGAPOLI-01-REP-2009-10, ISBN: 978-87-992924-1-7.

Costanza, R., d'Arge, R., deGroot, R., Farber, S., Grasso, M., Hannon, B., Limburg, K., Naeem, S., ONeill, R.V., Paruelo, J., Raskin, R.G., Sutton, P. and vanden Belt, M. (1997) The value of the world's ecosystem services and natural capital. *Nature* 387: 253–260.

Dandou, A., Tombrou, M. and Soulakellis, N. (2009) The Influence of the City of Athens on the Evolution of the Sea-Breeze Front. *Boundary-Layer Meteorology* 131(1): 35–51.

Dennison, W.C. (2008) Environmental problem solving in coastal ecosystems: A paradigm shift to sustainability. *Estuarine, Coastal and Shelf Science* 77: 185–196.

Derraik, J.G.B. (2002) The pollution of the marine environment by plastic debris: a review. *Marine Pollution Bulletin* 44: 842–852.

Dietzel, C., Herold, M., Hemphill, J.J. and Clarke, K.C. (2005) Spatio-temporal dynamics in California's central valley: Empirical links to urban theory. *International Journal of Geographical Information Science* 19: 175–195.

Dobson, J.E., Bright, E.A., Coleman, P.R., Durfee, R.C. and Worley, B.A. (2000) LandScan: A global population database for estimating populations at risk. *Photogrammetric Engineering and Remote Sensing* 66: 849–857.

Duckworth, F.A. and Sandberg, J.S. (1954) The effect on cities upon horizontal and vertical temperature gradients. *Bulletin of the American Meteorological Society*, 35, 198–207.

Duncan, R.P., Clemants, S.E., Corlett, R.T., Hahs, A.K., McCarthy, M.A., McDonnell, M.J., Schwartz, M.W., Thompson, K., Vesk, P.A. and Williams, N.S.G. (2011) Plant traits and extinction in urban areas: a meta-analysis of 11 cities. *Global Ecology and Biogeography* 20: 509–519.

Eckhardt, S., Cassiani, M. and Stohl, A. (2011) Influence of North American Megacities on European Atmospheric Composition. Deliverable D5.6, *MEGAPOLI Scientific Report 11-10*, MEGA-POLI-36-REP-2011-06.

Elton, C.S. (1958) *The Ecology of Invasions by Animals and Plants*, Methuen and Co. Ltd., London.

Emori, S. and Brown, S.J. (2005) Dynamic and thermodynamic changes in mean and extreme precipitation under changed climate. *Geophysics Research Letters*, 32.

Faerge, J., Magid, J. and de Vries, F.W.T.P. (2001) Urban nutrient balance for Bangkok. *Ecological Modelling* 139: 63–74.

Feddema, J., Oleson, K., Bonan, G., Mearns, L., Washington, W., Meehl, G., and Nychka, D. (2005) A comparison of a GCM response to historical anthropogenic land cover change and model sensitivity to uncertainty in present-day land cover representations. *Climate Dynamics* 25, 581–609

Fischer, E.M., Oleson, K.W. and Lawrence, D.M. (2012) Contrasting urban and rural heat stress responses to climate change, *Geophysics Research Letters* 39: L03705.

Flanner, M.G. (2009) Integrating anthropogenic heat flux with global climate models. *Geophysics Research Letters* 36: L02801.

Folberth, G.A., Rumbold, S., Collins, W.J. and Butler, T. (2011) Regional and Global Climate Changes due to Megacities using Coupled and Uncoupled Models. Deliverable D6.6, *MEGAPOLI Scientific Report 11-07*, MEGAPOLI-33-REP-2011-06.

Folke, C., Jansson, A., Larsson, J. and Costanza, R. (1997) Ecosystem appropriation by cities. *Ambio* 26: 167–172.

Folke, C., Kautsky, N., Berg, H., Jansson, A. and Troell, M. (1998) The ecological footprint concept for sustainable seafood production: A review. *Ecological Applications* 8: S63–S71.

Forman, R.T.T. (2008) *Urban regions: ecology and planning beyond the city*, Cambridge University Press, Cambridge.

Francis, R.A. and Chadwick, M.A. (2012) What makes a species synurbic? *Applied Geography* 32: 514–521.

Francis, R.A. and Hoggart, S.P.G. (2008) Waste not, want not: the need to utilise existing artificial structures for habitat improvement along urban rivers. *Restoration Ecology* 16, 373–381.

Francis, R.A. and Hoggart, S.P.G. (2012) The flora of urban river wallscapes. *River Research and Applications* DOI: 10.1002/rra.1497

Francis, R.A., Hoggart, S.P.G., Gurnell, A.M. and Coode, C. (2008) Meeting the challenges of urban river habitat restoration: developing a methodology for the River Thames through central London. *Area* 40: 435–445.

Freitas, E.D., Rozoff, C.M., Cotton, W.R. and Silva Dias, P.L. (2007) Interactions of an urban heat island and sea-breeze circulations during winter over the metropolitan area of São Paulo, Brazil. *Boundary-Layer Meteorology*, 122: 43–65.

Gauss, M., Myhre, G., Isaksen, I.S.A., Grewe, V., Pitari, G., Wild, O., Collins, W.J., Dentener, F.J., Ellingsen, K., Gohar, L.K., Hauglustaine, D.A., Iachetti, D., Lamarque, J.-F., Mancini, E., Mickley, L.J., Prather, M.J., Pyle, J.A., Sanderson, M.G., Shine, K.P., Stevenson, D.S., Sudo, K., Szopa, S. and Zeng, G. (2006) Radiative forcing since preindustrial times due to ozone change in the troposphere and the lower stratosphere. *Atmospheric Chemistry and Physics*, 6: 575–599.

Gedzelman, S.D., Austin, S., Cermak, R., Stefano, N., Partridge, S., Quesenberry, S., and Robinson, D.A. (2003) Mesoscale aspects of the urban heat island around New York City. *Theoretical Applied Climatology* 75: 29–42.

Gleick, P.H. (1996) Basic water requirements for human activities: meeting basic needs. *Water International*, 21, 83–92.

Goudey, C.A. and Moran, B.M. (2005) Four years of recirculating aquaculture in Urban Boston Harbor, USA. In Costa-Pierce, B., Desbonnet, A., Edwards, P. and Baker, D. (eds) *Urban Aquaculture*, CABI Publishing, Oxford.

Greer, K. and Stow, D. (2003) Vegetation type conversion in Los Penasquitos Lagoon, California: An examination of the role of watershed urbanization. *Environmental Management* 31: 489–503.

Grimm, N.B., Faeth, S.H., Golubiewski, N.E., Redman, C.L., Wu, J.G., Bai, X.M. and Briggs, J.M. (2008) Global change and the ecology of cities. *Science* 319: 756–760.

Grimmond, C.S.B., Blackett, M., Best, M., Barlow, J., Baik, J.J., Belcher, S., Bohnenstengel, S.I., Calmet, I., Chen, F., Dandou, A., Fortuniak, K., Gouvea, M.L., Hamdi, R., Hendry, M., Kawai, T., Kawamoto, Y., Kondo, H., Krayenhoff, E.S., Lee, S.H., Loridan, T., Martilli, A., Masson, V., Miao, S., Oleson, K., Pigeon, G., Porson, A., Ryu, Y.H., Salamanca, F., Steeneveld, G.J., Tombrou, M., Voogt, J., Young, D. and Zhang, N. (2010a) The International Urban Energy Balance Models Comparison Project: First results from Phase 1. *Journal of Applied Meteorology & Climatology*: 1268–1292.

Grimmond, C.S.B., Blackett, M., Best, M.J., Baik, J.-J., Belcher, S.E., Beringer, J., Bohnenstengel, S.I., Calmet, I., Chen, F., Coutts, A., Dandou, A., Fortuniak, K., Gouvea, M.L., Hamdi, R., Hendry, M., Kanda, M., Kawai, T., Kawamoto, Y., Kondo, H., Krayenhoff, E.S., Lee, S.-H., Loridan, T., Martilli, A., Masson, V., Miao, S., Oleson, K., Ooka, R., Pigeon, G., Porson, A., Ryu, Y.-H., Salamanca, F., Steeneveld, G.-J., Tombrou, M., Voogt, J.A., Young, D. and Zhang, N. (XXX) Initial Results from Phase 2 of the International Urban Energy Balance Comparison Project. *International Journal of Climatology* 31: 244–272.

Grimmond, C.S.B., Roth, M., Oke, T.R., Au, Y.C., Best, M., Betts, R., Carmichael, G., Cleugh, H., Dabberdt, W., Emmanuel, R., Freitas, E., Fortuniak, K., Hanna, S., Klein, P., Kalkstein, L.S., Liu, C.H., Nickson, A., Pearlmutter, D., Sailor, D. and Voogt, J. (2010): Climate & More Sustainable Cities: Climate Information for Improved Planning & Management of Cities (Producers/Capabilities Perspective). *Procedia Environmental Sciences*, 1, 247–274.

Holt, T. and Pullen, J. (2007) Urban canopy modeling of the New York City metropolitan area: A comparison and validation of single- and multilayer parameterizations. *Monthly Weather Review* 135: 1906–1930.

Holt, T., Pullen, J. and Bishop, C.H. (2009) Urban and ocean ensembles for improved meteorological and dispersion modeling of the coastal zone. *Tellus* 61A: 232–249.

Horton, R.M., Gornitz, V., Bader, D.A., Ruane, A.C., Goldberg, R. and Rosenzweig, C. (2011) Climate Hazard Assessment for Stakeholder Adaptation Planning in New York City. *Journal of Applied Meteorology and Climatology* 50(11) (November): 2247–2266.

Hu, D., Huang, S.L., Feng, Q., Li, F., Zhao, J.J., Zhao, Y.H. and Wang, B.N. (2008a) Relationships between rapid urban development and the appropriation of ecosystems in Jiangyin City, Eastern China. *Landscape and Urban Planning* 87: 180–191.

Hu, D., Li, F., Wang, B., Lei, K., Cao, A., Wang, Z. and Yin-Hua, L. (2008b) An effect analysis of changes in the composition of the water ecological footprint in Jiangyin City, China. *International Journal of Sustainable Development and World Ecology* 15: 211–221.

Hubacek, K., Guan, D., Barrett, J. and Wiedmann, T. (2009) Environmental implications of urbanization and lifestyle change in China: Ecological and Water Footprints. *Journal of Cleaner Production* 17(14): 1241–1248.

Huff, F.A. and Changnon Jr. S.A. (1972) Climatological assessment of urban effects on precipitation at St. Louis. *J. Appl. Meteor.*, 11, 823–842.

Hurtt, G., Chini, L.P., Frolking, S., Betts, R.A., Feddema, J., Fischer, G., Fisk, J.P., Hibbard, K., Houghton, R.A., Janetos, A., Jones, C.D., Kindermann, G., Kinoshita, T., Goldewijk, K.K., Riahi, K., Shevliakova, E., Smith, S., Stehfest, E., Thomson, A., Thornton, P., van Vuuren, D.P., Wang, Y.P. (2011) Harmonization of land-use scenarios for the period

1500–2100: 600 years of global gridded annual land-use transitions, wood harvest, and resulting secondary lands. *Climatic Change* 109, 117–161.

Iamarino, M., Beevers, S. and Grimmond, C.S.B. (2011) High-resolution (space, time) anthropogenic heat emissions: London 1970–2025. *International Journal of Climatology*. doi: 10.1002/joc.2390.

Ichinose, T., Shimodozono, K. and Hanaki, K. (1999) Impact of anthropogenic heat on urban climate in Tokyo. *Atmospheric Environment* 33: 3897–3909.

IEA, International Energy Agency (2008) *World Energy Outlook 2008*, International Energy Agency, Paris.

IPCC (2012) Summary for Policymakers. In: Field, C.B., Barros, V., Stocker, T.F., Qin, D., Dokken, D.J., Ebi, K.L., Mastrandrea, M.D., Mach, K.J., Plattner, G.-K., Allen, S.K., Tignore, M. and Midgley, P.M. (eds) *Managing the Risks of Extreme Events and Disasters to Advance Climate Change Adaptation*, A Special Report of Working Groups I and II of the Intergovernmental Panel on Climate Change, Cambridge University Press, Cambridge.

Jackson, T.L., Feddema, J.J., Oleson, K.W., Bonan, G.B. and Bauer, J.T. (2010) Parameterization of Urban Characteristics for Global Climate Modeling. *Annals of the Association of American Geographers* 100: 848–865.

Jacobson, M.Z. and Ten Hoeve, J.E. (2011) Effects of Urban Surfaces and White Roofs on Global and Regional Climate, *Journal of Climate*, 25: 1028–1044.

Jaing, Y., Kirkman, H. and Hau, A. (2001) Megacity development: managing impacts on marine environments. *Ocean and Coastal Management* 44: 293–318.

Jenerette, G.D., Marussich, W.A. and Newell, J.P. (2006a) Linking ecological footprints with ecosystem valuation in the provisioning of urban freshwater. *Ecological Economics* 59: 38–47.

Jenerette, G.D., Wu, W.L., Goldsmith, S., Marussich, W.A. and Roach, W.J. (2006b) Contrasting water footprints of cities in China and the United States. *Ecological Economics* 57: 346–358.

Jones, C.D., Hughes, J.K., Bellouin, N., Hardiman, S.C., Jones, G.S., Knight, J., Liddicoat, S., O'Connor, F.M., Andres, R.J., Bell, C., Boo, K.-O., Bozzo, A., Butchart, N., Cadule, P., Corbin, K.D., Doutriaux-Boucher, M., Friedlingstein, P., Gornall, J., Gray, L., Halloran, P.R., Hurtt, G., Ingram, W.J., Lamarque, J.-F., Law, R.M., Meinshausen, M., Osprey, S., Palin, E.J., Parsons, L., Chini, Raddatz, T., Sanderson, M.G., Sellar, A.A., Schurer, A., Valdes, P., Wood, N., Woodward, S., Yoshioka, M. and Zerroukat, M. (2011) The HADGEM2-ES implementation of cmip5 centennial simulations. *Geoscientific Model Development* 4: 543–570.

Kaye, J.P., Groffman, P.M., Grimm, N.B., Baker, L.A. and Pouyat, R.V. (2006) A distinct urban biogeochemistry? *Trends in Ecology and Evolution* 21: 192–199.

Khan, S.M. and Simpson, R.W. (2001) Effect of a heat island on the meteorology of a complex urban airshed. *Boundary-layer Meteorology* 100(3): 487–506.

Kharin, V.V., Zwiers, F.W., Zhang, X. and Hegerl, G.C. (2007) Changes in Temperature and Precipitation Extremes in the IPCC Ensemble of Global Coupled Model Simulations. *Journal of Climate* 20: 1419–1444.

Knutson, T.R., McBride, J., Chan, J., Emanuel, K.A., Holland, G., Landsea, C., Held, I.M., Kossin, J., Srivastava, A.K. and Sugi, M. (2010) Tropical cyclones and climate change. *Nature Geoscience*, 3, DOI:doi:10.1038/ngeo779

Lam, N.W.Y., Huang, R. and Chan, B.K.K. (2009) Variations in intertidal assemblages and zonation patterns between vertical artificial seawalls and natural rocky shores: A case study from Victoria Harbour, Hong Kong. *Zoological Studies* 48: 184–195.

Lamarque, J.F., Bond, T.C., Eyring, V., Granier, C., Heil, A., Klimont, Z., Lee, D., Liousse, C., Mieville, A., Owen, B., Schultz, M.G., Shindell, D., Smith, S.J., Stehfest, E., Van Aardenne, J., Cooper, O.R., Kainuma, M., Mahowald, N., McConnell, J.R., Naik, V., Riahi,

K. and van Vuuren, D.P. (2010) Historical (1850–2000) gridded anthropogenic and biomass burning emissions of reactive gases and aerosols: methodology and application. *Atmospheric Chemistry and Physics* 10: 7017–7039.

Landsberg, H.E. (1970) Man-made climatic changes. *Science* 170: 1265–1274.

Landscan (2004) High Resolution Global Population Data Set copyrighted by UT-Battelle, LLC, operator of Oak Ridge National Laboratory under Contract No. DE-AC05-00OR22725 with the United States Department of Energy, www.ornl.gov/sci/landscan/index.shtml (last accessed 3 April 2013).

Lanti, A. (2006) A Regulatory Approach to the Jakarta Water Supply Concession Contracts. *International Journal of Water Resources Development* 22(2): 255–276.

Lawrence, M.G., Butler, T.M., Steinkamp, J., Gurjar, B.R. and Lelieveld, J. (2007) Regional pollution potentials of megacities and other major population centers. *Atmospheric Chemistry and Physics* 7: 3969–3987.

Lebassi B., González, J., Fabris, D., Maurer, E., Miller, N., Milesi, C. and Bornstein, R. (2009) Observed 1970–2005 cooling of summer daytime temperatures in coastal California. J. Climate 22, 3558-3573. http://dx.doi.org/10.1175/2008JCLI2111.1.

Lebassi, B., González, J., Fabris, D., Maurer, E., Miller, N., Milesi, C., Switzer, P. and Bornstein, R. (2009) Observed 1970–2005 Cooling of Summer Daytime Temperatures in Coastal California. *Journal of Climate* 22: 3558–3573.

Leppäkoski, E., Gollasch, S. and Olenin, S. (2002) Introduction. In: Lepäkoski, E., Gollasch, S. and Olenin, S. (eds) *Invasive Aquatic Species of Europe: Distribution, Impacts and Management*, Kluwer Academic Publishers, Amsterdam.

Li, M.S. and Lee, S.Y. (1997) Mangroves of China: A brief review. *Forest Ecology and Management* 96: 241–259.

Lin, N., Emanuel, K., Oppenheimer, M. and Vanmarcke, E. (2012) Physically based assessment of hurricane surge threat under climate change. *Nature Geoscience* DOI: 10.1038/NCLIMATE1389

Liu, J.H. and Kueh, C.S.W. (2005) Biomonitoring of heavy metals and trace organics using the intertidal mussel *Perna viridis* in Hong Kong coastal waters. *Marine Pollution Bulletin* 51: 857–875.

Loose, T. and Bornstein, R.D. (1977) Observations of mesoscale effects on frontal movement through an urban area. *Monthly Weather Review* 105: 563–571.

Lowe, J.A. and Gregory, J.M. (2005) The effects of climate change on storm surges around the United Kingdom. *Philosophical Transactions of the Royal Society* (363(1831) (15 June): 1313–1328.

Mack, R.N. (2003) Global plant dispersal, naturalization, and invasion: Pathways, modes and circumstances. In: Ruiz, G.M. and Carlton, J.T. (eds) *Invasive Species: Vectors and Management Strategies*, Island Press, Washington, DC.

Martilli, A. (2003) A two-dimensional numerical study of the impact of a city on atmospheric circulation and pollutant dispersion in a coastal environment. *Boundary Layer Meteor* 108: 91–119.

Martinez, M.L., Intralawan, A., Vazquez, G., Perez-Maqueo, O., Sutton, P. and Landgrave, R. (1997) The coasts of our world: Ecological, economic and social importance. *Ecological Economics* 63: 254–272.

Matlack, G.R. (1997) Four centuries of forest clearance and regeneration in the hinterland of a large city. *Journal of Biogeography* 24: 281–295.

McCarthy, M.P. and Sanderson, M.G. (2011) Urban Heat Islands: Sensitivity of urban temperatures to climate change and heat release in four European cities. In: *Cities and climate change: responding to an urgent agenda*, World Bank, Washington, DC.

McCarthy, M.P., Best, M.J. and Betts, R.A. (2010) Climate change in cities due to global warming and urban effects. *Geophysical Research Letters* 37: L09705+.

McCarthy, M.P., Harpham, C., Goodess, C.M. and Jones, P.D. (2011) Simulating climate change in UK cities using a regional climate model, HadRM3. International Journal Climatology.

Meehl, G.A. and Tebaldi, C. (2004) More intense, more frequent, and longer lasting heat waves in the 21st century. *Science* 305(5686): 994–997.

Meehl, G.A., Arblaster, J.M. and Tebaldi, C. (2005) Understanding future patterns of increased precipitation intensity in climate model simulations, *Geophysical Research Letters* 32: L18719.

MegaPoli (2011) Megacities: Emissions, urban, regional and global atmospheric pollution and climate effects, and integrated tools for assessment and mitigation (EC FP7 project #212520). Project Final Report, http://megapoli.dmi.dk/publ/MEGAPOLI_FinalReport_Part1.pdf

Melas, D., Ziomas, I.C. and Zerefos Christos, J. (1995) Boundary layer dynamics in an urban coastal environment under sea breeze conditions. *Atmospheric Environment* 29(24): 3605–3617.

Mestayer, P.G., Durand, P., Augustin, P., Bastin, S., Bonnefond, J.M., Bénech, B., Campistron, B., Coppalle, A., Delbarre, H., Dousset, B., Drobinsk, P., Druilhet, A., Fréjafon, E., Grimmond, C.S.B., Groleau, D., Irvine, M., Kergomard, C., Kermadi, S., Lagouarde, J.P., Lemonsu, A., Lohou, F., Long, N., Masson, V., Moppert, C., Noilhan, J., Offerle, B., Oke, T., Pigeon, G., Puygrenier, V., Roberts, S., Rosant, J.M., Saïd, F., Salmond, J., Talbaut, M. and Voogt, J. (2005) The Urban Boundary-Layer Field campaign in Marseille (UBL/CLU-ESCOMPTE): Set-up and first results. *Boundary-Layer Meteorology* 114: 315–365.

Millennium Ecosystem Assessment, MEA (2005) Ecosystems and human well-being: wetlands and water. World Resources Institute, Washington, DC.

Mills, G. (2007) Cities as agents of global change. *International Journal of Climatology* 27: 1849–1857.

Mills, G., Cleugh, H.A., Emmanuel, R., Endlicher, W., Erell, E., McGranahan, G., Ng, E., Nickson, A., Rosenthal, J. and Steemer, K. (2010) Climate information for improved planning and management of mega cities (needs perspective). *Procedia Environmental Sciences* 1: 228–246.

Molle, F. (2004) Technical and Institutional Responses to Basin Closure in the Chao Phraya River Basin, Thailand. *Water International* 29(1): 70–80.

Molle, F. (2007) Scales and power in river basin management: the Chao Phraya River in Thailand. *Geographical Journal* 173(4): 358–373.

Molle, F. and Floch, P. (2008) Megaprojects and social and environmental changes: The case of the Thai water grid. *Ambio* 37(3): 199–204.

Moore, C.J., Lattin, G.L. and Zellers, A.F. (2011) Quantity and type of plastic debris flowing from two urban rivers to coastal waters and beaches of Southern California. *Journal of Integrated Coastal Zone Management* 11: 65–73.

Murray-Smith, C., Brummitt, N.A., Oliveira-Filho, A.T., Bachman, S., Moat, J., Lughadha, E.M.N. and Lucas, E.J. (2009) Plant diversity hotspots in the Atlantic coastal forests of Brazil. *Conservation Biology* 23: 151–163.

Murthy, R.C., Rao, Y.R. and Inamdar, A.B. (2001) Integrated coastal management of Mumbai metropolitan region. *Ocean & Coastal Management* 44: 355–369.

Myers, N., Mittermeier, R.A., Mittermeier, C.G., da Fonseca, G.A.B. and Kent, J. (2000) Biodiversity hotspots for conservation priorities. *Nature* 403: 853–858.

Nakayama, T., Watanabe, M., Tanji, K. and Morioka, T. (2007) Effect of underground urban structures on eutrophic coastal environment. *Science of the Total Environment* 373: 270–288.

Nelson, M.J., Pant, C.P., Self, L.S. and Usman, S. (1976) Observations on the breeding habitats of Aedes aegypti in Jakarta, Indonesia. *Southeast Asian Journal of Tropical Medicine and Public Health* 7: 424–429.

Nicholls, R.J., Wong, P.P., Burkett, V.R., Codignotto, J.O., Hay, J.E., McLean, R.F., Ragoonaden, S. and Woodroffe, C.D. (2007) Coastal systems and low-lying areas. In: Parry, M.L., Canziani, O.F., Palutikof, J.P., van der Linden, J.P. and Hanson, C.E. (eds) *Climate Change 2007: Impacts, Adaptation and Vulnerability. Contribution of Working Group II to the Fourth Assessment Report of the Intergovernmental Panel on Climate Change*, Cambridge University Press, Cambridge.

Nicholls, R.J., Hanson, S., Herweijer, C., Patmore, N., Hallegatte, S., Corfee-Morlot, J., Chateau, J. and Muir-Wood, R. (2008) Ranking Port Cities with High Exposure and Vulnerability to Climate Extremes: Exposure Estimates, OECD Environment Working Papers, No. 1, OECD Publishing.

Nixon, S.W. (1995) Coastal marine eutrophication: A definition, social causes and future concerns. *Ophelia* 41: 199–219.

Nixon, S.W., Granger, S.L. and Nowicki, B.L. (1995) An assessment of the annual mass balance of carbon, nitrogen, and phosphorus in Narragansett Bay. *Biogeochemistry* 31: 15–61.

Ohashi, Y. and Kida, H. (2002) Local circulations developed in the vicinity of both coastal and inland urban areas: Numerical study with a mesoscale atmospheric model. *Journal of Applied Meteorology* 41: 30–45.

Oke, T.R. (1982) The energetic basis of the urban heat island. *Quarterly Journal of the Royal Meteorological Society*, 108: 1–24.

Oleson, K.W., Bonan, G.B., Feddema, J., Vertenstein, M. and Grimmond, C.S.B (2008a) An urban parameterization for a global climate model: 1. Formulation and evaluation for two cities. *Journal of Applied Meteorology and Climatology* 47(4): 1038–1060.

Oleson, K.W., Bonan, G.B., Feddema, J. and Vertenstein, M. (2008b) An urban parameterization for a global climate model: 2. Sensitivity to input parameters and the simulated urban heat island in offline simulations. *Journal of Applied Meteorology and Climatology* 47(4): 1061–1076.

Oleson, K.W., Bonan, G.B., Feddema, J. and Jackson, T. (2011) An examination of urban heat island characteristics in a global climate model. *International Journal of Climatology*, 31: 1848–1865.

Olivier, J.G.J., van Aardenne, J.A., Dentener, F., Ganzeveld, L. and Peters, J.A.H.W. (2005): Recent trends in global greenhouse gas emissions: regional trends and spatial distribution of key sources. In: "Non-CO2 Greenhouse Gases (NCGG-4)", A. van Amstel (coord.), 325–330.

Pandey, V.P., Shrestha, S., Chapagain, S.K. and Kazama, F. (2011) A framework for measuring groundwater sustainability. *Environmental Science & Policy* 14(4): 396–407.

Parker, D.E. (2010) Urban heat island effects on estimates of observed climate change. WIREs Clim Change 1, 123–133. http://dx.doi.org/10.1002/wcc.21.

Pauchard, A., Aguayo, M., Pena, E. and Urrutia, R. (2006) Multiple effects of urbanization on the biodiversity of developing countries: The case of a fast-growing metropolitan area (Concepcion, Chile). *Biological Conservation* 127: 272–281.

Peng, S.T., Dai, M.X., Hu, Y.D., Bai, Z.P. and Zhou, R. (2009) Long-term (1996–2006) variation of nitrogen and phosphorus and their spatial distributions in Tianjin coastal seawater. *Bulletin of Environmental Contamination and Toxicology* 83: 416–421.

Porson, A., Clark, P.A., Harman, I.N., Best, M.J. and Belcher, S.E. (2010) Implementation of a new urban energy budget scheme in the MetUM. part i: Description and idealized simulations. *Quarterly Journal of the Royal Meteorological Society* 136: 1514–1529.

Pullen, J., Holt, T., Blumberg, A. and Bornstein, R. (2007) Atmospheric response to local upwelling in the vicinity of New York/New Jersey Harbor, *Journal of Applied Meteorology and Climatology* 46: 1031–1052.

Pullen, J., Ching, J., Sailor, D., Thompson, W., Bornstein, R. and Koracin, D. (2008) Progress toward meeting the challenges of our coastal urban future, *Bulletin of the American Meteorological Society*, 89 (11).

Quah, A.K.L. and Roth, M. (2012) Diurnal and weekly variation of anthropogenic heat emissions in a tropical city, Singapore, *Atmospheric Environment* 46: 92–103.

Ramaswamy, V., Boucher, O., Haigh, J., Hauglustaine, D., Haywood, J., Myhre, G., Nakajima, T., Shi, G.Y., Solomon, S., Betts, R., Charlson, R., Chuang, C., Daniel, J.S., Del Genio, A., van Dorland, R., Feichter, J., Fuglestvedt, J., de F Forster, P.M., Ghan, S.J., Jones, A., Kiehl, J.T., Koch, D., Land, C., Lean, J., Lohmann, U., Minschwaner, K., Penner, J.E., Roberts, D.L., Rodhe, H., Roelofs, G.J., Rotstayn, L.D., Schneider, T.L., Schumann, U., Schwartz, S.E., Schwarzkopf, M.D., Shine, K.P., Smith, S., Stevenson, D.S., Stordal, F., Tegen, I. and Zhang, Y. (2001) Radiative forcing of climate change. In: *Climate Change 2001: The Scientific Basis*. Contribution of Working Group I to the Third Assessment Report of the Intergovernmental Panel on Climate Change. Cambridge University Press, Cambridge.

Riahi, K., Gruebler, A. and Nakicenovic N. (2007) Scenarios of long-term socio-economic and environmental development under climate stabilization. *Technological Forecasting and Social Change* 74(7): 887–935.

Ribic, C.A., Sheavly, S.B., Rugg, D.J. and Erdmann, E.S. (2010) Trends and drivers of marine debris on the Atlantic coast of the United States 1997–2007. *Marine Pollution Bulletin* 60: 1231–1242.

Santi, L., Tavares, M. and Omena, E. (2006) Patterns of species richness and species density of sublittoral soft-bottom polychaetes in a grossly polluted urban bay: Guanabara Bay, Rio de Janeiro, Brazil. *Journal of Coastal Research* 39: 1127–1131.

Satterthwaite, D. (2008) Cities' contribution to global warming: notes on the allocation of greenhouse gas emissions. *Environment and Urbanization* 20: 539–549.

Sekovski, I., Newton, A. and Dennison, W.C. (2012) Megacities in the coastal zone: Using a driver-pressure-state-impact-response framework to address complex environmental problems. *Estuarine, Coastal and Shelf Science* 96: 48–59.

Seto, K.C., Fragkias, M., Guneralp, B. and Reilly, M.K. (2011) A Meta-Analysis of Global Urban Land Expansion, *PLOS One 6(8).*

Shepherd, J.M., Pierce, H. and Negri, A.J. (2002) Rainfall modification by major urban areas: Observations from space-borne rain radar on TRMM. *Journal of Applied Meteorology* 41: 689–701.

Smith, V.H. (2003) Eutrophication of freshwater and coastal marine ecosystems: a global problem. *Environmental Science and Pollution Research* 10: 126–139.

Strzepek, K. and Boehlert, B. (2010) Competition for water for the food system. *Philosophical Transactions of the Royal Society of London B*, 365: 2927–2940.

Sudol, M.F. and Ambrose, R.F. (2002) The US Clean Water Act and habitat replacement: Evaluation of mitigation sites in Orange County, California, USA. *Environmental Management* 30: 727–734.

Taniguchi, M., Shimada, J., Fukuda, Y., Yamano, M., Onodera, S., Kaneko, S. and Yoshikoshi, A. (2009) Anthropogenic effects on the subsurface thermal and groundwater environments in Osaka, Japan and Bangkok, Thailand. *Science of the Total Environment*, 407: 3153–3164.

Tamburri, M.N., Wasson, K. and Matsude, M. (2002) Ballast water deoxygenation can prevent aquatic introductions while reducing ship corrosion. *Biological Conservation* 103: 331–341.

Thompson, W., Holt, T. and Pullen, J. (2007) Investigation of a sea breeze front in an urban environment. *Quarterly Journal of the Royal Meteorological Society* 133: 570–594.

Trusilova, K. and Churkina, G. (2008) The response of the terrestrial biosphere to urbanization: land cover conversion, climate, and urban pollution. *Biogeosciences* 5: 1505–1515.

Trusilova, K., Jung, M., Churkina, G., Karstens, U., Heimann, M. and Claussen, M. (2008) Urbanization impacts on the climate in Europe: Numerical experiments by the PSU – NCAR mesoscale model (MM5). *Journal of Applied Meteorology and Climatology* 47: 1442–1455.

United Nations, Department of Economic and Social Affairs (UN-DESA), Population Division (2012a) "World Urbanization Prospects: The 2011 Revision – Highlights", http://esa.un.org/unpd/wup/index.htm [accessed 06/07/2012]

UN, United Nations (2009) World Urbanization Prospects: The 2009 Revision, UN Dept of Economic and Social Affairs, Population Division (http://esa.un.org/unpd/wup/)

UN Dept of Economic and Social Affairs 2005, Urban Agglomerations, Population Division: New York.

Valiela, I. and Martinetto, P. (2007) Changes in bird abundance in eastern North America: Urban sprawl and global footprint? *Bioscience* 57: 360–370.

Varis, O. (2006) Megacities, Development and Water. *International Journal of Water Resources Development* 22(2): 199–225.

Voogt, J.A. and Grimmond, C.S.B. (2000) Modeling surface sensible heat flux using surface radiative temperatures in a simple urban area. *Journal of Applied Meteorology* 39: 1679–1699.

World Water Assessment Programme (WWAP) (2009) The United Nations World Water Development Report 3: Water in a Changing World. Paris: UNESCO, and London: Earthscan.

World Wildlife Fund (2011) *Mega-Stress for Mega-Cities: A Climate Vulnerability Ranking of Major Coastal Cities in Asia*, http:*//assets.panda.org/downloads/mega_cities_report.pdf* (last accessed 6 November 2011).

Xie (2011) Evaluating Habitat Change and Boundary Adjustment of a Nature Reserve in Coastal Wetlands: Case Study of Beidagang Nature Reserve, China. *Journal of Coastal Research*

Yoshikado, H. (1992) Numerical study of the daytime urban effect and its interaction with the sea breeze. *Journal of Applied Meteorology and Climatology* 31: 1146–1164.

Yoshikado, H. (1994) Interaction of the sea breeze with the urban heat islands of different sizes and locations. *Journal of the Meteorological Society of Japan* 72: 139–142.

Yoshikado, H. and Tsuchida, M. (1996) High levels of winter air pollution under the influence of the urban heat island along the shore of Tokyo Bay. *Journal of Applied Meteorology and Climatology* 35(10): 1804–1813.

Zhang, J., Yang, S.L., Xu, Z.L. and Wu, Y. (2006) *Impact of human activities on the health of ecosystems on the Changjaing Delta region.* In: Wolanski, E. (ed.) *The Environment in Asia Pacific Harbours, Springer: Dordrecht.*

Zhang, Y., Peng, X.F., Sun, X. and Chen, Z.H. (2011) Microclimate effects and ecosystem services evaluation of urban wetlands in Guangzhou City. 5th International Conference on Bioinformatics and Biomedical Engineering, 1–7.

Zhao, S., Peng, C., Jiang, H., Tian, D., Lei, X. and Zhou, X. (2006) *Land use change in Asia and the ecological consequences. Ecological Research* 21: 890–896.

Chapter 5

# Reducing risk from natural hazard, pollution and climate change in megacities and associated networks

Lead author: *Jessica Lamond (University of the West of England, UK)*

Contributing authors: *Andreas Kannen (Helmholtz-Zentrum Geesthalcht, Germany), Rob Francis (King's College London), Harriet Bulkeley (University of Durham), Nassos Vafeidis (Christian-Albrechts University Kiel, Germany), Phillippa Mason (Imperial College London, UK), Chao-Yi Chung, Colin Booth (University of West England, UK), William Solecki (CUNY, USA), Kapil Gupta (Indian Institute of Technology, India), Justus Kithiia (Macquarie University, Australia), Tom Measham (CSIRO, Australia), Fabiana Barbi (UNICAMP), Magnus Moglia (CSIRO, Australia), Veronique Morin (Asian Institute for Technology, Thailand)*

## 1 Introduction

Coastal megacities face many challenges, as described in the preceding chapters. These challenges are related to the concentration of the urban environment, rate of land-use change and land-cover loss, the sheer scale of the population and its associated infrastructure needs that cause critical infrastructure networks and resource demands to spread well beyond the urban core. These are generic concerns for large cities but made more critical for cities on the coast where coastal location brings exposure to specific hazards (with associated implications for living with climate change) and intensifies vulnerability through the presence of international transport and trade centres and the populations they attract. However, cities are also well described as the engine room of a local economy, and can be the site of a large proportion of national and international product (gross world product) (Kamal-Chaoui and Robert, 2009; Bicknell *et al.*, 2009). Megacities on the coast therefore present huge exposures, challenges and the scope for focused efforts to reduce risk. In many ways as this book argues they are the fulcrum of global sustainability in the twenty-first century.

Given their size and resources, many coastal megacities have been the focus of attention for mitigation and more recently adaptation responses to climate risks, building on a considerable legacy of natural hazards and pollution reduction efforts such that there is a burgeoning resource of experience – successes and failures from coastal megacities. A recent report (Arup, 2011) on the climate actions undertaken by or planned in megacities, 15 of which are coastal, is subtitled '40 cities, 297 million residents and 4,734 climate actions', this alone illustrates the seriousness with which adaptation and mitigation has been viewed within megacities. Arup (2011) conclude that cities often lead the way in adaptation and that city mayors control many of the critical levers needed to dramatically reduce carbon emissions. For example the transport policies that have proved so successful for managing congestion and pollution including greenhouse gas emissions in London since the turn of the century are effective because they are largely within the control of the London mayor. However, not all problems associated with global environmental change can be tackled at city level. Different levels of governance need to cooperate across the scale and bring together the private and voluntary as well as state sector for interventions such as large-scale renewable energy generation to take hold (Bulkely and Betsill, 2003).

A critical aspect of urban governance in response to climate change is the availability and access to actionable scientific information. Cities serve as knowledge generators in both developed and developing countries due to the presence of leading universities and research institutions. The delivery of scientific knowledge to inform policy is essential to effective formulation and implementation especially when taken within the context of the complexity of the urban metabolism. The design and application of evidence-based policy relies on the ability to optimize opportunities present in cities to create an institutionalized nexus between knowledge and practice. Knowledge generation and utilization goes beyond the scientific. Recent work has explored the potential for experiments in inclusive, citizen-led urban planning and resource management to offer an alternative institutional architecture for urban development (Mitlin, 2012). The aim is not to find a single model for improved urban government to meet the multiple and interdependent challenges of the twenty-first century, but to see the coming decades as a prime opportunity to scope the range of experiments and embedded models that are confronting the challenges of sustainability for coastal megacities today and to ask how far such experiments are transferable, and how far they can contribute to a transformation of urban development.

This chapter sets out the existing framework for managing urban environmental risks with a focus on coastal megacities. It is followed in Chapter 6 by a more detailed critique of contemporary approaches and scoping of future demands. Chapter 5 lays out the existing landscape. First, we briefly revisit the primary hazards that come with a coastal location. This is followed in turn by accounts of contemporary experiences of urban planning and risk management, mitigation and adaptation.

## 2 Managing the Risks of a Coastal Location

The coast provides a wealth of opportunities for industry, urban development, residential activities, tourism, recreation, transport, fisheries, aquaculture, agriculture, trade and defence (military and natural) (Hadley, 2009). While the externalities of these developments lead to increased exposure of population and physical assets, they can lead to the deterioration and overexploitation of coastal resources, excessive urbanisation of the shore area, inappropriate disposal of solid, atmospheric and liquid waste and erosion or ecological degradation bringing loss of natural sea defences (Arthurton, 1998; Jiang *et al.*, 2001; Sekovski *et al.*, 2011). So much is well documented in the literature. Less clear, looking forward, is the extent to which such cities that are so adept at generating their own hazards and contributing to global environmental change will now also be on the receiving end of magnified risk, for example through sea-level rise and increased exposure to hazard, as outlined in Chapter 3 (Nicholls, 2002; Li, 2003). Uncertainty surrounding the likelihood and range of possible futures associated with climate variability and extreme events on the one hand and global economic and political change on the other pose wicked challenges for urban planners and risk managers, and especially those tasked with investing in fixed physical infrastructure solutions such as the design of critical infrastructure and flood or coastal defences.

### *2.1 Coastal erosion defences*

There is a pressing need to design, engineer and appropriately manage coastal defences (either manmade or natural) in a sustainable manner (King and Adeel, 2002), so as to protect those populations and neighbourhoods most at risk from coastal flooding (Wright, 2004). Though a wide variety of defensive structures (grouped as being either hard or soft engineering measures [French, 2001]) exists to address the issue, their implementation and management requires a thorough understanding and appreciation of coastal processes (Battjes, 2006). Otherwise, alleviating potential problems at one place can mean they are simply exacerbated elsewhere.

Hard engineering approaches include sea walls, breakwaters, revetments, groynes, gabions, offshore breakwaters and armourstones. These act to stabilize the shoreline position and are commonly applied to protect high-value or densely populated urban coasts. However, a static coastline can be problematic because of its inability to respond to sea-level changes in the medium- and long-term (coastal squeeze) and its cessation of interactions and of sediment inputs to the natural sediment budget (Doody, 2004; Wolters *et al.*, 2005). Thinking long-term and beyond the city has led to the growing appeal of working with nature and utilizing soft engineering approaches (Coy, 2000; Hartig *et al.*, 2011), which include beach nourishment feeding, dune building, increasing natural sedimentation and managed realignment. In a similar mode land-use up-river of urban centres is increasingly seen as playing a water retention role so that it is designed to provide space for river floodwaters to flow and slow, dissipating energy and smoothing flood peaks so to reduce damage in the city.

Coastal management decision making is not constrained by 'holding the line'. It considers three basic positions: (i) advance the existing shoreline seaward; (ii) hold the existing shoreline; or (iii) retreat the existing shoreline landward (Nur *et al.*, 2001; Tribbia and Moser, 2008; Brooks and Spencer, 2012). Each has its own technological and political as well as ecological and social consequences. In many locations, urban planning instruments attempt to monitor coastal shore development as part of everyday practice (O'Riordan and Ward, 1997; Milligan *et al.*, 2009). This can be a backbone for formulating and detailing a shoreline management plan. Implementation and maintaining a plan's relevance over time usually benefits from a partnership approach of engagement, and consideration of the opinions and concerns of all stakeholders (Tompkins *et al.*, 2008). The coming together of multiple interests and visions for the coast is championed by the Integrated Coastal Zone Management (ICZM) community (Shi *et al.*, 2001; Klinger, 2004).

Coastal wetlands are declining globally, partly in response to urbanisation (MA, 2005), though such ecosystems (e.g. mangroves, seagrasses, salt marshes, dunes, coral reefs) provide a useful regulating service against wave action and coastal flooding (Shuto, 1987; Danielsen *et al.*, 2005; Kandasamy and Narayanasamy, 2005). Their capacity to mitigate extreme events, while valuable, is constrained (Feagin *et al.*, 2010): the magnitude of wave energy that can be absorbed depends on the density and size of trees, shore slope and bathymetry as well as the characteristics of incident waves, and tidal stage (Algoni, 2008). The resilience of mangroves is impressive; mangroves have the capacity to regenerate rapidly following disturbance (Kamthonkiat *et al.*, 2011), and integral systems are a useful natural buffer against general flood disturbances. This said, disturbance is significant: Algoni (2008) argues that while climate change may lead to a maximum global loss of 10–15 per cent of mangrove forest, this must be considered of secondary importance compared with current average annual rates of 1–2 per cent deforestation driven by direct human pressure. Proactive planting and restoration of coastal vegetation in general has been performed in many coastal areas (see Mukherjee *et al.*, 2009; Feagin *et al.*, 2010) as a measure of coastal protection, though often erosion-resistant non-native species have been planted (sometimes in combination with habitat destruction, e.g. levelling of sand dunes), at cost to local native ecosystems (Feagin *et al.*, 2010). Tanaka *et al.* (2009) for example note that mixed layers of the trees *P. odoratissimus* and *C. equisetifolia* might be the most effective form of buffer for mitigating tsunami damage in Sri Lanka and Thailand. Coastal vegetation planting and restoration has taken place at several large coastal cities, e.g. Sydney, (McManus, 2006) and Tainan (Taiwan) (Fan, 2002), but there is limited literature on restoration in and around megacities. For example, Zamith and Scarano (2006) examine work on restoration in Rio de Janeiro, though not as a bioshield. An insightful exception is the experience of Bang Khun Tian, a coastal area in Bangkok province, which highlights the vulnerability of peri-urban and rural communities situated in urban jurisdictions. Here a range of initiatives including mangrove restoration have been applied in efforts to contain coastal erosion since the 1980s and protect local livelihoods (NTS, 2011).

### 2.2 *Flood/storm defences*

Coastal megacities are subject to multiple sources and causes of flood risk including specifically coastal risks such as sea surge and tsunami and also increased risk from pluvial flash flooding due to the lack of permeability of the dense urban environment, detailed in Chapter 2 and 4 (Fuchs *et al.*, 2011a). In addition many coastal megacities are the outlet point for large river basins or situated on deltas and therefore suffer from the accumulated fluvial flood risk of entire river catchments (Aerts *et al.*, 2009). Flood and storm defences are therefore a key part of the management of hazard risk for coastal megacities.

Hard engineered flood defence structures are intended to reduce the exposure of people and the built and natural environment to flooding. They are constructed to protect against flood events of a particular magnitude and provide limited protection for bigger floods; in fact these structures may worsen their impact when they trap flood water or fail causing catastrophic loss (MRC, 2009). Therefore it is important that flood defences are part of a strategic, integrated approach to flood risk management (Jha *et al.*, 2011), one that can be integrated into integrated coastal zone management (ICZM) and may need to take on a regional scale of application to integrate the management of environmental and urban processes (Kumar *et al.*, 2010). Inland flood management usually consists of systems of conveyance and storage at a catchment level, often upstream of the coastal megacity. This is supported by city-level drainage, flood defences and storage facilities and can be further backed up by community and individual property level schemes (Jha *et al.*, 2012). Within mega-cities watercourses are generally heavily channelized with restricted buffer zones as for example in Jakarta, Taipei and London. In a prosperous city like Taipei, economic and commercial pressures can result in urban land being developed in ignorance of or in spite of floodplain area restrictions. In such cases where the economic value of land is high and development pressures intense it can be difficult to protect land to build in flood retention ponds within the urban fabric. In developing countries, drainage channels and storage facilities may be attractive for informal settlement usage, and the chronic housing crisis in many coastal megacities often undermines spatial defence systems. In the Philippines, the loss of over 20 kilometres of drainage channels called *esteros*, is believed to have been a key factor to megacity-wide floods (Zoleta-Nantes, 2000).

Fluvial flood defences often consist of walls and embankments (levees or dykes), placed strategically around settlements or adjacent to water courses, to prevent ingress of water to inhabited areas. Examples are found across the Netherlands, Manila, Jakarta (Aerts *et al.*, 2009) as well as Tokyo and Kyoto (Kumagai and Nojima, 1999). The height of such structures is calculated to resist floods with a return period of up to 10,000 years in some areas of the Netherlands but total protection is not guaranteed. Most fluvial flood defences and many conveyance approaches effectively increase the capacity of channels and direct greater flows of water downstream and towards the coastal discharge points. As a result, coastal areas receive greater volumes of water, which may affect tidal recharge in the case of coastal areas below sea level. Redirected flows may have to be pumped out of channels or stored until tidal recess allows for gravitational flow. Hence, current research is directed towards methods for retaining flows elsewhere in the system, either outside the urban area or by environmentally enhancing urban retention basins.

Hard-engineered defences and levees are land-intensive solutions, they usually require the construction of permanent structures occupying valuable land. Many such systems are also very expensive. New developments include massive earth systems of multipurpose structures, which form the basis for other infrastructures or even residential or commercial development. In Japan, for example, super levees provide both flood defence and opportunities for community development (APFM, 2007), however, their applicability to existing megacities may be limited by resettlement issues.

Where urban space is limited and access to river spaces, roads, infrastructure and buildings is essential, temporary measures are attractive (Jha *et al.*, 2012). The advantage of such defence is that they can be installed just before or during a flood event, but under normal circumstances the space and access remains unchanged. Yet both demountable and temporary flood defences are only functional when in a closed position and both require operational procedures for putting them in place. Automatic flood barriers are also available that remain in situ and are raised by the force or presence of water. Developments continue in the technology of materials and control systems to ensure that such systems operate and can be relied upon. Recent examples include a European Union-funded consortium researching cost-effective solutions for flood risk reduction in urban areas worldwide (e.g. www.floodprobe.eu and www.corfu7.eu).

A flood barrier provides temporary protection from particularly high tides, or storm surges, at a point where water flow or shipping would otherwise be allowed to pass freely. It is usually just one component in an artificial or natural flood protection scheme. Flood barriers are major and often innovative civil engineering structures. Well-known examples include the Delta Works in the Netherlands (Aerts *et al.*, 2009) and the Thames Barrier in London. Cost is a major restriction on the wider installation of coastal barrier systems but they also have environmental impacts that may yet not be fully understood. For example, closure of the Jakarta bay area to prevent coastal flooding would cause pollution of the bay and necessitate changes in waste disposal and treatment (Jha *et al.*, 2012). Poorly designed and managed coastal barriers could potentially worsen flooding.

Wetlands (both terrestrial and coastal) are amongst the first ecosystems to be destroyed during urbanisation, partly because they have historically been seen as unproductive or actively harmful (e.g. unstable ground, sources for diseases like malaria) (Pauchard *et al.*, 2006). Forman (2008) records a very low abundance of wetlands in major cities, and notes that those with populations more than eight million had no major wetlands. Urban wetlands have utility for both flood control and pollutant removal (Jia *et al.*, 2011), and may also support biodiversity (Kim *et al.*, 2011); they are consequently often incorporated in sustainable urban drainage systems (SUDS). SUDS wetlands act by storing excess runoff and floodwaters, as well as associated pollutants such as sediments, heavy metals and nutrients that may cause freshwater or coastal eutrophication (Smith, 2003; Fan *et al.*, 2009; Scholz and Hedmark, 2010; Silva *et al.*, 2010). Such water may also then be utilized for non-potable applications in some cases, particularly where freshwater supplies are low. Smaller and more specific forms of 'bioretention' (Davis *et al.*, 2009), such as rain gardens or wastewater gardens, are also common at local scales.

Wetland restoration/construction has occurred in several large cities, including London, Seoul, Taipei, Nanjing, Seattle, Portland, Belo Horizonte and Kyoto (Fan *et al.*, 2009; Silva *et al.*, 2010; Kim *et al.*, 2011; Morimoto, 2011), though coastal megacity examples are not common in the literature. Much literature focusing on Europe and America examines planning and management of discrete wetland areas, while Asian cities have focused more on integrated wetland systems that allow space for flooding within residential areas, for example by raising ground floor housing levels (Morimoto, 2011). In some areas managed realignment of estuaries or salt marshes has been practised, but this is relatively uncommon in large cities due to habitation pressures along waterfront areas. Where this can be done, however, it reduces the risk of both riverine and tidal flooding in the urban area, and helps to retain pollutants.

### *2.3 Water supply, treatment and management*

Population growth and increases in climate variability such as heat waves and evapotranspiration is likely to put pressure on water resources in coastal megacities; excessive groundwater extraction and salinization is already a major problem for many Asian cities, as described in detail in Chapter 3 (Zaisheng, 1998; Bates *et al.*, 2009; Moglia *et al.*, 2009). Surface waters are also often affected by severe pollution, as it is in Jakarta, as are water levels by changing rainfall patterns, streamflows and extended periods of drought due to global environmental change impacts (Varis and Somlyódy, 1997; Bates *et al.*, 2009; Wilbers *et al.*, 2009). Sea level rise or flash floods constitute further hazard risks and impact on the availability of land and infrastructure (Satterthwaite, 2008). As urban water supply is also dependent on water systems in its surrounding regions and more remotely, water scarcity may escalate social tensions and cross-border water use regimes are already in place seeking to regulate access and distribution (Swyngedouw, 2004; Gleick *et al.*, 2005).

Many megacities have large areas with no access to piped water services (WHO/UNICEF, 2011) reliant upon unprotected wells, springs, commercial water provision, bottled water and tanker provided water. Slum settlement populations are most vulnerable to the health consequences of deterioration in the quantity and/or quality of freshwater resources (Moser and Satterthwaite, 2010). One response to water security in richer cities has been to consider desalination, though the energy intensity of this option combined with local environmental impacts has generated much debate on its potential for a sustainable solution (Tal, 2011). Other options, including wastewater recycling, may be more sustainable from an environmental perspective but can meet considerable community opposition (Hurlimann and Dolnicar, 2009). Decentralized water systems, which promote household and community level collection and distribution, are a challenge to existing centralized, or large-scale private sector management urban water distribution systems, but they represent a promising option (Sharma *et al.*, 2010; Biggs *et al.*, 2009) and include stormwater harvesting (Fletcher *et al.*, 2007), managed aquifer recharge (Dillon, 2004) or rainwater harvesting (Srinivasan *et al.*, 2010). Such options can enhance local governance as well as reduce costs of (i.e. leaks from) centrally managed urban water distribution systems. Options for enhanced conservation of existing resources includes demand management (Renwick and Green, 2000; Brooks, 2007), management of groundwater abstractions (Takizawa, 2008) and management of surface water quality (Kambole, 2003).

Climate change has wide-ranging implications for urban water planning, for example relating to the water–energy nexus, ensuring liveability of cities and finding appropriate strategies and planning approaches (Howe *et al.*, 2009). A key adaptation strategy may be to strive for the water sensitive city ideal, which involves adaptive, multi-functional infrastructure and urban design that reinforce water sensitive behaviours (Brown *et al.*, 2009). Likewise, Integrated Urban Water Management as a mainstream planning practice would foster adaptation (Maheepala *et al.*, 2010) by systematically identifying ways to supplement supplies in a way that meet existing social and environmental goals and constraints (Sharma *et al.*, 2009). Financing supports sustainability and is of critical importance for adaptation. In a drive towards ensuring the financial resources for systems, there has been a recent but sometimes problematic trend towards privatizing urban water (Bakker, 2003). In fact, there is limited knowledge about the drivers of the performance of urban water utilities (Berg and Marques, 2011).

Urban drainage systems need to be able to deal with both wastewater and stormwater whilst minimizing problems to human life and the environment, including flooding. Urbanisation both increases drainage flows demand and has a significant effect on the impact of drainage flows on the environment. Urban drainage systems can often move a local flooding problem to another location and pollution within

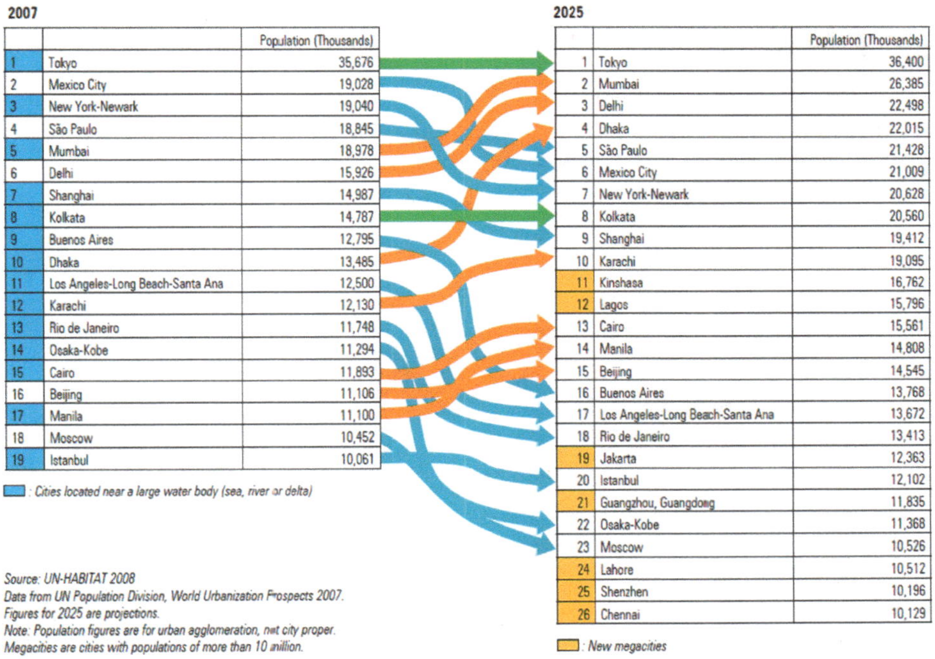


*Plate 4.1* Global mega-city populations, 2007 and projected for 2025

Source: UN-HABITAT (2008, pp.6)

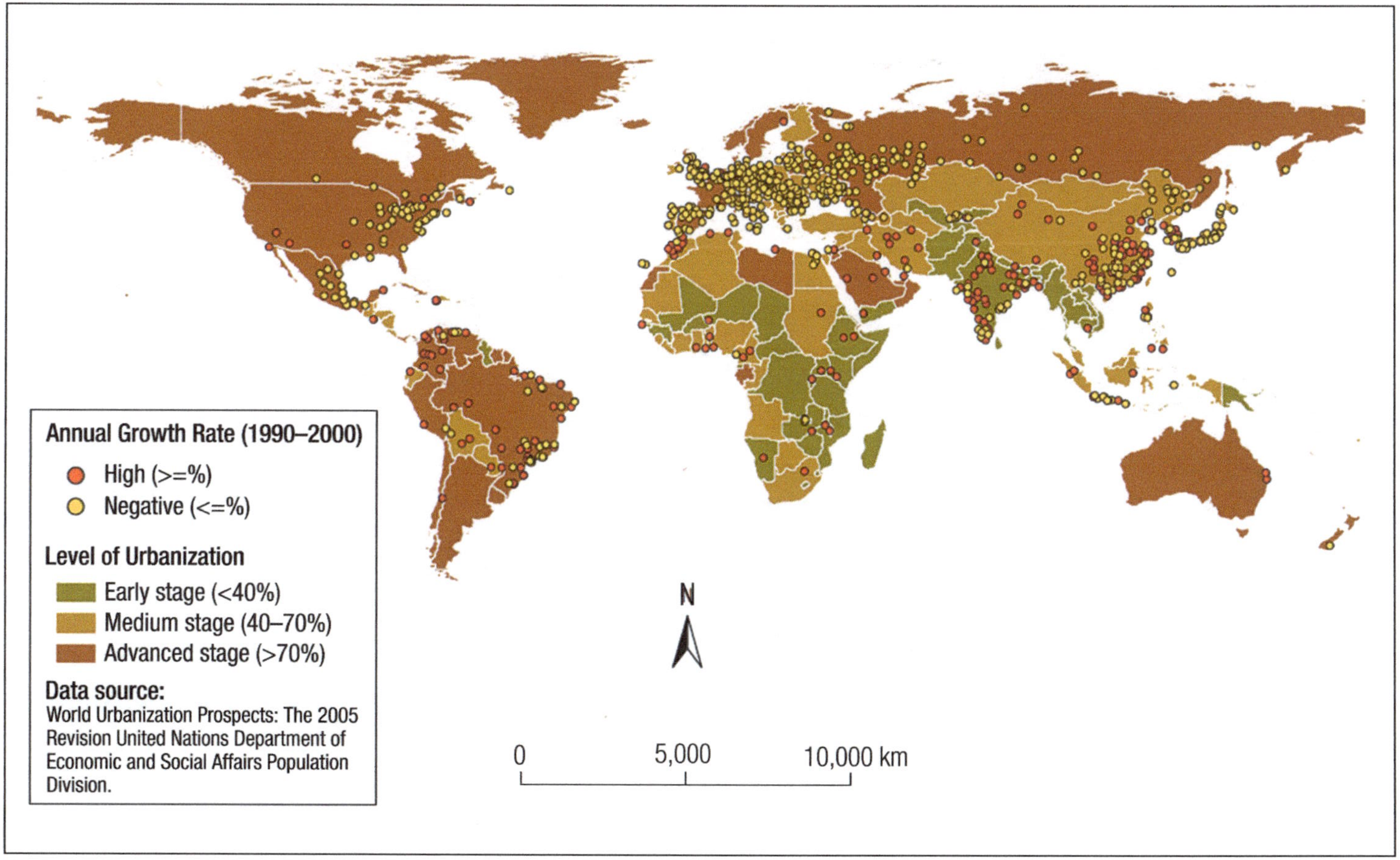


*Plate 4.2* Cities experiencing high and negative growth rates, 1990-2000

Source: UN-HABITAT Global Urban Observatory (UN-HABITAT 2008, pp.11)

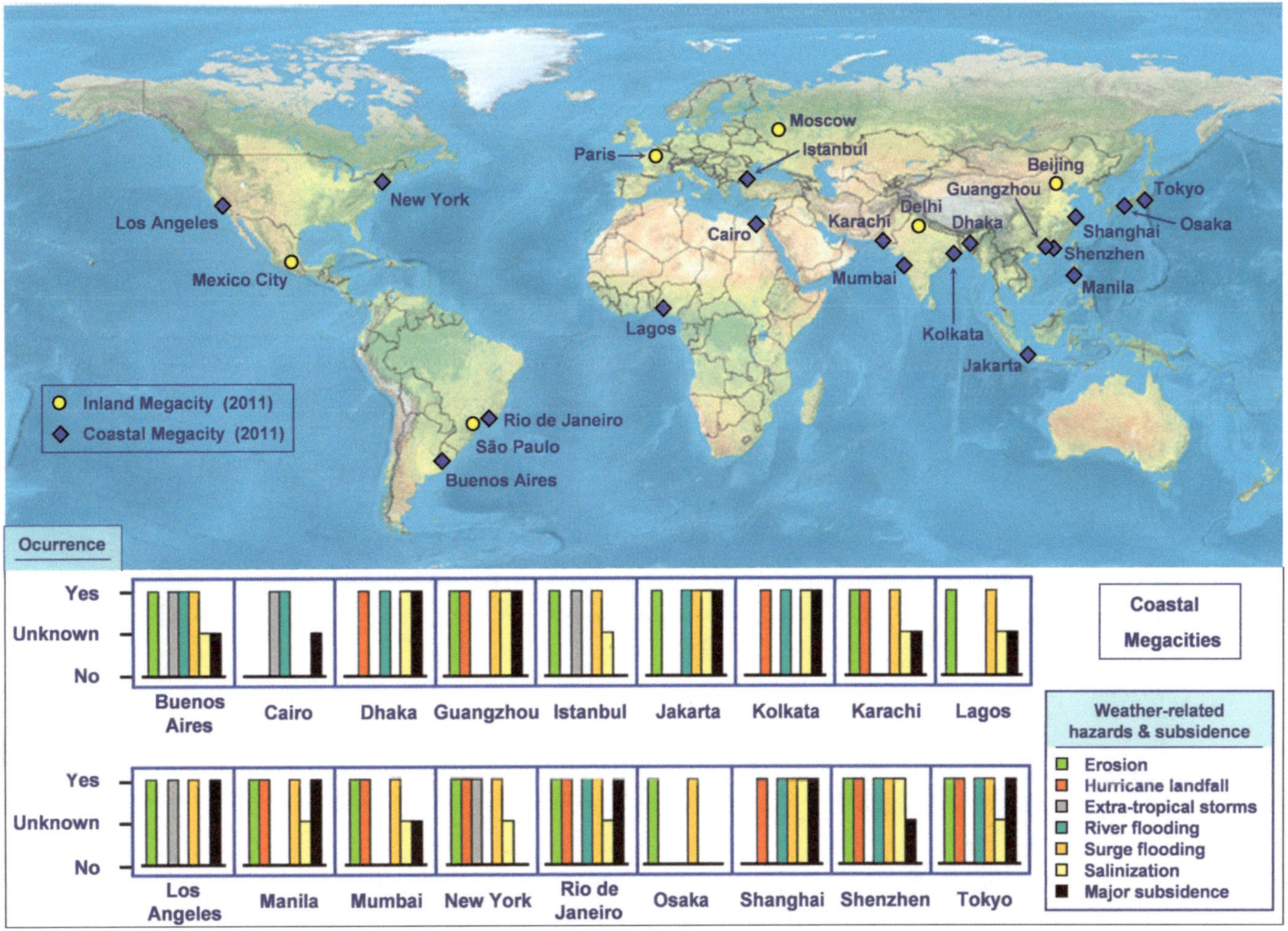


*Plate 4.3* Major weather-related hazards and subsidence occurrence for coastal megacities

Source: City population data sourced from UN-DESA (2012b) World Urbanisation Prospects 2011 table on the thirty largest urban agglomerations. Hazard data sourced from Klein et al (in Kreimer et al., World Bank, 2003, pp.104), Caljouw et al (2005), Li (2003), Nur et al (2001), Ward et al (2011) and expert judgement. World map provided by Natural Earth (available from www.naturalearthdata.com).

NB: Despite being more than 100km from the coastline, Dhaka and Cairo are included in this figure to reflect the strong atmospheric and geomorphological influence of the coast in these cities (see Chapter 1: section 3.2).

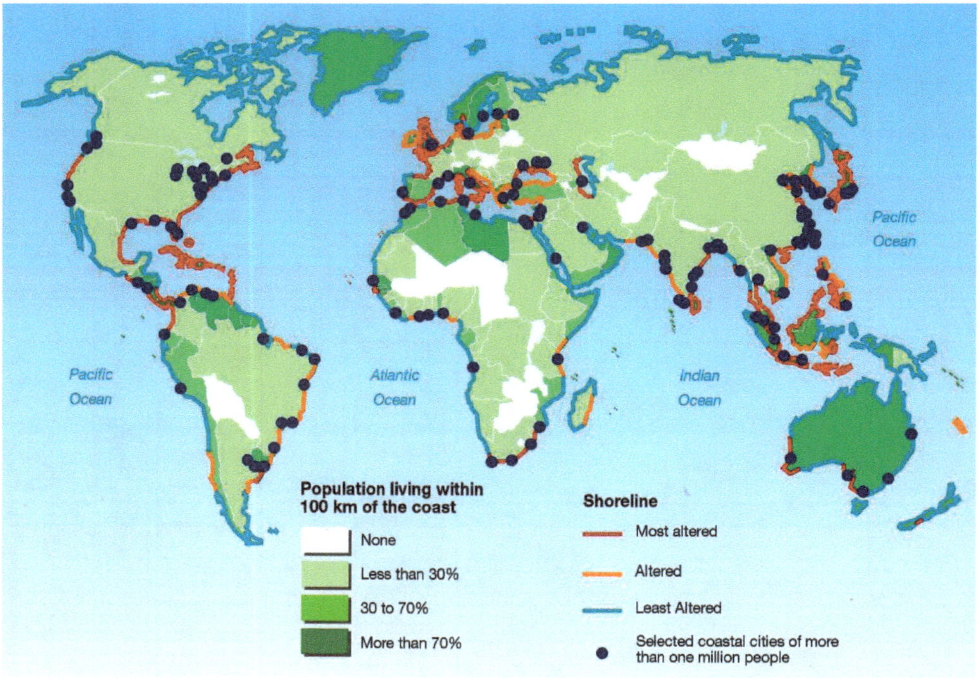


*Plate 4.4* Coastal population and shoreline degradation

Cartographer: P. Rekacewicz, 2006. Sources: Burke et al., 2001, Harrison et al., 2001. http://www.grida.no/graphicslib/detail/coastal-populations-and-shoreline-degradation_eba3

*Plate 4.5 (facing page)* Global earthquake distribution

Notes: (a) Cities with more than 5 million (small squares) or 10 million (large squares) population. (b) Earthquakes of the last 1000 years known to have killed more than 10 000 (red dots) or 100 000 (yellow dots) people. There are 113 earthquakes in this figure, many of their symbols overlying each other. A total of 34 of these earthquakes occurred in the last 100 years alone. (c) Histogram of the number of earthquakes killing more than 10 000 (grey) or 50 000 (red) people per century.

Source: Jackson (2006)

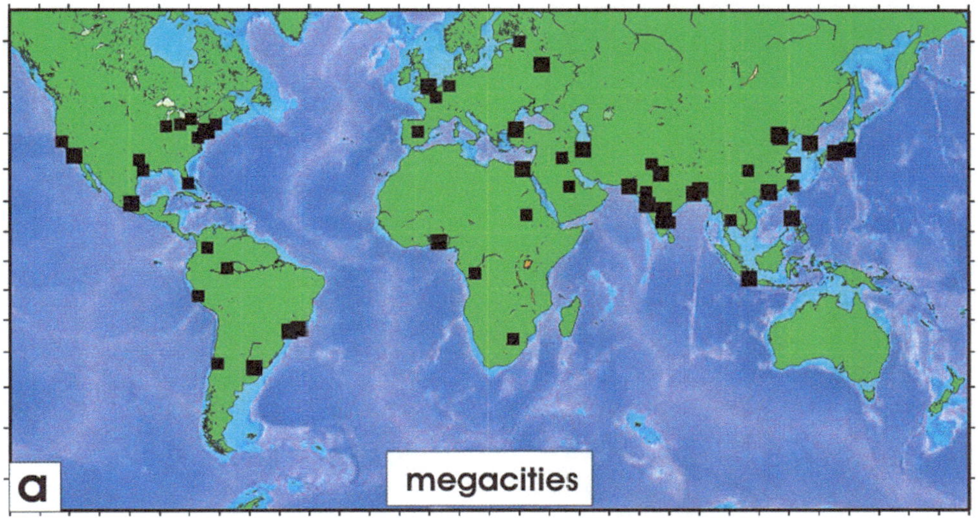
a
megacities

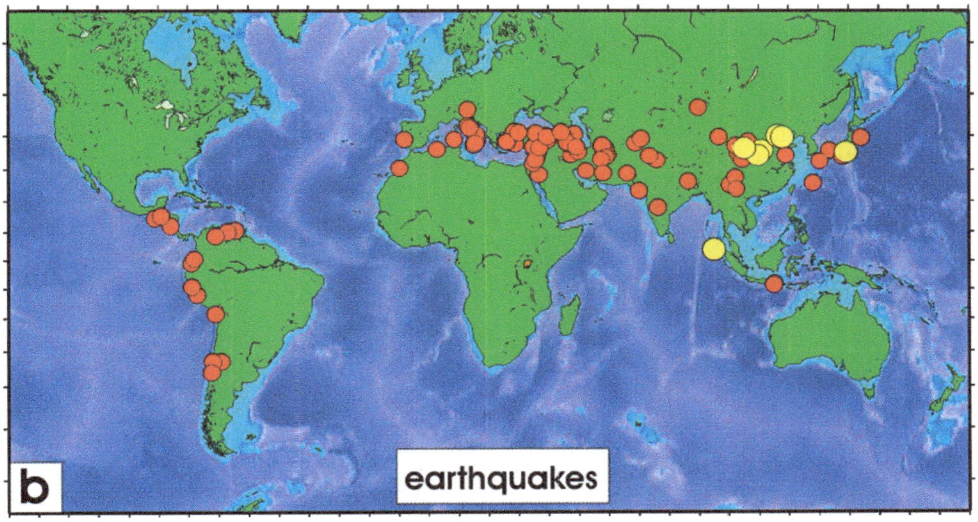
b
earthquakes

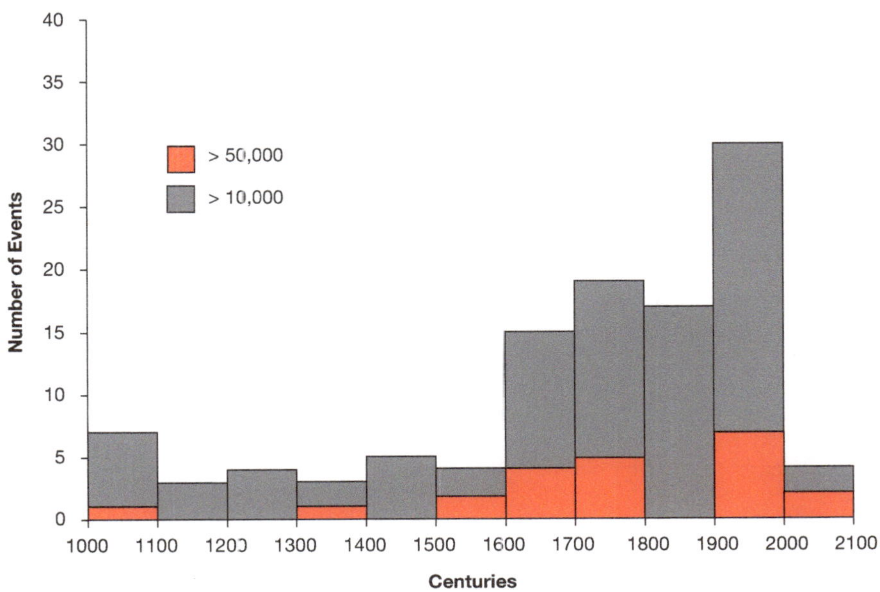
40
35
30
25
20
15
10
5
0
Number of Events
> 50,000
> 10,000
1000
1100
1200
1300
1400
1500
1600
1700
1800
1900
2000
2100
Centuries

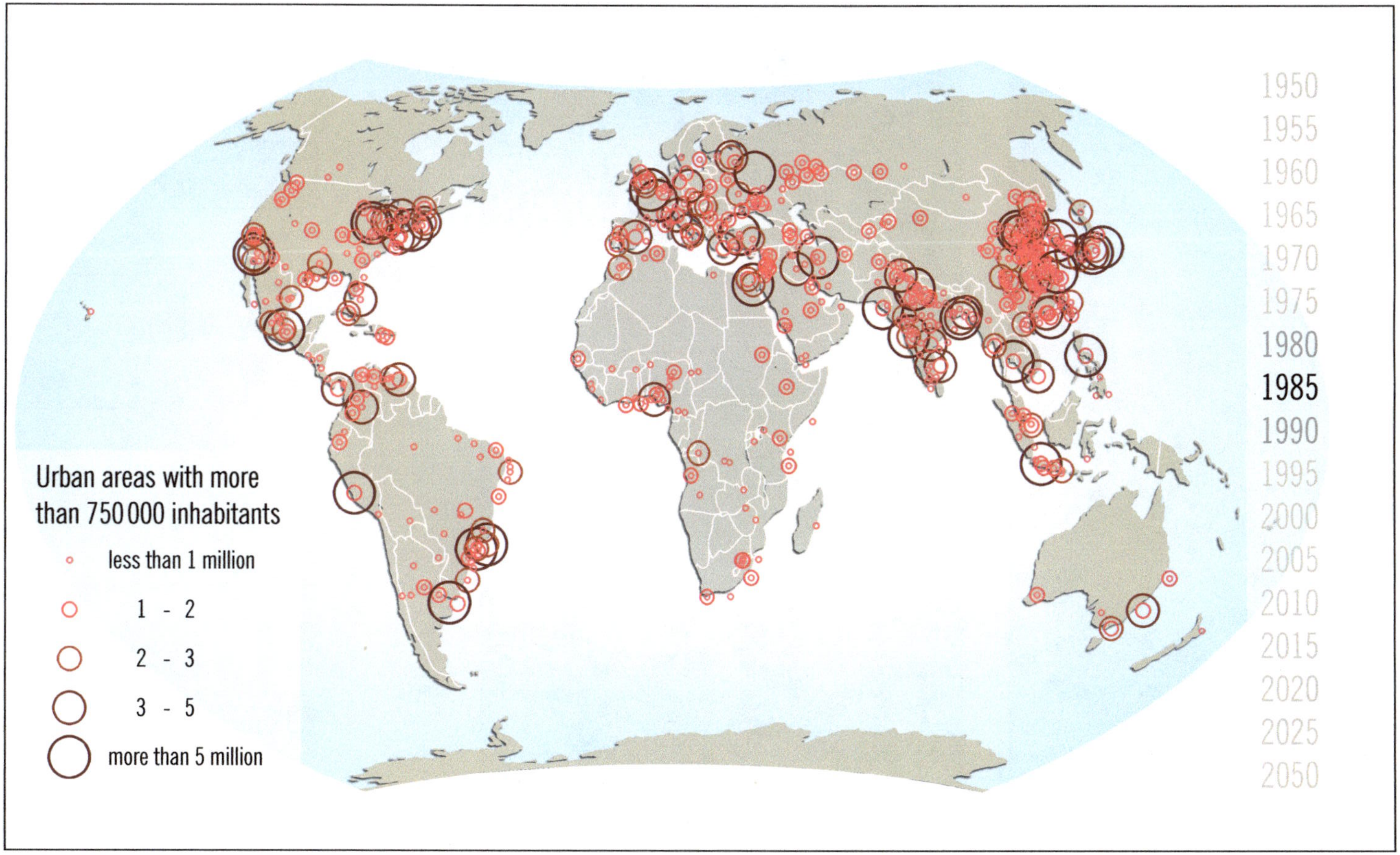


*Plate 4.6* Coastal megacities

Cartographer: Hugo Ahlenius, Nordpil

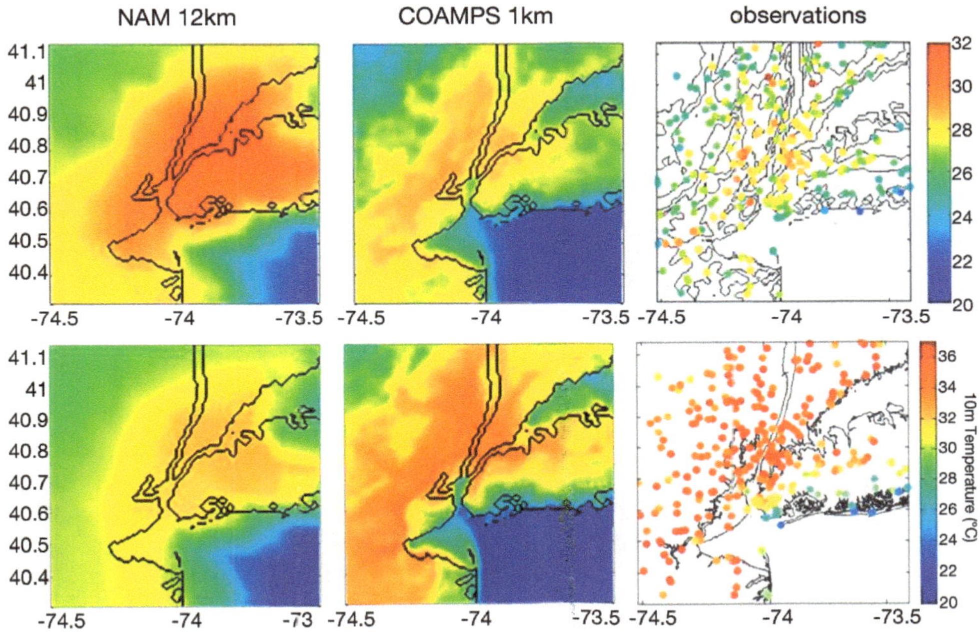


*Plate 4.7* New York City region mean 10-m air temperature during the 7-10 June 2011 heat wave modelled and observed

(left) National Weather Service's 12 km resolution North American Model (NAM) (not urbanized); (middle) ~1 km resolution urbanized Coupled Ocean-Atmosphere Mesoscale Prediction System (COAMPS); and (right) observations. (Top row) mean temperature for June 7-10, 2011 with the NYC UHI visible; and (bottom) 1500 h on June 8, when a sea breeze cooled southern Long Island, on the right side of the map. Data courtesy of NOAA (NOS-PORTS, NWS-ASOS, NWS-HADS, Urbanet), Rutgers NJ Weather and Climate Network, APRSWXNET, AWS Convergence Technologies, Inc. (WeatherBug), and Weatherflow, via Mark Arend (NYCMetNet).

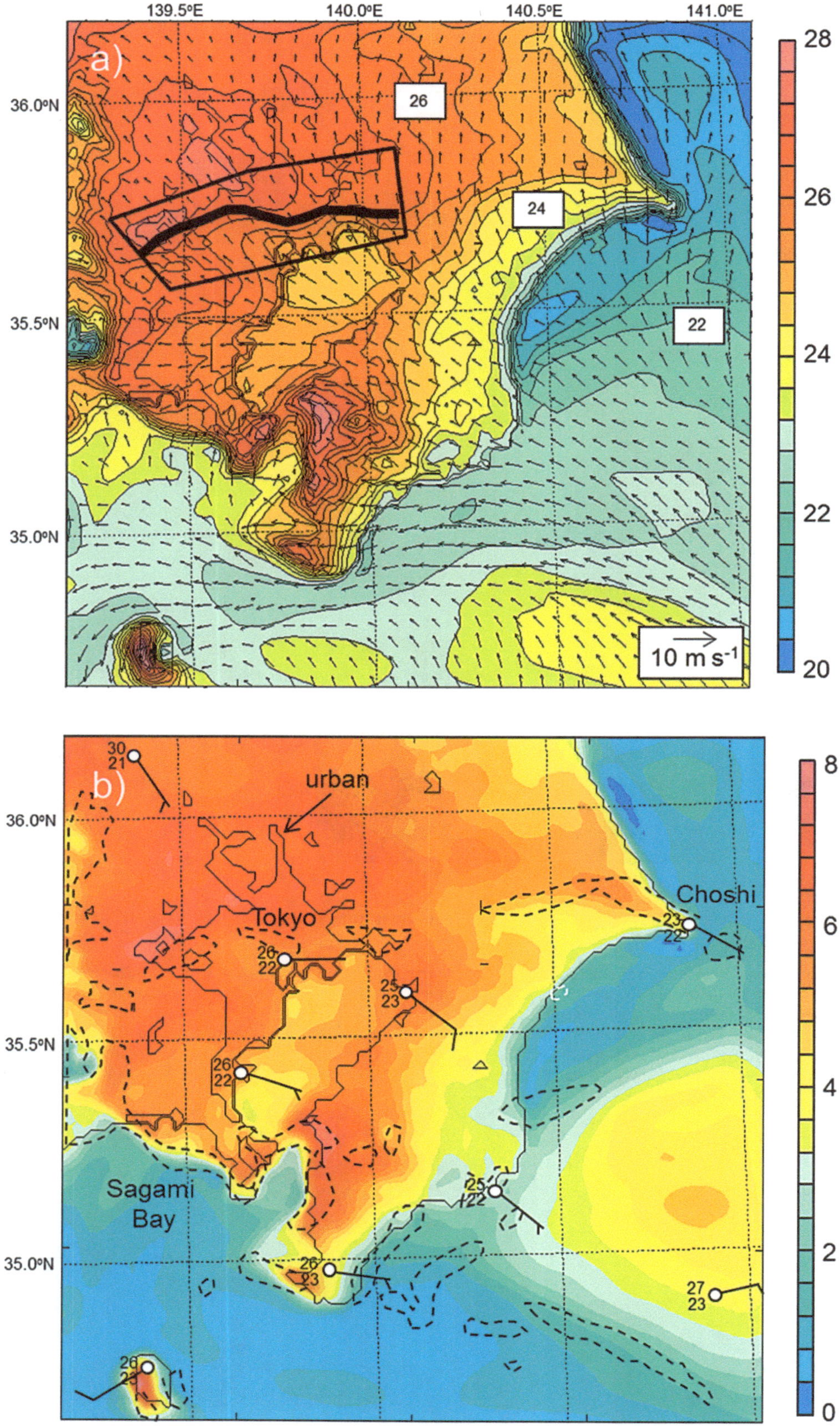

139.5°E
140.0°E
140.5°E
141.0°E
36.0°N
35.5°N
35.0°N
a)
26
24
22
10 m s-1
28
26
24
22
20
b)
urban
Tokyo
Choshi
Sagami
Bay
8
6
4
2
0

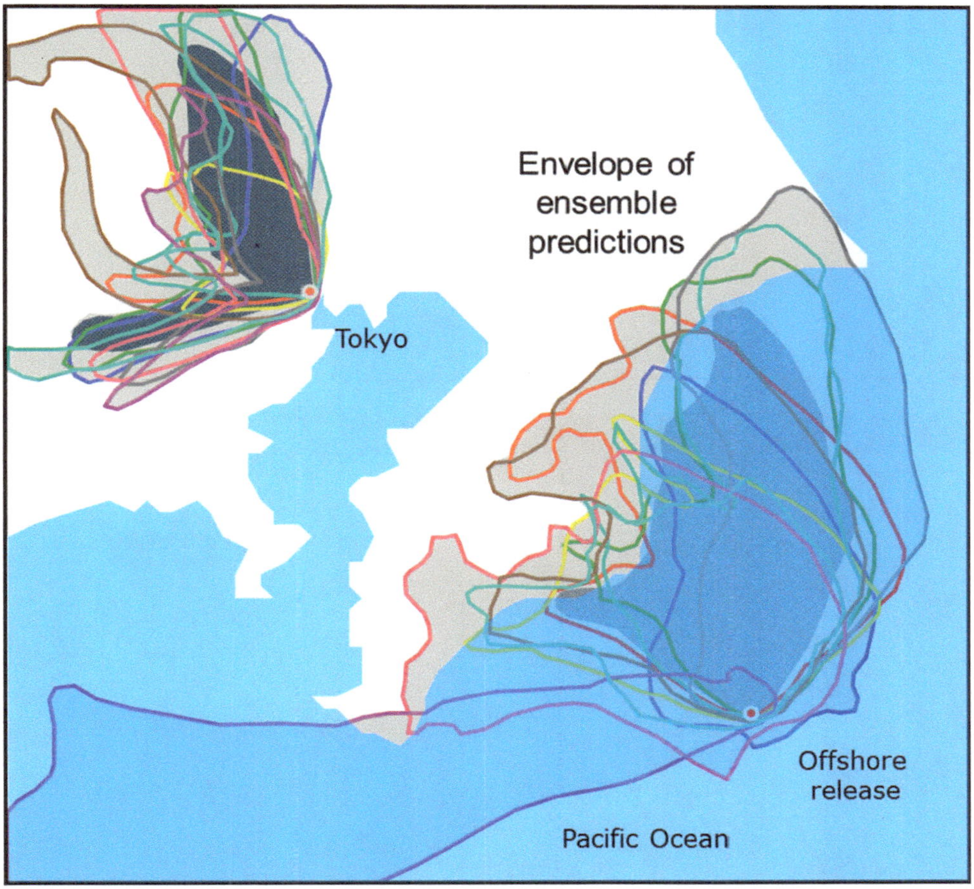


*Plate 4.9* Tokyo Metropolitan Area $10^3$ dosage contour from 12-hour COAMPS Nest 4 (1.67 km) ensemble members initiated June 30, 2005.

*Plate 4.8 (facing page)* COAMPS 6-h forecast of the Tokyo Bay region (06 UTC 30 June 2005) (1.67-km deterministic ensemble member) of condition at 10-m (a) wind speed and air temperature (contours every 0.4 °C), and b) dew point depression (contours every 0.4 °C) plus vertical velocity greater than 5 cm $s^{-1}$ at 210-m (black dashed contours) and station observations of air and dew point temperature (°C) and wind (full barb=5 m $s^{-1}$). Solid line shows urban extent. In (a) the heavy solid line is deterministic model location of sea breeze front, box – shows the ensemble spread of sea breeze front location (Holt et al. 2009).

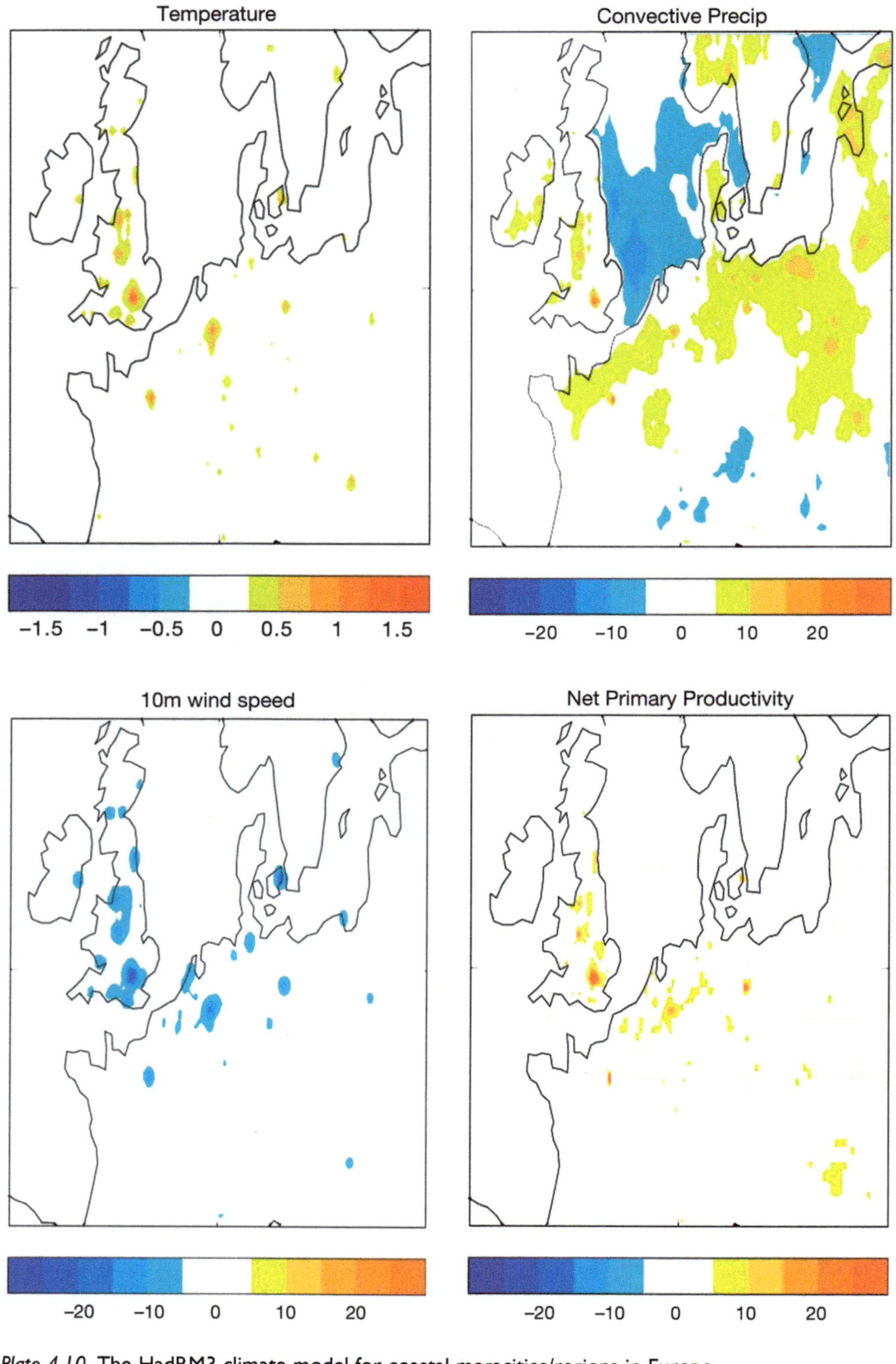


*Plate 4.10* The HadRM3 climate model for coastal megacities/regions in Europe

Notes: Difference between the HadRM3 climate model (coastal megacities of London, in addition Paris, Ruhr Valley region, Europe) with and without urban land surface for temperature (°C), convective precipitation (%), 10 m wind speed (%), and net primary productivity (NPP, %). The NPP data have been weighted to represent the productivity per unit area of vegetation, rather than total area.

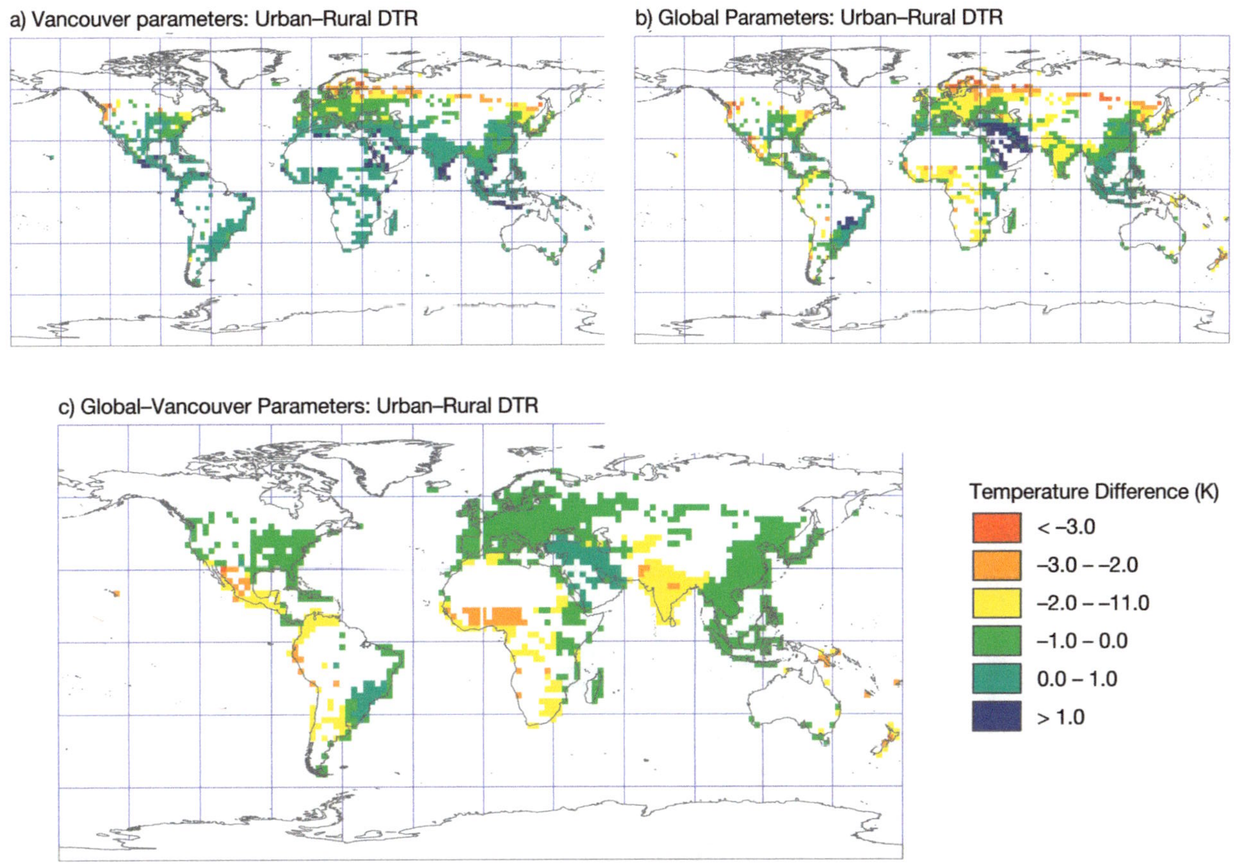


*Plate 4.11* The NCAR urban canyon model

Fig 4.6: Modelled annual temperature differences using NCAR CLMU (Oleson et al., 2008a, 2011; Jackson et al., 2010) a) the difference between urban and rural diurnal temperature range (DTR) in the Vancouver parameter simulation; b) the difference between urban and rural DTR in the global parameter simulation; and c) the difference between urban and rural DTR between the Global and Vancouver parameter simulations.

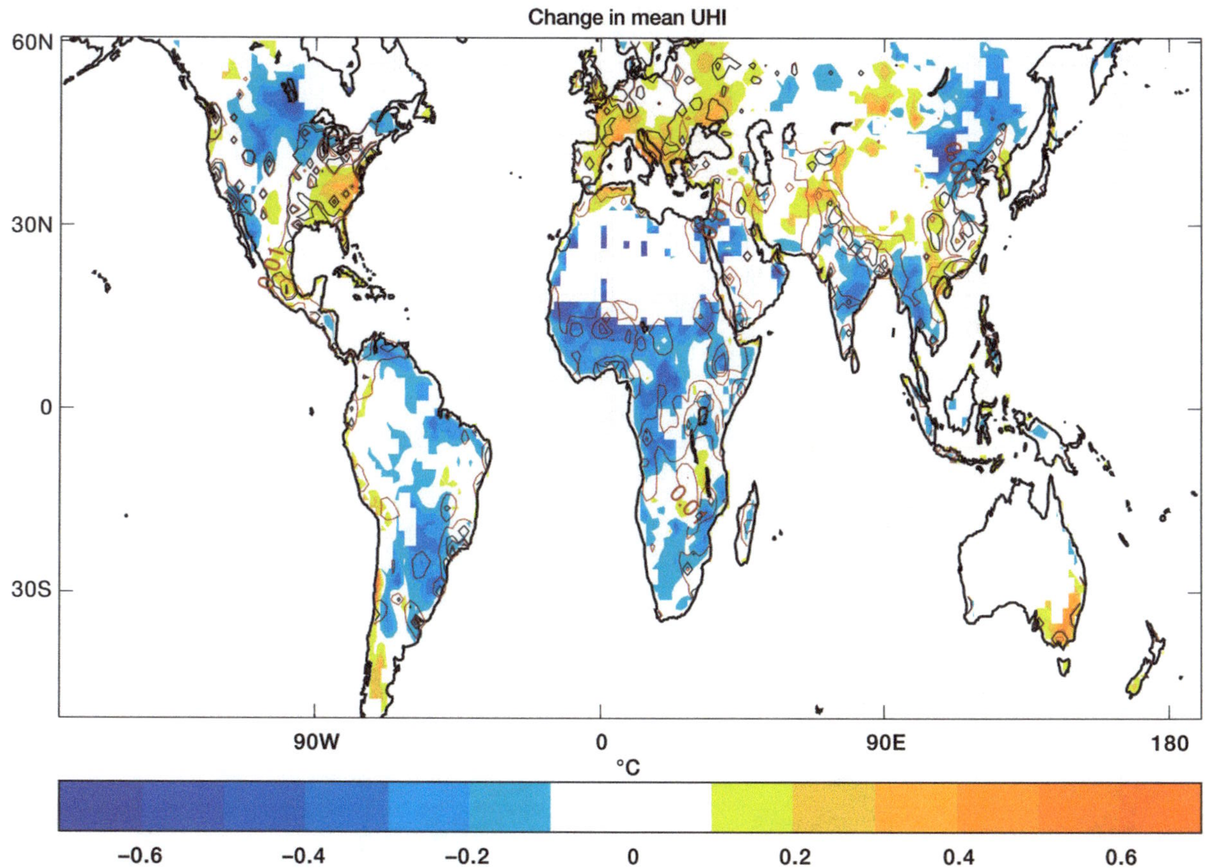


*Plate 4.12* Change in the UHI strength between the periods 1981-2000 and 2081-2100 simulated by HadGEM2-ES.

Notes: The 1%, 5%, and 10% urban land cover contours are overlain for (black) 2005 and (red) 2100. For this experiment the urban land cover remained fixed at 2005 levels.

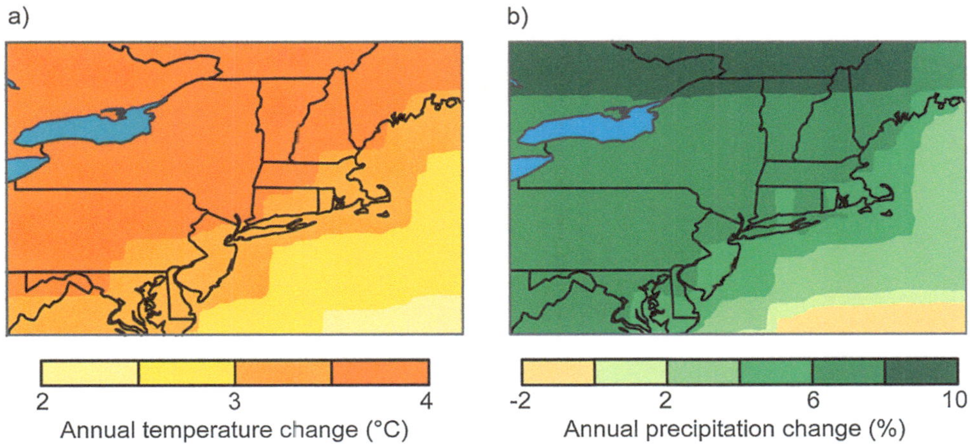


*Plate 4.13* Ensemble mean results for the North East Coast of North America (coastal megacity of New York City) from 16 GCMs for the A1B emission scenario for the 2030s relative to the 1970-99 base period for (a) temperature and (b) precipitation (Horton et al. 2011). A1B emission scenario (http://www.ipcc.ch/ipccreports/tar/wg1/029.htm#storya1) assumes rapid economic growth with the peak in global population mid-century and technological advances are efficient but balanced across energy sources.

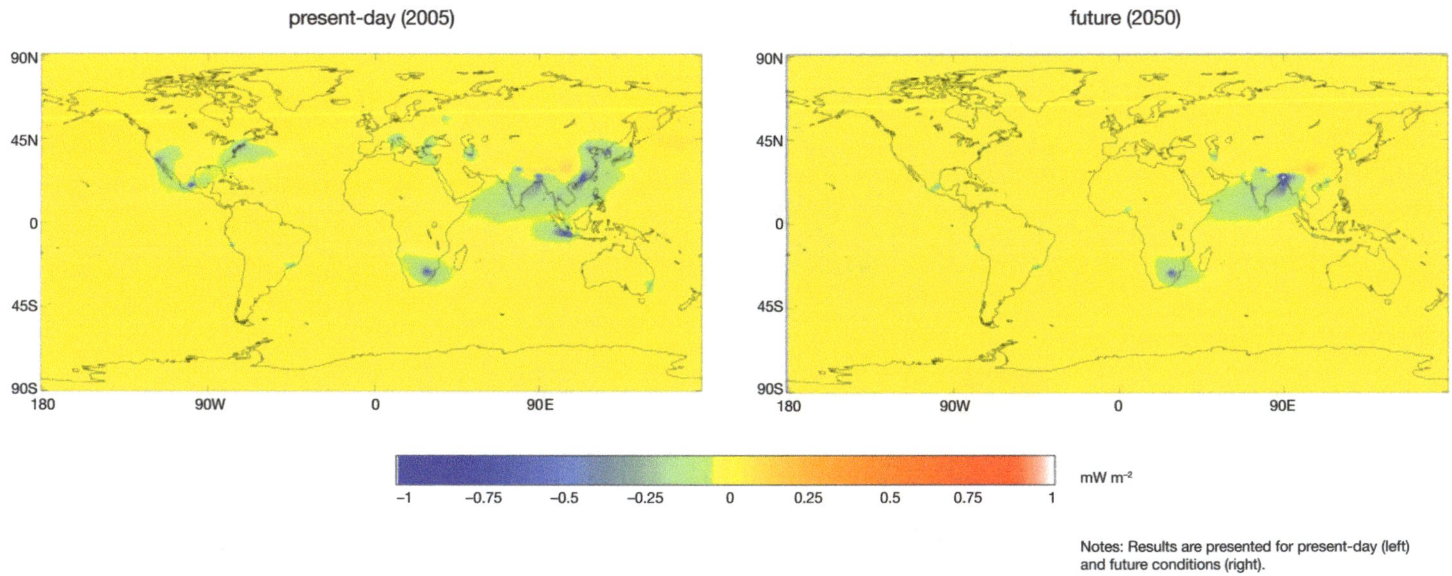


*Plate 4.14* Global distribution of TOA shortwave all-sky direct radiative forcing from megacity aerosols

Notes: Results are presented for present-day (left) and future conditions (right).

*Plate 4.15* Satellite image of August 2011 New Jersey coast algal bloom event

Notes: Courtesy of Mid-Atlantic Regional Association Coastal Ocean Observing System: http://seaandskyny.com/2011/07/22/where-is-the-sewage-spill-plume-going/

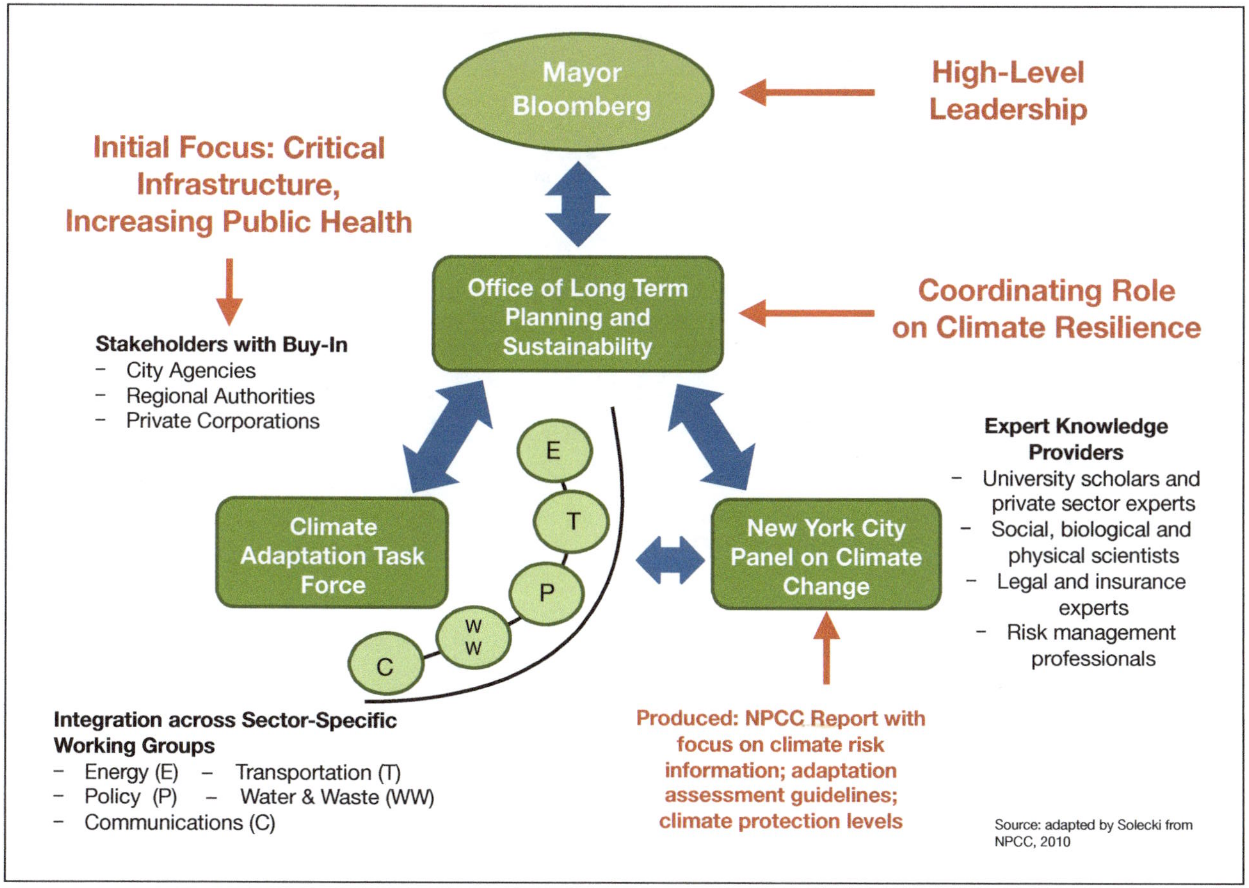


*Plate 4.16* Schematic of New York City Climate Change Action

Adapted by Solecki from NPCC (2010)

urban drainage flows can have major impacts on health downstream. In developed countries in recognition of the unsustainable nature of closed drainage, there is a move away from piped systems, towards more natural systems including open draining storm water (New York City Department of Environmental Protection, 2011). In many locations throughout the world, however, main drainage systems have never existed. Nevertheless, where they are present, many urban drainage systems are limited by political–administrative city boundaries and historical urban geography. Consequently storm water drainage may only be designed to meet a maximum capacity flow that is inadequate as a consequence of increased urbanisation (more rapid run-off) or increased rainfall intensity. Unregulated land use and land cover changes often lead to artificial damming and constrictions of the natural flow of water through the city. The phenomenon of fish kills in the Thames and flood water rising from drainage ducts are two manifestations of this challenge in London that have resulted in a major, new cross-city storm drain. More significantly, disregard for watershed boundaries and the orographic effect of steep mountains bounding coastal cities often result in institutional failures in preparedness, increasing disaster risk.

Both combined and separate sanitation and drainage systems can result in excess and polluted surface water flooding. The older parts of many cities such as New York are drained using the combined system, whereby wastewater and stormwater are mixed and carried together. Usually surcharges are allowed to overflow into watercourses bringing with it wastewater and pollution (Butler and Davies, 2011). In a separate system, wastewater and stormwater are drained by separate pipes, often constructed in parallel. The problem of combined sewer overflows is thereby avoided, but stormwater discharge is usually untreated, and this may also cause pollution particularly in megacities with high levels of particulates.

Surface water flows can be deliberately directed into design pathways that may have been created specifically to cope with exceedance flows. Design pathways include floodways, retention basins, or designated areas of public open space for temporary storage; pollution control can be built into such systems. On a small scale, some adaptation to existing urban features like road profiles and curb heights can improve the effectiveness of pathways for extreme events.

## 3 Urban planning and disaster risk management

Over the past two decades, a number of cities have sought responses to global environmental change and climate change impacts. There is a rapidly growing interest in urban conditions or contexts linked to particular sorts of responses. Nonetheless, urban policy agendas increasingly reflect adaptation and mitigation efforts and offer insights into where, and to what effect, these responses are taking place (Bai, 2007; Bulkeley and Betsill, 2003; Bulkeley *et al.*, 2010; Parnell *et al.*, 2007; Satterthwaite *et al.*, 2008; Schruers, 2008). Carmin *et al.* (2012) characterize urban adaptation policy formation as an act of bricolage – the stitching together of existing policy agendas and interests to form a new domain. Redclift *et al.* (2011) examining the formation of mitigation and adaptation policy on the rapidly urbanising coast of Qintana Roo, Mexico (from Cancun to the Belize boarder) identify a similar process from the perspective of civil society, with most purchase coming from those NGOs that can utilize mitigation or adaptation to push existing agendas, for example around improved urban solid waste and water management.

As with other issues, it is in those cities where climate change is framed as either a risk or an opportunity that engagement with the issue has been forthcoming. In some cases, this framing has been developed through the creation of scientific assessments of climate vulnerabilities, (Corfee-Morlot *et al.*, 2011; Roberts, 2010). In other cases, it has been the contribution that cities are making to GHG emissions that has provided the spur for action (Collier, 1997; Deangelo and Harvey, 1998). While scientific assessments, and the capacity to conduct and assess the implications of these processes, are important, this does not automatically lead to urban responses to the issue. Rather, a policy and wider public narrative giving significance to climate change for the city, and internalizing this as a problem for urban development – not only for international policy – needs to be created where this scientific evidence is connected to wider issues and concerns in the city.

Research suggests that three factors are particularly important in shaping the extent to which such narratives are developed and become established within the city, in turn serving to shape the responses that emerge in a particular urban context (Bai, 2007; Betsill and Bulkeley, 2007; Bulkeley and Kern, 2009; Bulkeley, 2010; Schreurs, 2008).

First, the role of municipal and political leadership. Many of the early pioneers in responding to climate change during the 1990s were members of transnational municipal networks that provided the opportunity for individual members of municipal authorities concerned with these agendas to draw on the knowledge and capacity of a broader community (Kern and Bulkeley, 2009). Indications are that some coastal megacities are now becoming more active members of such networks, including C40, Cities for Climate Protection and the Asian Cities Climate Change Resilience Network. In addition, the role of individual champions within cities has been critical in the development and pursuit of responses to climate change. Notable amongst coastal megacities have been Mayors Bloomberg, in New York, and Livingstone, in London, in providing political leadership, and the role of senior policy officials, for example in the case of Durban where the environmental manager, Deborah Roberts, has played a significant role in moving the agenda on climate change forwards (Roberts, 2010).

Second, the exploitation of strategic 'windows of opportunity'. Research has shown how cities such as Beijing, Cape Town and Durban have exploited the 'window of opportunity' created by global sporting events to promote action on climate change (Bulkeley *et al.*, 2009; Roberts, 2010). These opportunities are also structured by membership of municipal networks, specific conferences and events, and, unfortunately, climate-related disasters (Bulkeley *et al.*, 2009; Pelling, 2011). Disaster events can provide a physical opportunity to intervene in the urban landscape (e.g. transportation systems, housing, water and energy provision) they can also result in increased motivation to do so in an appropriate manner to address climate change.

Third, the complexity of the climate change issue has meant that cities frequently use a variety of different responses that depend upon forms of partnership working with private and civil society actors. In those cities where other actors are also engaged and committed to addressing climate change, new forms of political, community and policy response are forthcoming. For example, HSBC has worked with urban authorities in Shanghai, Mumbai, Hong Kong, London and New York to develop new initiatives, while a range of civil society organizations, from international NGOs to local community-based organizations, are also critical in delivering responses on the ground.

Taken together, these factors suggest that alongside sufficient scientific knowledge and opportunities to act, the critical variables shaping the different responses being developed to climate change in coastal megacities are related to issues of specific goals, capacity, political will and engagement.

### 3.1 Risk assessment

The range of possible hazard mitigation and vulnerability reduction measures for managing environmental risks are broad, but as noted above they succeed best when there is a shared view of the type and scale of the risk and the opportunity that can come from overcoming the risk. Assessment and communication of the risks faced by coastal megacities is therefore a critical step in generating positive responses that are targeted towards the most important risks rather than taking the easier step of following international agendas or adopting adaptation or mitigation measures that are easy or acceptable but do not engage with the key challenges associated with a coastal location.

Risk assessment as a development planning tool can be used to provide a comprehensive and interdisciplinary approach in assessing both the hazard and vulnerability components of risk (Pelling and Wisner, 2009). Additionally, the process of identifying and evaluating risk can bring increased awareness to key stakeholders and community members. Given the interdisciplinary nature of risk management, definitions of hazards, exposure, vulnerability, capacity and risk tend to vary across the various fields of study. For example: IPCC and UNISDR definitions differ with IPCC combining hazard and exposure whereas UNISDR/World Bank definitions see risk as the result of the compounding effect of hazard, exposure and vulnerability. While standardized definitions may be neither feasible nor desirable (Fuchs *et al.*, 2011a), there is general agreement that hazard focuses on the physical phenomenon with the potential to bring harm; exposure refers to the concentration of population, resources and infrastructure in the path of the hazard and the notion of vulnerability usually encapsulates the susceptibility of the physical built environment, social, economic, environmental and political components of society to harm; and capacity refers to the attributes within a system that help communities and individuals to cope with risk. It is the interaction of these components with each other that results in disaster risk, and it is well understood that for urban environments in particular, these interactions are complex and dynamic (Pelling, 2003), and improved understanding of such systems within the coastal zone is required to develop appropriate and effective responses to reduce risk.

The fact that coastal megacities potentially expose large populations and important centres of socioeconomic activities to hazards, combined with the growing recognition of the importance of city-scale studies for risk assessment (Hunt and Watkiss, 2011), has generated considerable interest in recent years in assessing the risks arising for the population and assets of coastal megacities from different types of hazards.

Already since the IPCC Third Assessment Report, Scott *et al.* (2001) have identified cities in coastal and riverine locations as the most likely to be affected by climate change impacts, and flooding is a primary source of risk. While several studies assessing risks or mapping hazards in coastal regions have focused on coastal flooding resulting from higher sea levels, storm surges, rainfall or tsunamis, few have looked specifically at coastal megacities. Studies focusing on coastal megacities include the studies of Gornitz *et al.* (2001), who assessed the impacts of sea-level rise in the New York Metropolitan area; Hanson *et al.* (2011), who ranked the world's largest port cities based on the exposure of population and assets to relative sea-level rise; and Ranger *et al.* (2011) who quantified current and future flood risk due to increased rainfall in Mumbai.

Sea-level rise will act to exacerbate erosion and flooding conditions along coastal areas, and is therefore an important consideration when assessing tropical cyclone risk (Nicholls and Cazenave, 2010). While climate-related components of sea-level rise have received much attention, locally driven components such as land subsidence warrant more systematic study. Rodolfo and Siringan (2006) demonstrate this with the example of Manila, the Philippines, where groundwater extraction and the subsequent land subsidence has resulted in a local sea-level rise of more than ten times that attributed to climate change. Their study further points to the use of ineffective infrastructural projects that are both poorly designed and fail to address the root causes of increasing flood risk. In addition to sea-level rise, climate change is expected to have some impacts on various tropical cyclone parameters, such as intensity, frequency and storm track. There remains, however, a large amount of uncertainty in how such parameters will be impacted (Grossmann and Morgan, 2011) and in what way those changes will act to either reduce and/or increase risk for local areas.

The Wenchuan earthquake in 2008, in China, had a devastating effect on the local communities but for more complex reasons than ground shaking alone. Whilst the disaster did not affect a megacity, more than 70,000 people were killed and more than 25 million homes were destroyed. The disaster stimulated a great deal of research and many publications have arisen on wide-ranging topics such as the geometry and mechanism of fault that moved during the earthquake (Hashimoto *et al.*, 2010; Jia *et al.*, 2010; Kusky *et al.*, 2010; Huang *et al.*, 2011); the quantification of landslide hazards and risks in the complex mountain environment (Lin *et al.*, 2009; Chen *et al.*, 2009; Chigira *et al.*, 2010; Ouimet, 2010; Cui *et al.*, 2011; Dai *et al.*, 2011; Liu *et al.*, 2011); as well as the development of rapid stereo-mapping, using unmanned airborne vehicles, as a disaster mitigation and management tool. Some of the research revealed that conditioning factors, preparatory to the earthquake trigger, were the reason for the destruction of the county town of Beichuan. The published seismic intensity map for the event has been greatly improved as a result of the work (Liu *et al.*, 2011) and building regulations will accommodate this new information and be improved as a result. A direct response on the part of the Chinese has subsequently been the relocation of the town tens of kilometres away to a much safer location.

In most of the above studies, risk assessment and hazard mapping at city level have been primarily based on geographic information systems analysis, employing global or regional datasets (e.g. population, elevation, surge levels, etc.) and a wide range of sea-level rise and socioeconomic scenarios. These assessments usually address a single type of hazard, or in some cases, combinations of hazards (e.g. Hanson *et al.*, 2011, who consider sea-level rise, surges and subsidence). In fewer cases, more detailed datasets and methods for mapping and assessing risks (such as those described by Kaiser *et al.*, 2011) have been employed when trying to assess the risks associated with single or multiple hazards in coastal megacities (e.g. World Bank, 2008; Environment Agency, 2010). However, detailed studies are rare as they require large amounts of resources in order to address limited data availability and high computational requirements. Moreover, incorporating adaptation in risk assessment and hazard mapping constitutes a major challenge in risk assessment and hazard mapping. Although assessments that consider adaptation have been carried out at regional scales (Hinkel *et al.*, 2010), very few examples of studies on coastal megacities exist where a limited range of responses to risks have been incorporated in the analysis (Kirshen *et al.*, 2008; Environment Agency, 2010).

Assessing the risks that arise from multiple hazards, such as sea-level rise, storm floods, river floods, natural and human-induced subsidence and tsunamis, while considering a range of potential adaptation responses, currently constitutes a major challenge for developing improved risk assessments in coastal megacities.

## 3.2 Spatial planning at all scales

If the economic benefits of growing coastal megacities continue to be felt, the question is not whether to let cities grow but how to do so to ensure the positive features (Jha *et al.*, 2011). Urbanisation is accompanied by increasingly larger-scale urban spatial expansion to accommodate population increases. This expansion exacerbates hazard pressures associated with changing climates and development legacy (such as subsidence or inadequate land drainage) as it changes the natural landscape, land uses and land cover, for example by changing water flows and increasing impermeable areas, thereby adding to the flood hazard problem (Satterthwaite, 2011), or by the intensification of urban heat islands (see Chapter 4). Denser population and concentration of assets makes urban settings more vulnerable to natural hazard impacts (Corfee-Morlot *et al.*, 2009). Conversely, lower density expansion can lead to higher rates of emissions as transport costs escalate (Glaeser and Kahn, 2008). The Working Group on Climate Change and Development (WGCCD, 2009) suggests that, in developing countries, increased urban resilience in the face of climate change will only be achieved by tackling the two issues of inadequate infrastructure and urban expansion.

In the largest cities and urban areas, therefore, the balance of prevention may need to shift from individual measures to collective action (World Bank and United Nations, 2010). Rapid urbanisation means that cities will need to invest in infrastructure services, given the increase in the demand for these (Jha and Brecht, 2011), and presents the opportunity to make those investments in sustainable ways as economies of scale and proximity can provide more cost-efficient infrastructure and services. Risk assessment also can help locate infrastructure and populations in areas at low risk. In order for cities to take advantage of their potential, however, good and robust governance and spatial planning are prerequisites.

The need for, and opportunity created by, good spatial planning to ensure that this expansion is managed with minimum impact on the environmental risks and still yield the best living and working conditions is clear. The optimum urban form, however, remains an open question. The debate about urban planning in general centres around, on the one hand, the undesirability of urban sprawl, which is characterized by inefficient land use and planning policies (World Bank 2008), and increased transportation demands against the observation that higher densities tend to lead to an increase in non-permeable surfaces and the occupation of vulnerable terrains generating hazard risk, and levels of congestion which can compromise or even overwhelm the operation of infrastructure services such as solid waste and air quality control. In a coastal megacity this debate can be compounded by the complex physical topography, sometimes steep surrounding terrain or the confluence of multiple river basins, extensive wetlands and other low-lying areas, and the limitations of coastal and land interactions such as subsidence.

Spatial planning needs to be addressed at all scales from the city to the region to be successful, however, the mechanisms for integrating scales may be inadequate. It has become increasingly clear that much meaningful adaptation within coastal megacities adaptation can be realized at the regional scale, however, in most cases effective governance structures are not sufficient to successfully implement such efforts. Past planning efforts involving watershed or water basin management illustrate that region-wide strategies situated at varying spatial scales down to the local level can have some measure of success. It will be important to examine these lessons learned and translate them in the contemporary urban spatial planning arena (UNHABITAT, 2007).

Under the wider definition of spatial planning cities need not only to consider the coastal land use zones of acceptable development but also the type of acceptable landscape and building that can occupy these zones. This must be seen in the context that occupation of zones with acknowledged associated hazards does exist and will continue to exist in the foreseeable future given the current location of major coastal megacities.

Spatial planning responses to environmental hazards, generally, and climate hazards, specifically, in the coastal megacity therefore encompasses urban master plans; climate resilient infrastructure planning; marine spatial planning; hazard zoning; poverty reduction, resettlement, and relocation; definition of acceptable development and building codes; greening of urban environments.

The approach to long-term planning of coastal megacities can include these overlapping concepts. With many coastal megacities undertaking risk assessments that are designed to generate master plans, which often include hazard zone definition and management, and infrastructure planning. The main limitation of these measures lie in their implementation, whereby goals are often not achieved and a new plan is conceived before the main benefits of the previous one are realized. Enforcement of more passive features of the urban master plan such as zoning restrictions can also be highly problematic in the face of rapid population growth and competing economic priorities (Jha *et al.*, 2011). Infrastructure planning is also limited by the massive uncertainties associated with predicted manifestations of climate change such that robust and no regrets options need to be developed (Ranger *et al.*, 2010).

Coastal megacities – as many urbanised coastal areas in general – face a multitude of human activities not only on land, but also in the adjacent sea areas. In particular coastal megacities usually encompass large port areas and necessary access for shipping. Furthermore, spatial demand for marine areas comes from recreation, fisheries, nature protection, coastal defence, land reclamation, construction activities (e.g. for bridges), sand and gravel mining, oil and gas industry (as for example in the Bohai Sea in China) and today in renewable energies such as offshore wind farming, which is one of the largest growing activities, e.g. in the North Sea. The result is an increasingly complex spatial pattern of marine uses with some of them being in conflict with each other. Using the German North Sea as a case study such developments have been investigated in detail in Gee *et al.* (2004) and Kannen *et al.* (2008, 2010). As a new instrument for managing the increasing diversity of marine uses Marine/Maritime Spatial Planning (MSP) is emerging, particularly in Europe and North America, but also for example in China. According to the UNESCO-IOC Marine Spatial Planning Initiative, MSP is defined as a 'public process of analyzing and allocating the spatial and temporal distribution of human activities in marine areas to achieve ecological, economic and social objectives that have been specified through a political process' (UNESCO, undated). In this sense MSP is understood as a normative approach to the development, ordering, and securing of space (Douvere and Ehler, 2009). As marine spatial planning often (and particularly in Europe) goes beyond internal waters and territorial seas, international legal frameworks such as the UN Law of the Sea (UNCLOS), international conventions such as the Convention on Biological Diversity (CBD) and international policies, e.g. on fisheries, also need to be considered (Maes, 2008). In Europe MSP is promoted by the EU Integrated Maritime Policy (IMP, European Commission, 2007). Subsequently, in 2008, the Commission adopted a Roadmap for Maritime Spatial Planning: Achieving Common Principles in the EU (European Commission, 2008). However, the BaltSeaPlan Vision 2030 for the Baltic Sea MSP notes that in practice 'MSP is pro-active and future oriented. It delivers the desired outcome of sustainable socio-economic development within a healthy marine environment by balancing all relevant interests in a fair and unbiased manner' (Gee *et al.*, 2011). It should be noted, however, that among countries with competing claims on territorial waters, MSP is difficult to practice and may sometimes lead to open conflict. MSP therefore is nevertheless an integrated planning measure, which aims for 'optimizing sea use and ensuring the integrity of the ecosystem at the same time' (Gee *et al.*, 2011) based on the recognition of environmental, economic, and social goals formulated by society. The integration of MSP with climate change mitigation is in the developmental phase with notable exceptions such as the coral triangle initiative (Fernandes *et al.*, 2012) and guidance prepared by UNESCO and the IOC (Ehler and Douvere). There is much to be offered here for developing coastal planning for megacities and extensively urbanised coastal regions.

The preservation and creation of green space with pervious surfaces and the planting of street trees and urban vegetation in general all help to ameliorate surface runoff, flash flooding, aquatic and atmospheric urban pollution, and heat island effects (e.g. Escobedo and Nowak, 2009). Most ecological engineering has focused on living roofs and walls, which have the advantage of actively supporting the structure on which they are based (rather than abandoning land as in the case of some wetland construction or managed realignment) while at the same time providing a very effective water storage and water/air filtration service (e.g. Oberndorfer *et al.*, 2007; Francis and Lorimer, 2011). Much of this efficacy varies depending on the form of construction (substrate type and depth, vegetation type, etc.), and this remains an area of much research; most have focused on northern temperate areas but there is now an increase in tropical and arid/semi-arid studies (e.g. Jim and Tsang, 2011). Many cities have encouraged green roofs and living walls and there has been notable expansion in research studies in recent years. Some key examples from coastal megacities include London (Grant, 2006), Hong Kong (Jim and Tsang, 2011), Chicago (Yang *et al.*, 2008), Washington, DC (Niu *et al.*, 2010), Shanghai (Roehr and Kong, 2010), New York City (Gaffin *et al.*, 2009), Singapore (Wong *et al.*, 2003) and Toronto (Van Seters *et al.*, 2009), among others. Francis and Lorimer (2011) have suggested that such structures may be tailored to replicate a range of different ecosystems at the landscape scale, and the potential therefore exists to provide both water and pollutant retention and species habitat under different climate scenarios.

Relocation away from areas at risk and resettlement of populations is highly problematic in coastal megacities (Fuchs *et al.*, 2011b). Urban density and scale and associated infrastructure often makes large-scale resettlement prohibitively expensive. Even the clearance of highly localized areas can take years to negotiate due to resistance from local populations as experienced by authorities in Jakarta and Metro Manila (Jha *et al.*, 2012) in attempting to make space to improve flood risk management infrastructure. Resettlement after total demolition of an area by natural or technological hazard may seem logical but is rarely successful (Gupta, 2012). In very large coastal cities where informal settlement presents as a consistent pattern of urban growth, resettlement requires approaches to physical and socioeconomic security that are both innovative and integrated (Twigg, 2004).

Adaptation of buildings and landscape to cope with current and plan for future possible hazards can reduce risks significantly either in rapidly expanding new sectors or in retrofitting existing stock (Salagnac, 2012). In Kobe after the 1995 earthquake, regulations on seismic resistance were strengthened and may have saved lives during the recent Tsunami (Glanz and Onishi, 2011). Developments in flood-proof housing includes floating and elevated structures (BRE, 2009), and resilient and adaptive building technologies suitable to address many risks exist. However, the political will may be lacking and developing the skills necessary to upscale pilot projects and schemes to the scale of coastal megacities will involve major investment in training and communication programmes (Salagnac, 2012). While artisan and community-based approaches to retrofitting and reconstruction are well documented, these tend to be practiced in low-income communities. It is important to have better understanding of these approaches and determine the limits to scaling up such activities to have an impact in megacities. Other barriers to be overcome range from financial constraints to emotional resistance by property owners and occupiers (Lamond and Proverbs, 2010). Uncertainties associated with climate change and the issues with downscaling of climate predictions to property level also hamper efforts to proactively adapt (Nielsen, 2006). Moreover, downscaling climate scenarios will need to be coupled with multi-temporal assumptions regarding engineering design and the behaviour of physical properties of building materials in coastal settings.

### *3.3 Environmental regulation*

The sheer scale of human activity within megacities inevitably generates a vast quantity of pollutants. For a coastal megacity the advantage of the proximity of coastal waters has often been seen as a convenient disposal mechanism. For example in the

Bangkok greater urban area an estimated 50–100 tonnes of garbage every day are dumped into rivers and canals and can be washed out to sea (Jiang, Kirkman, and Hua 2001). Coastal location can also be seen as an aid to the control of air quality within the city without concern for the eventual destination of pollutants. Coastal megacities can contribute hugely to international levels of air-quality management, indicated by the fact that particulates from megacities can be traced inter-continentally (Molina and Molina, 2004). Without adequate regulation, therefore, coastal megacities may contribute massively to global environmental degradation while avoiding some of the full impacts. Conversely, other coastal megacities may suffer from pollutants generated in their hinterland and become the last point of possible treatment before discharge to the sea.

Poor urban air quality has long been recognized as an issue, and of all environmental pollutants air pollution is of the most concern within megacities as it is highly visible and experienced by all residents and visitors (Mage *et al.*, 1996; UNHABITAT, 2011). Therefore many lessons have been derived from air-quality management that will be equally applicable in other areas of environmental regulation.

Often the most cost-effective route to reduce air pollution is to stem the emissions at source via emission control (Mage *et al.*, 1996). This can be approached using technology to improve fuel efficiency, use less polluting fuels (e.g. lead free or low sulphur), or pre-treat emissions. Development of such technology can be encouraged through regulatory controls such as regulated emission limits as has been seen in respect of vehicle emissions (Lloyd and Cackette, 2001). Another alternative is to regulate land-use planning and permit only well-planned infrastructure and transport growth such as the movement of infrastructure to outside the urban area. This has been achieved in Los Angeles (Molina and Molina, 2004).

Regulation needs to be backed by economic instruments that can provide the incentive for adoption of expensive technological measures and cleaner but more costly fuels. Examples can include positive incentives such as grant schemes and emissions trading or penalties such as fines and taxes. Equally, the legislative framework to measure and enforce standards relies on good governance, including appropriate technical capacity within regulators (Olowoporoku, 2011). There is no question that such regulation can be very effective as the stabilization and decline of air pollution in many developed cities can attest (Mage *et al.*, 1996). Evidence from Jakarta and Manila also suggest that tighter environmental legislation has contributed to a lowering of heavy-metal contamination in Jakarta and Manila Bays (Hosono, 2011). The fact that this is likely to be reproducible in other coastal megacities warrants the call for tough environmental regulations in all coastal megacities. However, further research into the most effective economic and regulatory structures to tackle the issue in rapidly growing coastal megacities is needed.

### 3.4 Disaster risk management

Great efforts have, in the past, been directed at predicting the occurrence of natural hazards in megacities (Stein *et al.*, 1997). We are not, however, really any closer to reliably predicting when such events might occur. Even if a natural disaster could be predicted with some precision, the large numbers of people and sprawling infrastructure within large cities make evacuation impractical. As a consequence, the emphasis has switched away from prediction and forecasting and towards increasing understanding of vulnerability, exposure, and preparedness for natural disasters.

The responses in Japan to the Kobe earthquake in 1995 include the development of a widely adopted and effective earthquake readiness programme. In Hong Kong a highly effective traffic light system is in place all over the city, to warn of imminent typhoons and their likely severity. When such systems are activated, local people then carry out well-drilled patterns of behaviour, which ensure that they are more likely to survive the event that is about to happen.

Any increase in understanding of the complexities of a disaster can improve the resilience and preparedness for subsequent events, provided that knowledge is made available to those on the ground and contexts are not greatly changed. The internet has provided us with an ideal method of making such knowledge and understanding rapidly available to anyone with online access. For example, an online disaster response website with live information including onsite photographs (in Flikr), anecdotal evidence (Twitter and Facebook), maps, satellite images, and live information on the positions of the various hazardous entities (earthquake epicentres, volcanoes, typhoons, etc.) can be found at www.esri.com/services/disaster-response/global-viewer.html (last accessed 8 April 2013). Virtual discussions sponsored by institutional stakeholders such as the World Bank, UNISDR, and others (at www.understandrisk.org/ur/ (last accessed 9 April 2013)) present new opportunities to learn about risk-reduction policies and practices. Such tools are commonplace now and provide valuable resources to increase awareness and resilience. The 2004 Indian Ocean Tsunami disaster prompted a change in attitude toward the availability and delivery of Earth observation satellite image data to the public; it was decided that such information should be made available, via the internet, for download by anyone who needed it, within hours of the event. Several programmes were developed to achieve this delivery of information and which led to and include the launch and exploitation of the Disaster Monitoring Constellation of satellites (www.dmcii.com), and the development of the Global Monitoring for Environment and Security or GMES programme (www.gmes.info), which is contributed to by multiple organizations at regional and national levels.

State-of-the-art tools for developing appropriate responses include high-detail terrain mapping and geographical information systems modelling of landslides (e.g. Liu *et al.*, 2011), tsunamis (e.g. Ioualalen *et al.*, 2010), and floods (e.g. Taubenböck *et al.*, 2010), or the use of Differential Interferometry (DInSAR) for volcano monitoring and for earthquake deformation mapping to better constrain the likely damage through ground shaking (e.g. Liu *et al.*, 2011). Such scientific effort is vital to help build capacity and resilience. Responses may then be formal and include the development of early warning systems, or the creation of new organizations, or the modification of building regulations to assist in better design to withstand earthquakes or avoid unstable slopes. The responses may also be informal and include such unprompted behaviour as the formation of local cooperatives to share risk or the gradual movement of properties and businesses to higher ground, or away from vulnerable slopes and tsunami-prone coast lines.

Beyond technological development, better knowledge and understanding of the ways in which risk information, including early warning, is used and acted upon is now gaining priority as the most likely focus for research and development to reduce human loss to disaster. While not a megacity, the millionaire city of Padang, Indonesia has attracted much attention as a city at high risk from earthquakes and tsunamis. Work here has shown the potential for combining satellite imagery of hazard and land use with agent-based modelling and grounded social research to better model and advise government and communities on evacuation paths, possible bottlenecks and ways of maximizing early warning time (Lämmel *et al.*, 2008).

## 4 Climate change mitigation

Clearly coastal megacities can make a significant contribution to reducing total emissions and are in a strong position to utilize offshore energy. Another emerging approach is that of carbon sequestration in recognition that the capacity of the biosphere to store carbon dioxide either directly, via injection into geological features, or indirectly by planting of trees is vast (Coley, 2012). Advanced methods such as coal gasification also offer potential. Storage in the oceans may be an obvious sequestration option for coastal megacities but further research is needed into the impacts of such strategies on ocean processes and on their interaction with adjacent coastal zones (ibid., 2012).

### *4.1 Renewable energy*

Renewable energies have gained increasing prominence as a means of reducing greenhouse gas emissions (GHG) and thereby mitigating climate change. Technically and commercially, wind energy is, beside hydropower, one of the most advanced renewable energy technologies and forms a growing industrial sector. The potential to mitigate climate change by generating electricity from large grid-connected wind turbines is acknowledged and they may be sited either on- or offshore (Wiser *et al.*, 2011) with offshore generation increasing in popularity. While the current capacity of installed wind farms is still small (approximately 1.8 per cent of global electricity demand by 2009) growth in turbine installations is fast with China having become the world leader in 2010 with more than 44,000 MW totally installed capacity by the end of 2010, of which nearly 19,000 MW have been installed within the year 2010 (GWEC, 2011). The globally installed total capacity in 2010 amounts to 197,000 megawatts (ibid., 2011). The contribution could grow to more than 20 per cent by 2050 with the right policy and legislative developments to address impediments to increased wind-energy deployment (Wiser *et al.*, 2011).

Offshore wind is of increasing interest to developers because of several advantages compared to the conditions onshore, including higher yields: wind speeds are higher and more predictable, fewer conflicts with other types of land use are to be expected (e.g. nature protection or tourism), and public resistance against new wind farm developments seems less likely. However, reality can be a bit more difficult and in particular visual impacts, impacts on fisheries, and ecosystem impacts can be sensitive issues from the perspective of local coastal communities (Gee, 2010). Offshore wind farms currently drive wind energy growth particularly in Europe with large installations planned for the Doggerbank in the North Sea for example. In 2010 EWEA estimate that a total of 1,136 turbines were installed and grid connected; this represents almost 3,000 MW new capacity in Europe (EWEA, 2011). For Germany, in the long term, offshore wind farms are expected to provide a capacity of 20,000 to 25,000 megawatts by 2030, meeting about 15 per cent of the German electricity demand (and 30 per cent by on- and offshore wind farms). This might – according to today's technology – cover an estimated sea area of 2,000 to 4,000 square kilometres (BMU, 2002). The potential for offshore wind energy to mitigate the future emissions from energy demand in China has been demonstrated by the Shanghai Donghai Bridge Offshore wind farm (Gail, 2010), sufficient to supply 200,000 households and used to power the Cities Expo in 2010.

Wind farms are typically situated about ten kilometres away from the coast on the continental shelf area with associated technical challenges that result in a higher cost of installation than land-based systems (Zhixin *et al.*, 2009). The attraction of constructing intertidal or floating wind farms (Henderson and Witcher, 2010) lies in their lower cost but little experience or research is available on intertidal generation. Research to improve efficiency of wind technology, grid integration, and security of installations to coastal conditions is still in early stages (Gail, 2010). The considerable spatial requirements of offshore wind farming meet sea areas that are already much contested, in particular in sea areas adjacent to coastal megacities. Shipping and port development, land reclamation and construction projects, marine aggregate extraction, nature protection, and fisheries are just some examples of marine uses that also demand their share of space. Therefore, despite its potential for electricity generation without GHG emissions and its current growth trends, the introduction of offshore wind farming in large scales and near highly contested sea areas requires careful planning and new forms of governance for marine space (Kannen and Burkhard, 2009; Kannen *et al.*, 2010).

Wave energy is also a potential area for renewable offshore energy development; so far it is less developed than wind energy but the capacity of wave energy could be much greater due to the higher energy densities achieved (Coley, 2012b). Wave energy sites would be placed at some distance from the shore and therefore issues arise as to ownership and also the economic viability of installations in harsh marine environments. Tidal power is another coastal resource that has yet to be fully exploited with few installations for example in Canada and France. Two major routes are worthy of further investigation, storage of tidal waters at high tide coupled with traditional hydropower generation and offshore turbines to capture tidal streams (Coley, 2012b). Finally, geothermal energy is the third most utilized renewable source and is prevalent in the US, Mexico, Japan, New Zealand, and Italy. Offshore geothermal installations may make geothermal energy a practical consideration for more nations and coastal megacities in particular. Research is required into efficient and robust offshore generation technologies and also grid integration and energy storage techniques in order to realize the full potential of these renewable energy sources.

### *4.2 Sustainable transport*

Coastal megacities are, as noted above, geographically large and economically important. It follows that their transport needs will be equally intensive. The function of transport systems is to provide connections and access between people and the goods and services they require. The form of coastal megacities present particular challenges as they are likely to be conduits through which goods and services flow to international destinations, as well as centres of industrial and service industries concentrated in proximity to the coast and estuary with extended commuting networks into the hinterland. Coastal megacities subject to threat from natural hazard also require a secure evacuation system and that can be severely constrained by the physical restrictions imposed by coastal and estuarine morphology. There is a growing need to understand how these systems operate and are used and to find ways to manage them to minimize their impact on GHG emissions yet enhance productivity and increase security. In many cities there is an emphasis on shortening travel distances, making the city compact and moving facilities closer to where people live. Another alternative is to switch journeys to less carbon intensive modes of transport such as cycling or more efficient transport systems. The other major development that can reduce transport needs for people is the ability to connect virtually via telecommunications, however, their users are heavily dependent upon the system's infrastructure.

Transport is one of the areas over which mayors of megacities have a large measure of control (Arup, 2011); many cities own much of the public transport infrastructure and can designate cycle lanes and road charging if they wish. Therefore sustainable transportation policy is a way in which such cities can make a major contribution to the reduction of emissions. However, in many cases sustainable transport policies fail because of public opposition to the necessary behavioural change (Georgeakis and Nwagaboso, 2012) and the difficulty in quantifying many of the benefits of sustainable transport. There is very little research on the linkages between spatial distribution of poverty, structure of incomes and sustainable mobility options. This analysis would be of particular significance in coastal cities located in developing countries where sections of the population make their living through various forms of formal and informal employment.

Electric vehicles are seen as a natural replacement to the higher pollutant petrol or diesel alternatives. Clearly this reduces the pollutants at point of journey and improves the air quality of a megacity. However, the total emissions reduction benefit is only realized where the electricity grid is not carbon intense (Arup, 2011). Other low emissions options for public transport include the use of gas powered vehicles as was implemented in Delhi to improve the air quality (Gurjar *et al.*, 2007) and, in view of the improvements in Delhi (Ravindra *et al.*, 2006), has led to similar programmes in Tehran, Los Angeles, Bangkok, Santiago, Cairo, Beijing, and many other cities.

Traffic limitation has been implemented in London (in 2003) via the congestion charging scheme and this has been shown to have an impact on emissions via the speeding up of traffic and the number of miles travelled (Beevers and Carslaw, 2005). Increased public transport emissions were mitigated via the introduction of particle traps to the bus fleet. Effective car pooling can also reduce journeys by reducing the number of single occupant vehicles (DFT, 2004). In Los Angeles a scheme requiring companies to incentivize car pooling eliminated 90,000 trips a day (Wisner, 1999).

A more positive approach to limit traffic movements is to increase the desirability, safety, and availability of alternative transport such as cycling, walking, and public transport. With the notable exception of the Amsterdam Randstad, provision of improved cycling facilities has been a relatively recent development after the high-profile success of the Paris 'Velib' cycle hire scheme, which started in 2007. Bogota, Chicago, London, Mexico City, and Toronto have introduced large-scale schemes (Arup, 2011).

GDP and carbon emissions are usually positively correlated and, yet, because of their high density and extensive public transport some of the cities with the highest per capita GDPs in the world (e.g. Hong Kong, New York, and London) record relatively low per capita carbon emissions (Arup, 2011). Retrofitting more extensive rail and underground systems may be impractical in many coastal megacities because of the disruption caused and the land subsidence issues they face. However alternatives include Bus Rapid Transit as implemented in many Latin American megacities with high speed, segregated bus lanes and sheltered bus stops.

Increasing the use of public transport depends on speed and ease of use for the traveller, high-quality travel information (Lodden and Brechan, 2003), and integrated tariff and ticketing systems (Abrate *et al.*, 2009) can exert a positive impact on passenger demand. In London and Hong Kong 'smart tickets' are already popular and transformative actions have been taken in integrated ticketing in other megacities such as Tokyo.

Sea transportation of freight and passengers can result in a concentration of emissions due to freight vehicles both on land and in the sea in major coastal megacity ports. Possible responses include improved logistics to reduce congestion for container vehicles and the recent development of an Environmental Ship Index (ESI), which indicates that the differences in emissions can be very high for the oceangoing vessel (World Ports Climate Initiative, no date).

### *4.2 Energy efficient buildings*

There is no doubt that the construction and occupation of buildings is an energy-intensive activity that represents a significant proportion of global carbon emissions (ARUP, 2011). Increased energy efficiency can start at the power station by reclaiming the energy lost to heat during generation (Atkins, 1986; ARUP, 2011) through to efficient construction methods (Kibert, 2007) the use of recycled and recyclable materials (Khatib, 2012), and including efficiency standards for homes, offices, and industrial buildings. Recent developments in energy-efficient buildings include technologies such as micro generation, nanotechnology, advanced materials, and sophisticated control devices, so-called smart buildings, that promise to reduce energy needs without loss of functionality. The adoption of these relatively expensive innovations will depend on legal, financial, and social pressures to reduce emissions, and cities can choose to instigate local schemes to encourage and facilitate adaptation. Retrofitting energy efficient appliances, light bulbs and insulation are popular city-scale responses to climate mitigation, for example heating and cooling efficiency programmes in Mumbai cover residences (public and private) and offices (commercial and municipal) (Arup, 2011). The introduction of buildings ratings and minimum expected codes, particularly for new buildings, is another popular approach as this may have legal and economic leverage to improve efficiency. In Hong Kong, one of the highest carbon contributors, the Carbon Audit Campaign provides a systematic approach to identify potential emissions improvements and is proving highly successful (Arup, 2011).

## 5 Risk governance for coastal megacities

It is clear from the foregoing sections that there are multiple technological and policy responses suitable to address the future challenges of coastal megacities. Examples above and the case studies below demonstrate that these responses can be instrumental in improving the urban environment, limiting the global impact of megacities and protecting the residents of coastal megacities from hazards. Wider adoption and implementation of technology and policy responses is dependent on the character of urban governance and wider policy narratives. Frequently, especially in rapidly expanding coastal megacities, development agendas continue to emphasize economic growth as development. This sets the context and bounds what is possible for risk management. Recent discussions on adaptation as resilience and transformation build on established critiques of development from the risk management community and call for disaster risk management (and by implication other environmental management regimes) to focus not only on the proximate causes of risk and hazard, but to recognize the root causes that drive vulnerability – and, importantly, to also consider these for action. This takes risk management away from being a purely technical or administrative domain into mainstream political discourse. In this way risk governance is increasingly seen as being set within, and needing to complement, wider urban/national development goals. Megacities bring their own governance challenges and quickly resolving risk at the political as well as technical levels will require multi-level and cross-sector engagement with leaders from civil society, the private sector, and government agencies that have important contributions to make as well (Pelling, 2012; Johnson, 2012).

### *5.1 Urban governance*

The challenges of environmental governance for megacities are most often considered within a framework that considers governance to be a matter of institutional design – of achieving the right combination of resources, competencies, and mandate for public institutions and of garnering appropriate private sector and civil society resources towards collective goals. Within such a framework, the challenges of environmental governance are related to the scope of such institutions, how they are resourced, on relations between institutions and actors of different kinds, and of various facets of their capacity to respond across scales (Betsill and Bulkeley, 2007; Bulkeley and Betsill, 2005; Corfee-Morlot *et al.*, 2011; Gustavasson *et al.*, 2009; Puppim de Oliveira, 2009). Multilevel urban governance recognizes that while municipalities are relatively coherent and autonomous entities, the relationship between international, national, regional, and local governments is critical in shaping the capacity to govern climate change (O'Brien *et al.*, 2012). Multilevel governance approaches seek to influence urban and national policy to shape (lead, influence, or circumvent) the framing of environmental management, especially for climate change (including rural studies of water management regimes), determine the extent of the duties and powers that municipalities have, and enable or constrain forms of policy integration across and between levels of government.

Multilevel governance highlights the problem of 'fit' in regimes – the lack of coincident boundaries between the scale of the issues that need to be addressed and municipal authority (Bai, 2007; Romero Lankao, 2007). Emergence of horizontal forms of coordination and cooperation in the form of transnational municipal networks has also provided a new form of 'multilevel' governance within which urban responses have been developed. Such networks can provide significant support to their members, but may also function to narrow down the range of actions regarded as appropriate and in some cases to exclude those who have insufficient resources to take part in network activities (Kern and Bulkeley, 2009).

Membership of multinational networks has been especially important for medium-sized cities (Bulkeley and Betsil, 2003) in elevating political influence and access to technical insight. Megacities have been less prominent in such networks than they might have been indicating the capacities already enjoyed by the very largest of cities, and in comparison to national government. Work in Sao Paulo and Rio de Janeiro (Martins and Ferreira, 2012) shows the advantages that membership in a transnational network offers local governments in smaller cities is also of use to local government in megacities. These include support for leadership in the early stages of lobbying for climate change mitigation policy. Adaptation as a discrete policy area has proven less easy to promote, perhaps reflecting the amount of everyday risk management work that has been described as adaptation and the challenge of fundamental changes to the procedure of decision making that moving beyond this to integrate adaptation into urban development planning requires. In the Asian region, CITYNET is an established network of urban governments that has championed disaster risk reduction and climate change alongside poverty alleviation and governance capacity amongst and for its members, which include Bangkok, Dhaka, Jakarta, and Mumbai (www.citynet-ap.org/, last accessed 9 April 2013).

The ability to implement and enforce policies and measures is also a governance factor that is key in tackling climate change challenges. Such challenges range from issues of the importation of inappropriate planning models, poor instrument design to corruption, and from challenges of enforcing measures to those of measuring and accounting for progress with implementation, a complex issue where what counts as mitigation and adaptation are in themselves difficult to determine (Bulkeley *et al.*, 2009; Satterthwaite, 2008). A third set of factors that critically affect urban environmental governance relate to institutional capacity, in the form of access to appropriate knowledge, resources, and personnel (Holgate, 2007; Granberg and Elander, 2007). In particular, financial resources can be a barrier to implementing urban responses to climate change. Given the difficulty in financing even basic services for their constituents, and the many other competing issues on urban agendas, municipal authorities are unlikely to invest in climate change response. At the same time, research has also found that even where resources are available, the institutional structures through which they are allocated and invested can serve to impede action (Bulkeley *et al.*, 2009; Bulkeley and Castan Broto, 2011). Alternative financial mechanisms that can overcome these challenges have been important in shaping municipal responses to climate change, but are arguably harder to achieve in relation to adaptation and wider risk management – including for example responding to hazards of subsidence – than climate change mitigation.

Frequently, the response to such issues of institutional challenges is to generate more capacity through more knowledge, the provision of resources, the creation of new institutions, for 'good' governance, or through giving further autonomy to municipalities (Bulkeley, 2010). Important as these issues of institutions and capacity may be, they do not, however, provide a full picture of the ways in which governance is conducted and, therefore, of the challenges that are encountered.

A second set of important dynamics relates to the political economy of urban responses to climate change, and especially mitigation. The ways in which urban climate change agendas are manifest and contested relates to underlying dynamics of urban political economy, which serve to shape and reinforce relations of power and vulnerability in cities. At the most fundamental level, it is debatable whether city authorities should be addressing climate change at all. On the one hand, climate change has been resisted by politicians more concerned with immediate challenges, and the arguments 'not on my turf' and 'not in my term' are prevalent in many cities particularly in the global south where resources are limited and other concerns are more pressing (Bai, 2007). On the other hand, the absence of political concern for issues that shape the possibilities of addressing climate change, particularly those that relate to urban vulnerability and the lack of basic service provision for the urban poor, can also hamper responses (Satterthwaite, 2011). Strategies to mobilize political leadership, to create and utilize windows of political opportunity (such as international events or conferences), and to reframe climate change in relation to issues of local concern, from air pollution to economic development, have been important in overcoming some of these challenges (Betsill and Bulkeley, 2007; Koehn, 2008). However, such approaches have to date made little inroads on normal patterns of urban development. This is because the factors that drive urban growth, including the availability of land at the urban fringe, short payback periods on capital investment, increased mobility, the consumption of energy and resource-intensive goods and services, are also those that contribute to rising GHG emissions and vulnerability to environmental hazards (Bulkeley and Castan Broto, 2011; Pelling, 2011).

These issues may be particularly pressing for megacities and other urban centres in developing countries where reducing GHG emissions is associated with reduced opportunities for growth and development (Lasco *et al.*, 2007). However, the challenges that emerge between 'economy as usual' climate change and wider risk management concerns are not confined to such contexts. These major blockages point to the need to consider alternative approaches that identify the need for transformation in urban development vision, management structure, and process. Mitlin (2012), for example, describes the opportunities for reinvigorating governance opened through local collective action supported by membership of the international network Slum and Shack Dwellers International.

A final set of challenges facing urban environmental governance relates to the ways in which the (re)production of urban societies and economies is continually maintained through the material fabric and infrastructure – of housing, energy provision, water systems, and so on – that support them (Bulkeley *et al.*, 2010; Monstadt, 2009). These *socio-technical* networks, webs of technologies, artefacts, institutions, and the everyday practices that they constitute and that maintain them are often obdurate and resistant to change. The challenges this poses for governing climate change in the city are two-fold. First are issues of urban design and morphology. There are very real practical differences in seeking to reduce GHG emissions from transportation in compact, dense cities or sprawling urban areas, or in addressing flooding in delta cities from other coastal locations. Second are the traditional means through which infrastructure systems are developed and maintained to supply services, such as energy, water, and waste collection, as well as existing building stock, within cities.

Cultural and institutional practices, as well as political and economic interests, accrue around such networks, making them resistant to change and contestation. Rather than being a matter of overt political interest, here the challenges lie in the ways in which particular issues and alternatives are kept off urban agendas by accepted norms and operating conditions that fail to consider different directions for urban development. Of critical importance here is the infrastructure deficit experienced in many urban contexts, particularly in cities in developing economies, where established cultures concerning what sorts of infrastructures should be provided in different urban areas persist and serve to sustain existing vulnerabilities (Satterthwaite, 2011).

The challenges of urban environmental governance can therefore be thought of as three-fold – as institutional, political, and socio-technical. While research has tended to focus on the former, identifying critical areas where interventions may improve capacity to govern, in the absence of consideration of the latter it is unlikely that these will have a significant effect. At the same time, it is important to recognize that the importance of these different dimensions will shift in relation to the nature of urban governance being undertaken. Municipal responses to climate change can be regarded as involving multiple modes of governance, including self-governing, regulation, provision, and enabling (Bulkeley and Kern, 2006). While governing climate change in a self-governing mode may raise institutional challenges, modes of provision and enabling may be more constrained by challenges of a political economy and socio-technical nature.

### *5.2 Government actions*

A number of initiatives have recently emerged from megacities seeking to develop community-based resilience in order both to achieve a low-carbon transition and to protect those most vulnerable to the potential implications of the rising costs of energy in a carbon-constrained future. In London, for example, the Transition Town movement has led to the development of local community-based groups that focus on the challenges of peak oil and climate change to promote a 'relocalization' of the economy through local food provision, decentralized energy services, and reengagement with local economies (Bulkeley, 2012). In Los Angeles, local community groups are mobilizing around the issue of climate change to protect local green spaces and create alternative discourses about the ways in which water should be created, stored, and used in the city (Bulkeley and Schroeder, 2012). Analysis of the 2005 Mumbai floods has identified failure in everyday development and risk management: failure to manage solid waste collection and maintain city drains and the development of coastal mangrove that prevented natural drainage from the city. Both are symptoms of inappropriate development vision and priority and point to the need for urban development, and, especially in megacities where the stakes are so high, to protect environmental assets surrounding cities and the infrastructure within (Red Cross, 2004). Rapid population growth in Mumbai, especially in the slum settlements exceeded the capacity of risk management agencies (and also infrastructure providers) to keep up. This led to confusion. Elsewhere the merging of city districts as urban areas coalesce in megacities has a similar effect where disaster management agencies are not integrated and coordinated (Chafe, 2006).

In Jakarta, increased run-off, partly caused by deforestation and urbanisation upstream, has increased flood risk in the city. Responding to such risks – generated by development – requires national and local actions. In 2004 the government began construction of a 24-kilometre canal to carry flood water away from the city's northern and eastern districts. This was combined with headwater reforestation programmes, improved flood early warning systems, and emergency response planning alongside volunteer-led efforts to keep city drains clear from solid waste. Locally hazards are also being mitigated, for example in the Bidara Cina neighbourhood, which has improved sanitation, roads, and dykes. Elsewhere, though, low-income communities encroaching on drainage reserves are identified as a problem and many have been relocated. Criticism of this approach has argued that flood management has focused on moving low-income households and acting on local polluters rather than formal neighbourhoods and corporate industry. This case study shows the importance of acting at scale and in an integrated approach to reduce risk, but also the need for government to reflect on its priorities and how closely these match interest groups in society, including the urban and rural poor (Nagara, 2006).

### *5.3 Community level and civil society actions*

The importance of stakeholder engagement has been emphasized throughout the literature on managing vulnerability to natural hazards and climate change (e.g. Field *et al.*, 2012). This calls for urgent and practical decisions, not least those that involve harnessing and catalyzing synergies found within and amongst the array of stakeholders, for the successful implementation of adaptation and mitigation options.

In the face of climate change, mitigation and adaptation measures will most likely require decisions that fall outside the usual paradigm of economic efficiency and rationalization. This is mainly because most climate change response measures are slow, multi-scale, involve multiple players, and have long timescales. Consequently, community and civil society involvement in public discourse, if not in policy decision making, can serve to hold government (and the private sector) to account, for example in pressing for decentralized risk-management structures to be supported by adequate finance and legal weight. The literature has many examples of low-income communities in megacities and other urban centres that not only possess strong traditions of collaboration, sustained by bonds of trust, reciprocity, and connectedness, but are also organized, active, and influential in risk governance (see Kithiia and Dowling, 2010; Rakodi, 2006; Pelling and High, 2005). These capacities can be leveraged to facilitate the launch of effective city-wide adaptation strategies. This will not only reduce costs associated with adaptation but will also help utilize local innovativeness and achieve legitimacy.

Engaging an array of interest groups in large coastal conurbations is seen as necessary to ensure success and establish legitimacy. These are often well organized in megacities where levels of education and politicization are amongst the highest in a country. Chatterjee (2011) comments on the important basic contribution to risk management played by networks of self-help groups and religious organizations acting both at the local and city levels in Mumbai. However, building consensus can be a challenge, it means coordinating interests in city management with other, often conflicting, perspectives; such as the use of the coastal zone resources or valuable real estate. Mitigation and/or adaptation measures that involve change of livelihoods, abandonment of investments, relocation, retrofitting, and so on are likely to face opposition from politicians, developers, and other stakeholders and residents themselves.

Decision-making tools are only just able to approach the trade-offs required when risk management (including climate change mitigation) is mainstreamed into development. Research suggests that as a starting point, development planning that recognizes and addresses the impacts of climate change will require a more inclusive approach that seeks input from non-traditional sources, including academia, research organizations, the poor, and the vulnerable. An approach that emphasizes flexibility in organizational positioning to move beyond a political and ideological stalemate and the continuous incorporation of lessons learnt into future plans (see Kithiia and Lyth, 2011).

Recently, there has been a turn in the climate change community towards a 'no regrets' approach to vulnerability reduction (Heltberg *et al.*, 2009). This is an approach that is intentionally forward looking and opens scope for deliberative choices between transformation, resilience, and resistance approaches. Instead of targeting specific sectoral impacts of a hazard, this approach supports social policies that simultaneously address climatic risk, issues of poverty and sustainable development. Microfinance, weather-based index insurance, and social funds for community-based adaptation, for instance, are all examples of the multi-sectoral, multi-level policies typical of the 'no regrets' approach (Heltberg *et al.*, 2009). While the 'no regrets' approach is not cost free, involving real and opportunity costs and trade-offs (Wilby cited in Heltberg 2009), it does offer distinct advantages. Unlike sector-specific policies, this approach is not as susceptible to the uncertainty of climate projections or scenarios; policies will be beneficial regardless of whether a shock occurs or occurs differently than expected (Barnett cited in Heltberg 2009).

Including a diverse range of civil society actors (such as community residents and local experts) in adaptation activities has important benefits. The inclusion of these actors in the governance of climate-related risks increases the knowledge base that can inform adaptation initiatives (Renn and Schweizer, 2009; Measham and Preston, 2012). Risk theorists have demonstrated that when risks are associated with uncertainty, science is best considered a first step towards dialogue over the likelihood of impacts, and, importantly, the significance of impacts to the affected communities (Klinke and Renn, 2002). Therefore, engaging civil society actors can result in better responses to coastal challenges. Furthermore, the involvement of community members helps to build their understanding and their 'ownership' of the problems and solutions (Measham *et al.*, 2011). This can translate to a higher degree of community acceptance of responses and more effective implementation of actions (Renn and Schweizer, 2009; Preston *et al.*, 2011; Mukheibir and Ziervogel, 2007).

It is important to recognize the limitations and strengths of community participation in civil society actions. The first is that local communities rarely reflect a discrete set of interests, so much as multiple and frequently conflicting interests that extend across spatial scales (Lane and McDonald, 2005). Therefore community participation is better seen in relative rather than absolute terms. Furthermore, the policy context for community-level civil society actions is usually dependent upon the resources and acceptance of higher levels of governance, such as states or provinces. This support is important to ensure that the initiatives of one locale don't negatively impact upon those of another community, either directly e.g. by shifting flood water, or indirectly through the introduction of new social inequalities (Measham *et al.*, 2011).

## 5.4 Market responses

Coastal megacities often command large resources, social, natural, and financial. With a large population, commercial and industrial base is associated the potential to generate large local revenues in order to invest in physical and institutional capacity. Directing investment of these resources can utilize existing metrics such as cost-benefit analysis where sufficient quantitative data is available (Hunt and Watkiss, 2011). However, with the increased emphasis on policy and regulation rather than hard engineering and technological measures, the distribution of costs and benefits are increasingly complex, novel sophisticated economic, financial and fiscal instruments are required, and new economic models need to be developed (Webster, 2005). For example market imperatives towards risk reduction could include land price adjustment. The threat of lower resale value can provide incentives for the installation of mitigation and adaptation measures (Lamond, 2011). Conversely the lower price of undeveloped land may lead to the settlement within hazard zones of the poorest and most vulnerable. Evidence for the rational response of property markets to hazard zoning is mixed (World Bank and the United Nations, 2010; Lamond, 2005) and confused by the intervention of hazard insurance. Another example is the provision of alternative employment routes to reinforce environmentally sound practice and reduce the impact of overfishing in coastal waters (UNESCO, 2000) the impact of which will be felt locally but the wider costs and benefits may not be apparent for years to come. Climate change also brings with it the challenge of hugely increased uncertainty into decision mathematics (Lamond and Bateman, 2012; Barbier, 2012) stimulating a search for alternative decision metrics (Ranger *et al.*, 2010; Floodsite, 2007).

The importance of the private sector and the national and international markets in incentivizing and implementing risk reduction and responses to climate change is increasingly recognized, although often international policy is moving ahead of theory and the evidence of effectiveness of instruments is lacking (Stern, 2008). The complexity of global interdependency in finance and trade and the link between the prosperity and security of coastal megacities, often major financial and industrial centres, and the operations of governments and multinational companies is an area yet to be fully explored. Insurers are starting to recognize the importance of megacities and the issues associated with the massive probable maximum loss in a coastal disaster involving potentially millions of people (Munich Re, 2004). For example the most expensive individual catastrophe in history (prior to the Japanese Tsunami) was the earthquake of 17 January 1995 in the Japanese city of Kobe, causing an economic loss of US$140 billion at 2007 values (Spranger, 2008). The potential for even greater harm exists if a similar quake were to hit on a winter day at 6 pm (ibid.).

The probability of a strong earthquake hitting Tokyo in the next 30 years with potentially catastrophic losses is about 35–40 per cent (Stein *et al.*, 2006). Megacities continue to be a focal point of observation and concern for the insurance and global reinsurance industries as part of both disaster management and mitigation and adaptation processes. Insurers and in particular reinsurers have therefore recognized a key role of driving mitigation and adaptation and have been instrumental in assisting the development of global information systems such as the Global Earthquake Model www.globalquakemodel.org/.

The local distribution of risk is also shaped by market relations, and in particular those between employer and employee. This may be especially important in megacities with multiple, widespread, and complicated labour markets. In Mumbai, for example, research has identified the practice of employers that assist individual workers in recovering from city flood events in preference to the potentially larger and less focused assistance to communities at large (Chatterjee, 2011). Certainly it is advantageous for the employer to support workers and so limit exposure to the indirect impacts of disaster (i.e. by workers being forced to stay away from work to assist in repairs and local recovery) (Basu, 1997).

## 6 Conclusion

Innovation in responding to natural hazards and climate and pollution risks in the twenty-first century have been identified in this chapter to lean towards mitigation of climate change. Control of emissions and air quality are principal mechanisms as they are seen to have direct benefit to the city as a whole and can contribute to global goals. In this there appears to be little differentiation between coastal and other megacities of the world, albeit the opportunity to access offshore resources such as wave and tidal energy may be more immediate for the coastal city.

Coastal megacities have specific responses to their own particular risks, which are related to coastal erosion and storm and flood risk. These are seen to be increasing over time by the dual action of climate change and increased urbanisation. Many coastal megacities are facing sea-level rise, coastal erosion, and subsidence, representing major threats to their populations.

Further, coastal megacities are often centres of international trade, intrinsically linked to global financial and economic markets with complex transport and communications infrastructure to support physical and virtual connections with the wider world.

These cities have drawn on a strong science base to implement change but many, particularly the rapidly expanding urban centres in the developing world, have been hampered by a lack of coordination in urban governance and planning and difficulties in bring vested interests to address trade-offs in risk reduction and development. Constraints on resources, competing priorities, and coping with the impact of natural disasters and massive urban growth on existing infrastructure and systems complicate risk management capacity for large cities.

Priority areas for ongoing research to support further environmental management and risk-reduction activities, including climate change adaptation and mitigation, have been identified in technological, social, economic, and policy realms as follows:

1. Research into cost-effective and robust offshore renewable technology that reduces the demands of coastal megacities on traditional power generation while minimizing the environmental impact of such generation on the coastal zone.
2. Evaluation of the effectiveness of city-scale economic, financial, and fiscal instruments in incentivizing climate-sensitive behaviour and accelerating adaptation within the broader national and international market.
3. Urban governance research that addresses the complexity of urban form within a coastal megacity and the particular challenges associated with the scale and heterogeneity of these unique locations.
4. Identifying and modelling probable changes in coastal processes and sediment budget regimes so as to identify potential flood risks and, therefore, inform coastal protection management strategies.
5. Development of techniques for the downscaling of climate change parameters for urban flood forecasting.
6. Experimentation with ecological engineering techniques to ensure that hard flood engineering structures support a range of ecological communities, as a way of reconciling the use of urban coastal zones by both humans and marine/estuarine communities.
7. Better understanding of the political economy of water use in the city, and of the consequences beyond the city – in rural and coastal environments – of burgeoning water demand.
8. Social research to develop ways to make transparent and balance the selfish impacts of vested interests while encouraging inclusive modes of decision making at all scales from the local up.
9. Exploration of how extreme events – such as coastal storms – can be used as opportunities for increasing adaptive capacity and resilience. Building on work that highlights the importance of pre-disaster social relations, science, and popular discourse for post-disaster trajectories, and that shows the potential for working before disaster to shape policy directions post disaster.
10. Extreme events need to be studied as tipping points for coupled human and natural transitions and larger scale transformations for enhanced urban sustainability.

## References

Abrate, G., Piacenza, M., and Vannoni, D. (2009) The impact of Integrated Tariff Systems on public transport demand: Evidence from Italy. *Regional Science and Urban Economics*, 39, pp. 120–127.

Aerts, J.C.J.H., Major, D.C., Bowman, M.J., Dircke, P., and Marfai, M.A. (2009) *Connecting Delta Cities, coastal cities, flood risk management and adaptation to climate change*, Amsterdam: VU University Press.

Algoni, D.M. (2008) Mangrove forests: Resilience, protection from tsunamis, and responses to global climate change. *Estuarine, Coastal and Shelf Science*, 76, pp. 1–13.

APFM (2007) Associated Programme on Flood Management, WHO, www.affm.info (last accessed 9 April 2013).

Arthurton, R.S. (1998) Marine-related physical natural hazards affecting coastal megacities of the Asia–Pacific region – awareness and mitigation. *Ocean & Coastal Management*, 40, pp. 65–85.

Atkins, G. (1986) The advantages of CHP systems. In: *Proc I Mech E Symposium on CHP*, Sheffield.

Bai, X. (2007) Integrating Global Environmental Concerns into Urban Management: The Scale and Readiness Arguments. *Journal of Industrial Ecology*, 11(2), pp. 15–29.

Bakker, K., (2010) *Privatizing water: governance failure and the world's urban water crisis*, Cornell University Press: New York.

Barbier, E.B. (2012) Progress and Challenges in Valuing Coastal and Marine Ecosystem Services. *Rev Environmental Economics and Policy*, 6(1), pp. 1–19.

Basu, S. (1997) Why institutional credit agencies are reluctant to lend to the rural poor: a theoretical analysis of the Indian rural credit market. *World Development*, 25 (2), pp. 267–280.

Bates, B.C., Z.W. Kundzewicz, Wu, S., J.P., and Palutikof, E. (2008) *Climate Change and Water. Technical Paper VI of the Intergovernmental Panel on Climate Change*, IPCC Secretariat: Geneva.

Battjes, J.A. (2006) Developments in coastal engineering research. *Coastal Engineering*, 53, pp. 121–132.

Berg, S. and Marques, R. (2011) Quantitative studies of water and sanitation utilities: a benchmarking literature survey. *Water Policy*, 13(5), pp. 591–606.

Betsill, M. and Bulkeley, H. (2007) Looking Back and Thinking Ahead: A Decade of Cities and Climate Change Research. *Local Environment: The International Journal of Justice and Sustainability*, 12(5), pp. 447–456.

Beevers, S.D. and Carslaw, D.C. (2005) The impact of congestion charging on vehicle emissions in London. *Atmospheric Environment*, 39, pp. 1–5.

Bicknell, J., Dodman, D., and Satterthwaite, D. (2009) *Adapting Cities to Climate Change: Understanding and Addressing the Development Challenges*, London and Sterling, VA: Earthscan.

Bodin, Ö. and Crona, B.I. (2008) Management of Natural Resources at the Community Level: Exploring the Role of Social Capital and Leadership in a Rural Fishing Community. *World Development*, 36, pp. 2763–2779.

Booth, C., Hammond, F.N., Lamond, J., and Proverbs, D.G. (2012) *Solutions for Climate Change Challenges in the Built Environment: Innovation in the Built Environment*, Wiley-Blackwell: London.

Betsill, M. and H. Buke (2007) 'Looking Back and Thinking Ahead: A Decade of Cities and Climate Change Research', *Local Environment: The International Journal of Justice and Sustainability*, 12(5) pp. 447–456

Biggs, C., Ryan, C., Wiseman, J., and Larsen, K. (2009) *Distributed Water Systems: A networked and localised approach for sustainable water services.* Victorian Eco-Innovation Lab: Melbourne, Australia.

Brooks, D.B. (2006) An Operational Definition of Water Demand Management. *International Journal of Water Resources Development*, 22(4), pp. 521–528.

Brooks, S.M. and Spencer, T. (2012) Shoreline retreat and sediment release in response to accelerating sea level rise: Measuring and modelling cliffline dynamics on the Suffolk Coast, UK. *Global and Planetary Change*, 80–81, pp. 165–179.

Brown, R.R., Keath, N., and Wong, T.H.F. (2009) Urban water management in cities: historical, current and future regimes. *Water Science & Technology*, 59(5), pp. 847–855.

Building Research Establishment (BRE) (2009) *Life handbook. Long-term initiatives for flood-risk environments*, Bracknell: BRE.

Bulkeley, H. (2010) Cities and the Governing of Climate Change. *Annual Review of Environment and Resources*, 35, pp. 229–253.

Bulkeley, H. (2012) Climate change and urban governance: a new politics? In: Lockie, S., Sonnenfeld, D., and Fisher, D. (eds) *International Handbook of Social and Environmental Change*, Abingdon: Routledge.

Bulkeley, H. and Betsill, M.M. (2003) *Cities and Climate Change: Urban Sustainability and Global Environmental Governance*, London: Routledge.

Bulkeley, H. and Betsill, M. (2005) Rethinking Sustainable Cities: Multilevel Governance and the 'Urban' Politics of Climate Change. *Environmental Politics*, 14(1), pp. 42–63.

Bulkeley, H. and Kern, K. (2006) Local Government and the Governing of Climate Change in Germany and the UK. *Urban Studies*, 43(12), pp. 2237–2259.

Bulkeley, H. and Schroeder, H. (2012) Beyond state/non-state divides. *European Journal of International Relations*, forthcoming 2012.

Bulkeley, H., Schroeder, H., Janda, K., Zhao, J., Armstrong, A., Chu, S.Y., and Ghosh, S. (2009) *Cities and Climate Change: The role of institutions, governance and urban planning*, Marseille, World Bank Urban Research Symposium Cities and Climate Change.

Bulkeley, H., Castan Broto, V., Hodson, M., and Marvin, S. (eds) (2011) *Cities and Low Carbon Transitions*, London: Routledge.

Bundesministerium für Umwelt, Naturschutz und Reaktorsicherheit (BMU) (2002) *Strategie der Bundesregierung zur Windenergienutzung auf See*, Berlin: BMU.

Butler, D. and Davies J.W. (2011) *Urban Drainage*, London: Spon Press.

Carmin, J., Anguelovski, I., and Roberts, D. (2012) Urban Climate Adaptation in the Global South: Planning in an Emerging Policy Domain. *Journal of Planning Education and Research*, published online.

Chafe, Z. (2006) Reducing natural disaster risk in cities. In: *The Worldwatch Institute, State of the World 2007*, Washington, DC: Worldwatch Institute.

Chatterjee, M. (2012) Flood loss redistribution in a third world megacity: the case of Mumbai. In: Brauch, H.G., Spring, Ú.O., Mesjasz, C., Grin, J., Kameri-Mbote, P., Chourou, B., Dunay, P., and Birkmann, J. (eds.) *Coping with Global Environmental Change, Disasters and Security – Threats, Challenges, Vulnerabilities and Risks*, Hexagon Series on Human and Environmental Security and Peace, vol. 5, Berlin: Springer-Verlag.

Chigira, M., Wu, X., Inokuchi, T., and Wang, G., (2010) Landslides induced by the 2008 Wenchuan earthquake, Sichuan, China. *Geomorphology*, 118, pp. 225–238.

Coley, D. (2012) Energy efficiency: Alternative routes to mitigation. In: Booth, C., Hammond, C.A., Lamond, J., and Proverbs, D. (eds) *Climate change challenges for the built environment*, London: Blackwell.

Corfee-Morlot, J., Kamal-Chaoui, L., Donovan, M.G., Cochran, I., Robert, A., and Teasdale, P.-J. (2009) *Cities, Climate Change and Multilevel Governance*, OECD Environmental Working Papers N° 14, OECD.

Corfee-Morlot, J., Cochran, I., Hallegate, S., and Teasdale, P.J. (2011) Multilevel risk governance and urban adaptation policy. *Climatic Change*, 104, pp. 169–197.

Coy, A.C. (2000) Best value for coastal defence services. *Proceedings of Institution of Civil Engineers: Municipal Engineer*, 139, pp. 91–96.

Cui, P., Chen, X., Zhu, Y., Su, F., Wei, F., Han, Y., Liu, H., and Zhuang, J., (2011) The Wenchuan Earthquake (May 12, 2008), Sichuan Province, China, and resulting geohazards. *Natural Hazards*, 56, pp. 19–36.

Dai, F.C., Xu, C., Yao, X., Xu, L., Tu, X.B., and Gong, Q.M. (2011) Spatial distribution of landslides triggered by the 2008 Ms 8.0 Wenchuan earthquake, China. *Journal of Asian Earth Sciences*, 40, pp. 883–895.

Danielsen F. *et al.* (2005) The Asian tsunami: a protective role for coastal vegetation. *Science*, 320(5748), pp. 643.

Davis, A.P., Hunt, W.F., Traver, R.G., and Clar, M. (2009) Bioretention technology: overview of current practice and future needs. *Journal of Environmental Engineering (ASCE)*, 135, pp. 109–117.

Dean, R.G. (2003) Beach Nourishment: theory and practice. World Scientific Publishing Company: London.

De Sherbinin, A., Schiller, A., Pulsipher, A., (2007) The vulnerability of global cities to natural hazards. *Environment & Urbanization* 19(1) pp. 39–64.

DFT (2004) *Making Car Sharing and Car Clubs Work* – Final Report.

Dillon, P. (2004) Future management of aquifer recharge. *Hydrogeology Journal*, 13(1), pp. 313–316.

Doody, J.P. (2004) Coastal squeeze – an historical perspective. *Journal of Coastal Conservation*, 10, pp. 129–138.

Dray, A., Perez, P., Le Page, C., D'Aquino, P. & White, I. (2007). Who wants to terminate the game? The role of vested interests and metaplayers in the ATOLLGAME experience, *Simulation and Gaming* 38(4), pp. 494–511.

Douvere, F. and Ehler, C. (2009) New perspectives on sea use management: Initial findings from European experience with marine spatial planning. *Journal of Environmental Management*, 90, pp. 77–88.

Ehler, C., and Douvere, F. (2009) *Marine Spatial Planning: a step-by-step approach toward ecosystem-based management.* Intergovernmental Oceanographic Commission and Man and the Biosphere Programme. IOC Manual and Guides No. 53, ICAM Dossier No. 6. Paris: UNESCO.

Environment Agency (2010) Thames Estuary 2100 project (TE2100), www.thamesweb.com/page.php?page_id=60&topic_id=9

Escobedo, F.J. and Nowak, D.J. (2009) Spatial heterogeneity and air pollution removal by an urban forest. *Landscape and Urban Planning*, 90, pp. 102–110.

European Commission (2007) Communication from the Commission to the European Parliament, the Council, the European Economic and Social Committee and the Committee of the Regions. An Integrated Maritime Policy for the European Union. COM(2007) 575.

European Commission (2008) Roadmap for Maritime Spatial planning: achieving common principles in the EU. COM(2008)791.

European Wind Energy Association (EWEA) (2011) *The European offshore wind industry – key trends and statistics 2010*, January, www.ewea.org/fileadmin/ewea_documents/documents/00_POLICY_document/Offshore_Statistics/110120_Offshore_stats_Exec_Sum.pdf (last accessed 9 April 2013).

Fan, C.H., Chang, F.C., Ko, C.H., Sheu, Y.S., Teng, C.J., and Chang, T.C. (2009) Urban pollutant removal by a constructed riparian wetland before typhoon damage and after reconstruction. *Ecological Engineering*, 35, pp. 424–435.

Fan, K.-C. (2002) Mangroves in Taiwan: Current status and restoration projects. *Bois et Forets des Tropiques*, 273, pp. 43–54.

Feagin, R.A., Mukherjee, N., Shanker, K., Baird, A.H., Cinner, J., Kerr, A.M., Koedam, N., Sridhar, A., Arthur, R., and Forman, R.T.T. (2008) *Urban regions: ecology and planning beyond the city*, Cambridge: Cambridge University Press.

Fernandes, L., Green, A., Tanzer, J., White, A., Alino, P.M., Jompa, J., Lokani, P., Soemodinoto, A., Knight, M., Pomeroy, B., Possingham, H., and Pressey, B. (2012) *Biophysical principles for designing resilient networks of marine protected areas to integrate fisheries, biodiversity and climate change objectives in the Coral Triangle.* Report prepared by The Nature Conservancy for the Coral Triangle Support Partnership.

Field, C.B., Barros, V., Stocker, T.F., Qin, D., Dokken, D.J., Ebi, K.L., Mastrandrea, M.D., Mach, K.J., Plattner, G.-K., Allen, S.K., Tignor, M., and Midgley, P.M. (eds.) *Managing the Risks of Extreme Events and Disasters to Advance Climate Change Adaptation A Special Report of Working Groups I and II of the Intergovernmental Panel on Climate Change (IPCC)*, Cambridge: Cambridge University Press.

Fletcher, T.D., Mitchell, V.G., Deletic, A., Ladson, T., and Seven, A. (2007) Is stormwater harvesting beneficial to urban waterway environmental flows? *Water Science and Technology*, 55(4), pp. 265–272.

Floodsite (2007) *Development of DSS for Long-term Planning, Review Of Existing Tools – T18-06-01*, Wallingford, FLOODsite.

Forman, R.T.T. (2008) *Urban Regions: Ecology and Planning Beyond the City*, Cambridge: Cambridge University Press.

French, P.W. (2001) *Coastal Defences: Process, Problems and Solutions*, London: Routledge.

Fuchs, S., Kuhlicke, C., and Meyer, V. (2011a) Editorial for the special issue: vulnerability to natural hazards – the challenge of integration. *Natural Hazards*, 58, pp. 609–619.

Fuchs, R., Conran, M., and Louis, E. (2011b) Climate change and Asia's Coastal Urban Cities. Can they meet the challenge? *Environment and Urbanization ASIA*, 2(1), pp. 13–28.

Gaffin, S.R., Khanbilvardi, R., and Rosenzweig, C. (2009) Development of a green roof environmental monitoring and meteorological network in New York City. *Sensors*, 9, pp. 2647–2660.

Gail, R. (2010) China gets serious on offshore wind. *Renewable Energy Focus*, 11(5), pp. 16–19.

Gee, K. (2010) Offshore wind power development as affected by seascape values on the German North Sea coast. *Land Use Policy*, 27(2), pp. 185–194.

Gee, K., Kannen, A., and Heinrichs, B. (2011) BaltSeaPlan Vision 2030 for Baltic Sea Space. Hamburg, Autumn.

Gee, K., Kannen, A., Glaeser, B., and Sterr, H. (2004) National ICZM strategies in Germany: A spatial planning approach. In: Schernewski, G. and Loeser, N. (eds) Managing the Baltic coast. Coastline Reports 2, www.eucc-d.de/coastline-reports-2-2004.html (last accessed 9 April 2013).

Georgeakis, P. and Nwagoboso, C. (2012) Sustainable transportation. In: Booth, C.A., Hammond, F.N., Lamond, J.E., and Proverbs, D.G. (eds) *Climate change challenges for the built environment*, London: Blackwell.

Glaeser, E. and Kahn, M. (2008) *The greenness of cities*, Policy Brief, Boston, Rappaport institute for greater Boston, Taubman center for state and local government.

Glanz, J. and Onishi, N. (2011) Japan's Strict Building Codes Saved Lives, NYT Asia Pacific 11 March.

Gleick, P.H., Cooley, H., and Groves, D. (2005) *California Water 2030: An Efficient Future*, Oakland, CA: Pacific Institute.

Gornitz, V., Couch, S., and Hartig, E.K. (2001) Impacts of sea level rise in the New York City metropolitan area. *Global and Planetary Change*, 32, pp. 61–88.

Granberg, M. and Elander, I. (2007) Local Governance and Climate Change: Reflections on the Swedish Experience. *Local Environment: The International Journal of Justice and Sustainability*, 12(5), pp. 537–548.

Grant, G. (2006) Extensive green roofs in London. *Urban Habitats*, 4, pp. 51–65.

Grossmann, I. and Morgan, M.G. (2011) Tropical cyclones, climate change, and scientific uncertainty: what do we know, what does it mean, and what should be done? *Climatic Change*, 108(3), pp. 543–579.

Gupta, M. (2012) Settlement and shelter reconstruction. In: Wisner, B., Gaillard, J.C., and Kelman, I. (eds) *Routledge Handbook of Hazards and Disaster Risk Reduction*, London: Routledge.

Gurjar, B.R., Butler, T.M., Lawrence, M.G., and Lelieveld, J. (2008) Evaluation of emissions and air quality in megacities. *Atmospheric Environment*, 42(7), pp. 1593–1606.

Gustavsson, E., Elander, I., and Lundmark, M. (2009) Multilevel governance, networking cities, and the geography of climate-change mitigation: two Swedish examples. *Environment and Planning C: Government and Policy*, 27, pp. 59–74.

GWEC (Global Wind Energy Council) (2011) *Global Wind Report*. Annual Market Update 2010. April, www.gwec.net/fileadmin/images/Publications/GWEC_annual_market_update_2010_-_2nd_edition_April_2011.pdf .

Hadley, D. (2009) Land use and the coastal zone. *Land Use Policy*, 26, pp. 198–203.

Haiqing, Li (2003) Management of coastal mega-cities – a new challenge in the 21st century. *Marine Policy*, 27, pp. 333–337.

Hanson, S., Nicholls, R., Ranger, N., Hallegatte, S., Corfee-Morlot, J., Herweijer, C., and Chateau, J. (2011) A global ranking of port cities with high exposure to climate extremes. *Climatic Change*, 104, pp. 89–111.

Hartig, J.H., Zarull, M.A., and Cook, A. (2011) Soft shoreline engineering survey of ecological effectiveness. *Ecological Engineering*, 37, pp. 1231–1238.

Hashimoto, M., Enomoto, M., and Fukushima, Y. (2010) Coseismic Deformation from the 2008 Wenchuan, China, Earthquake Derived from ALOS/PALSAR Images. *Tectonophysics*, 491, pp. 59–71.

Heltberg, R., Siegel, P., and Jorgensen, S. (2009) Addressing Human Vulnerability to Climate Change: Toward a 'No-Regrets' Approach. *Global Environmental Change*, 19, 89–99.

Henderson, A. and Witcher, D. (2010) Floating Offshore Wind Energy – A Review of the Current Status and an Assessment of the Prospects, *Wind Engineering*, 34(1), pp. 1–16.

Hinkel, J., Nicholls, R.J., Vafeidis, A.T., Tol, R.S.J., and Avagianou, T. (2010) Assessing risk of and adaptation to sea-level rise in the European Union: an application of DIVA. *Mitigation and Adaptation Strategies for Global Change*, 15(7), pp. 703–719.

Hogan, D.M. and Walbridge, M.R. (2007) Best management practices for nutrient and sediment retention in urban stormwater runoff. Journal of Environmental Quality 36: 386–395.

Holgate, C. (2007) Factors and Actors in Climate Change Mitigation: A Tale of Two South African Cities. *Local Environment: The International Journal of Justice and Sustainability*, 12(5), pp. 471–484.

Hosono, T., Su, C., Delinom, R., Umezawa, Y., Toyota, T., Kaneko, S., and Taniguchi, M. (2011) Decline in heavy metal contamination in marine sediments in Jakarta Bay, Indonesia due to increasing environmental regulations. *Estuarine, Coastal and Shelf Science*, 92(2), pp. 297–306.

Howe, C., Henderson, J., and Smith, J. (2009) *Climate Change and Water*, London: IWA Publishing.

Hunt A. and Watkiss P. (2011) Climate change impacts and adaptation in cities: a review of the literature. *Climatic Change*, 104, pp. 13–49.

Hurlimann, A. and Dolnicar, S. (2009) When Public Opposition Defeats Alternative Water Projects – the Case of Toowoomba Australia. *Water Research*, 44(1), pp. 287–297.

Hurlimann, A., Dolnicar, S. & Meyer, P., (2009) Understanding behaviour to inform water supply management in developed nations – A review of literature, conceptual model and research agenda. *Journal of Environmental Management*, 91(1), pp. 47–56.

Ioualalen, M., Migeon, S., and Sardoux, O. (2010) Landslide tsunami vulnerability in the Ligurian Sea: case study of the 1979 October 16 Nice international airport submarine landslide and of identified geological mass failures, *Geophysical Journal International*, 181, pp. 724–740.

Jayatissa, L.P., Lo Seen, D., Menon, M., Rodriguez, S., Shamsuddoha, M., Dahdouh-Guebas, F. (2010) Shelter from the storm? Use and misuse of coastal vegetation bioshields for managing natural disasters. *Conservation Letters* 3 pp. 1–11.

Jenamani, R.K., Bhan, S.C., and Kalsi, S.R. (2006) 'Observational/ forecasting aspects of the meteorological event that caused a record highest rainfall in Mumbai', *Current Science*, 2006, 90, pp. 1344–1362.

Jha, A., Lamond, J., and Bloch, R. (2012) Cities and Flooding: A Guide to Integrated Urban Flood Risk Management for the 21st Century, Washington, DC: GFDRR/World Bank.

Jha, A.K. and Brecht, H. (2011) *Building Urban Resilience in East Asia. An Eye on East Asia and Pacific*, East Asia and Pacific. Economic Management and Poverty Reduction, No 8, Washington, DC: World Bank.

Jha, A., Lamond, J., Bloch, R., Bhattacharya, N., Lopez, A., Papachristodolou, N., Bird, A., Proverbs, D., Davies, J., and Barker, R. (2011) Five Feet High and Rising – Cities and Flooding in the 21st Century. *Policy Research Working Paper 5648*, Washington, DC: World Bank.

Jia, H.F., Ma, H.T., and Wei, M.J. (2011) Urban wetland planning: A case study in the Beijing central region. *Ecological Complexity*, 8, pp. 213–221.

Jia, D., Li, Y., Lin, A., and Wang, M. (2010) Structural model of 2008 Mw 7.9 Wenchuan earthquake in the rejuvenated Longmenshan thrust belt, China. *Tectnophysics*, 491, pp. 174–184.

Jiang, Y., Kirkman, H., and Hua, A. (2001) Megacity development: managing impacts on marine environments, *Ocean & Coastal Management*, 44, 293–318.

Jim, C.Y. and Tsang, S.W. (2011) Ecological energetics of tropical intensive green roof. *Energy and Buildings*, 43, pp. 2696–2704.

Johnson, C. (2012) Urban and regional planning and disaster. In: Wisner, B., Gaillard, J.C., and Kelman, I. (eds) *Routledge Handbook of Hazards and Disaster Risk Reduction*, London: Routledge, pp. 641–651.

Kaiser, G., Scheele, L., Kortenhaus, A., Lovholt, F., Romer, H., and Leschka, S. (2011) The influence of land cover roughness on the results of high resolution tsunami inundation modeling. *Natural Hazards and Earth System Sciences*, 11, pp. 2521–2540.

Kamal-Chaoui, L. and Robert, A. (ed.) (2009) *Competitive Cities and Climate Change*. OECD Regional Development Working Papers No. 2, OECD publishing.

Kambole, M.S. (2003) Managing the water quality of the Kafue River. *Physics and Chemistry of the Earth, Parts A/B/C*, 28(20-27), pp. 1105–1109.

Kamthonkiat, D., Rodfai, C., Saiwanrungkul, A., Koshimura, S., and Matsuoka, M. (2011) Geoinformatics in mangrove monitoring: damage and recovery after the 2004 Indian Ocean tsunami in Phang Nga, Thailand. *Natural Hazards and Earth Systems Sciences*, 11, pp. 1851–1862.

Kandasamy, K. and Narayanasamy, R. (2005) Coastal mangrove forests mitigated tsunami. *Estuarine, Coastal and Shelf Science*, 65(3), pp. 601–606.

Kannen, A., Gee, Licht-Eggert, K. (2008) Managing changes in sea use across scales: North Sea and North Sea coast of Schleswig-Holstein. In: Krishnamurthy, R.R. *et al.* (eds) *ICZM – The global challenge*, Chennai: Research Publishing.

Kannen, A. and Burkhard, B. (2009) Integrated Assessment of Coastal and Marine Changes Using the Example of Offshore Wind Farms: the Coastal Futures Approach. *GAIA*, 18(3), pp. 229–238.

Kannen, A., Gee, K., and Bruns, A. (2010) Governance Aspects of Offshore Wind Energy and Maritime Development. In: Lange, M., Burkhard, B., Garthe, S., Gee, K., Lenhart, H., Kannen, A., and Windhorst, W. (eds) *Analysing Coastal and Marine Changes – Offshore Wind Farming as a Case Study: Zukunft Kueste – Coastal Futures Synthesis Report*. LOICZ R & S Report No. 36, www.loicz.org/products/publication/reports/index.html.en (last accessed 9 April 2013).

Kern, K. and Bulkeley, H. (2009) Cities, Europeanization and Multi-level Governance: Governing Climate Change through Transnational Municipal Networks. *JCMS: Journal of Common Market Studies*, 47(2), pp. 309–332.

Khatib, J. (2012) Progress in eco and resilient construction materials development. In: Booth, C.A., Hammond, F.N., Lamond, J.E., and Proverbs, D.G. (eds) *Climate change challenges for the built environment*, London. Blackwell.

Kibert, C. (2007) The next generation of sustainable construction. *Building Research and Information*, 35(6), pp. 595–601.

Kim, K.G., Lee, H., and Lee, D.H. (2011) Wetland restoration to enhance biodiversity in urban areas: a comparative analysis. *Landscape and Ecological Engineering*, 7, pp. 27–32.

King, C. and Adeel, Z. (2002) Strategies for sustainable coastal management in Asia and the Pacific – perspectives from a regional initiative. *Global Environmental Change*, 12, pp. 139–142.

Kirshen, P., Knee, K., and Ruth, M. (2008) Climate change and coastal flooding in Metro Boston: impacts and adaptation strategies. *Climatic Change*, 90 pp. 453–473.

Kithiia, J. and Dowling, R. (2010) An integrated city-level planning process to address the impacts of climate change in Kenya: The case of Mombasa. *Cities*, 27, pp. 466–475.

Kithiia, J. and Lyth, A. (2011) Urban wildscapes and green spaces in Mombasa and their potential contribution to climate change adaptation and mitigation. *Environment and Urbanization*, 23(1), pp. 251–265.

Klinger, T. (2004) International ICZM: in search of successful outcomes. *Ocean & Coastal Management*, 47, pp. 195–196.

Klinke, A. and Renn, O. (2002) A New Approach to Risk Evaluation and Management: Risk-Based, Precaution-Based, and Discourse-Based Strategies. *Risk Analysis*, 22, pp. 1071–1094.

Koehn, P.H. (2008) Underneath Kyoto: Emerging Subnational Government Initiatives and Incipient Issue-Bundling Opportunities in China and the United States. *Global Environmental Politics*, 8(1), pp. 53–77.

Kumagai, Y. and Nojima, Y. (1999) Urbanization and disaster management in Tokyo. In Mitchell, K. (ed.) *Crucibles of Hazard: Megacities and disasters in transition*, New York: UN Press.

Kumar, T.S., Mahendra, R.S., Nayak, S., Radhakrishnana, K., and Sahu, K.C. (2010) Coastal vulnerability assessment for Orissa state, east coast of India. *Journal of Coastal Research*, 26(3), 523–534.

Kusky, M.T., Ghulam, A., Wang, L., Liu, J.G., Li, Z., and Chen, X. (2010) Focusing seismic energy along faults through time-variable rupture modes: Wenchuan earthquake, China. *Journal of Earth Science*, 21, pp. 910–922.

Lämmel, G., Rieser, M., Nagel, K., Taubenböck, G.S., Gosberg, N., Schlurmann, T., Klupfel, H., Setiadi, N., and Birkmann, J. (2008) Emergency preparedness in the case of tsunami: evacuation analysis and traffic optimisation for the Indonesian city of Padang. *Pedestrian and Evacuation Dynamics*, 1, pp. 171–182.

Lamond, J. (2011) Financial implications of flooding and the risk of flooding on households. In: Lamond, J.E., Proverbs, D.G., Booth, C.A., and Hammond, F.N. (eds) *Flood hazards, impacts and responses for the built environment*, New York: Taylor CRC Press.

Lamond, J.E. and Bateman, I.J. (2012) Methods for Valuing Preferences for Environmental and Natural Resources: An overview. In: Booth, C.A., Hammond, F.N., Lamond, J.E., and Proverbs, D.G. (eds) *Climate change challenges for the built environment*, London: Blackwell.

Lamond, J. and Proverbs, D. (2009) Resilience to flooding: learning the lessons from an international comparison of the barriers to implementation. *Urban Design and Planning*, 162, pp. 63–70.

Lamond, J., Proverbs, D., and Antwi, A. (2005) The effect of floods and floodplain designation on the value of property: an analysis of past studies. *2nd Probe Conference*, 16–17 November, Glasgow, www.gcal.ac.uk/probe2007/probe2005.pdf.

Lane, M.B. and McDonald, G. (2005) Community-based Environmental Planning: Operational Dilemmas, Planning Principles and Possible Remedies. *Journal of Environmental Planning and Management*, 48, pp. 709–731.

Lasco, R., Lebel, L., Sari, A., Mitra, A.P., Tri, N.H., Ling, O.G., and Contreras, A. (2007) *Integrating carbon management into development strategies of cities – establishing a network of case studies of urbanisation in Asia Pacific*, Final Report APN project 2004-07- CMY-Lasco.

Li, H. (2003) Management of coastal mega-cities – a new challenge in the 21st century. *Marine Policy*, 27, pp. 333–337.

Liu, J.G., Mason, P.J., Yu, E., Wu, M.-C., Chuan, T., Huang, R., and Liu, H. (2012) GIS modelling of earthquake damage zones using satellite remote sensing and DEM data, *Geomorphology*, 139, pp. 518–535.

Lloyd, A.C. and Cackette, T.A. (2001) Critical Review-Diesel Engines; Environmental Impact and Control. *JAWMA*, 51(6), pp. 809–847.

Lodden, U.B. and Brechan, I. (2003) Reiseinformasjonens betydning for bruk av kollektivtrafikk, Effekten av tjenestetilbudet til Trafikanten [The importance of travel information for the use of public transport, the effects of the travel information services of Trafikanten]. In: Booth, C., Hammond, F.N., Lamond, J., and Proverbs, D.G. (2012) *Solutions for Climate Change Challenges in the Built Environment: Innovation in the Built Environment*, London: Wiley-Blackwell.

Maes, F. (2008) The international legal framework for marine spatial planning. *Marine Policy*, 32, pp. 797–810.

Mage, D., Ozolins, G., Peterson, P., Webster, A., Orthofer, R., Vandeweerd, V., and Gwynne, M. (1996) Urban air pollution in megacities of the world. *Atmospheric Environment* 30(5), pp. 681–686.

Maheepala, S., Blackmore, J., Diaper, C., Moglia, M., Sharma, A., and Kenway, S. (2010) *Integrated Urban Water Management Planning Manual*, Denver, CO: Water Research Foundation.

Martins, R.D.A., Ferreira, L.C., (2011) Climate change action at the city level: tales from two megacities in Brazil. *Management of Environmental Quality: An International Journal* 22(3) pp. 344–357.

McManus, P. (2006) Mangrove battlelines: culture/nature and ecological restoration. *Australian Geographer*, 37, pp. 57–71.

Measham, T.G. and Preston, B.L. (2012) Vulnerability analysis and community engagement: A case study of the Sydney Climate Change Adaptation Initiative. In: Measham, T.G. and Lockie, S. (eds) *Risk and Social Theory in Environmental Management*, Melbourne: CSIRO Publishing.

Measham, T.G., Preston, B., Smith, T., Brooke, C., Gorddard, R., Withycombe, G., and Morrison, C. (2011) Adapting to climate change through local municipal planning: barriers and challenges. *Mitigation and Adaptation Strategies for Global Change*, 16, pp. 889–909.

Mekong River Commission (MRC) (2009) Best Practise Guidelines for Structural Measures and Flood Proofing. The Flood Management and Mitigation Programme, Component 2: Structural Measures & Flood Proofing in the Lower Mekong Basin.

Moglia, M., Grant, A.L., and Inman, M.P. (2009) Estimating the effect of climate on water demand: Towards strategic policy analysis. *Australian Journal of Water Resources*, 13(2), pp. 81–94.

Mesquita, A.R., (2004), On Sea Level Along the Brazilian Coast. *Afro America GLOSS News*, 8(1): 20–35.

Millennium Ecosystem Assessment (2005) Ecosystems and human well-being: wetlands and water. World Resources Institute, Washington, DC.

Milligan, J., O'Riordan, T., Nicholson-Cole, S.A., and Watkinson, A.R. (2009) Nature conservation for future sustainable shorelines: Lessons from seeking to involve the public. *Land Use Policy*, 26, pp. 203–213.

Mitlin, D. (2012) Lessons from the urban poor: collective action and the rethinking of development. In: Pelling, M., Manuel-Navarrete, D., and Redclift, M. (eds) (2011) *Climate Change and the Crisis of Capitalism: a change to reclaim self, society and nature*, Routledge Series in Human Geography, London: Routledge.

Molina, M.J. and Molina, L.T. (2004) Megacities and atmospheric pollution. *Journal of Air and Waste Management Association*, 54, pp. 644–680.

Monstadt, J. (2009) Conceptualizing the political ecology of urban infrastructures: insights from technology and urban studies, *Environment and Planning A*, 41, pp. 1924–1942.

Moreland Solar City (2011) Energy Hub, www.morelandsolarcity.org.au/index.php?nodeId=23 (last accessed 9 April 2013).

Morimoto, Y. (2011) Biodiversity and ecosystem services in urban areas for smart adaptation to climate change: "Do you Kyoto"? *Landscape and Ecological Engineering*, 7, pp. 9–16.

Moser, C. and Satterthwaite, D. (2010) Towards Pro-Poor Adaptation to Climate Change in the Urban Centres of Low- and Middle-Income Countries. In: Mearns, R., Norton, A. (eds.), *Social dimensions of climate change: equity and vulnerability in a warming world*, World Bank Publications.

Mukheibir, P. and Ziervogel, G. (2007) Developing a Municipal Adaptation Plan (MAP) for climate change: the city of Cape Town. *Environment and Urbanization*, 19, pp. 143–158.

Mumbai Metropolitan Development Authority Available: http://www.mmrdamumbai.org/

Munich Re (2004) *Megacities – Megarisks Trends and challenges for insurance and risk management*. Munich, Münchener Rückversicherungs-Gesellschaft.

Municipal Corporation of Greater Mumbai (MCGM), (1993) *BRIMSTOWAD (Brihanmumbai Stormwater Drainage) project, final report.*

Municipal Corporation of Greater Mumbai (MCGM), (2011) *Flood preparedness guidelines.*

Murthy, R.C. (1999) 'Integrated coastal management of Mumbai (Bombay), India: a coastal city report', IOC-SOA International Workshop on Coastal Megacities, Hangzhou, China, 27–30 September 1999.

Murthy, R.C., Rao, Y.R. and Inamdar, A B. (2001) 'Integrated coastal management of Mumbai metropolitan region', *Ocean & Coastal Management*, 44, pp. 355–369.

Nagara, B. (2006) Jakarta River Management. In: The Worldwatch Institute, *State of the World 2007*, Washington, DC. Worldwatch Institute.

Neves, C.F., Muehe, D., (2008) Vulnerabilidade, impactos e adaptação a mudanças do clima: a zona costeira. *Parcerias Estratégicas* 27, pp. 217–296.

New York City Department of Environmental protection (2011) *NYC Green Infrastructure Plan: A sustainable strategy for clean waterways*, New York City, www.nyc.gov/html/dep/pdf/green_infrastructure/NYCGreenInfrastructurePlan_ExecutiveSummary.pdf (last accessed 9 April 2013).

Nicholls, R.J. (1995) Coastal megacities and climate change. *GeoJournal* 37, pp. 369–379.

Nicholls, R.J. (2002) Analysis of global impacts of sea-level rise: a case study of flooding. *Physics and Chemistry of the Earth*, 27, pp. 1455–1466.

Nicholls, R.J. and Cazenave, A. (2010) Sea-level rise and its impact on coastal zones. *Science*, 328(5985), pp. 1517–1520.

Nielsen, J. (2006) *On the design of buildings in relation to climate change.* World Conference for Accelerating Excellence in the Built Environment, Birmingham.

Niu, H., Clark, C., Zhou, J., and Adriaens, P. (2010) Scaling of economic benefits from green roof implementation in Washington, DC. *Environmental Science and Technology*, 44 pp. 4302–4308.

NTS (2011) Hidden Vulnerabilities in Asian Megacities: the case of the Bang Khun Tian community in Bangkok, Thailand. *NTS Alert*, 1/

Nur, Y., Fazi, S., Wirjoatmodjo, N., and Han, Q. (2001) Towards wise coastal management practice in a tropical megacity – Jakarta. *Ocean & Coastal Management*, 44, pp. 335–353.

Oberndorfer, E., Lundholm, J., Bass, B., Coffman, R.R., Doshi, H., Dunnett, N., Gaffin, S., Köhler, M., Liu, K.K.Y., and Rowe, B. (2007) Green roofs as urban ecosystems: ecological structures, functions and services. *BioScience*, 57, pp. 823–833.

O'Brien, G., Bhatt, M., Saunders, W., Gaillard, J.C., and Wisner, B. (2012) Local government and disaster. In: Wisner, B., Gaillard, J.C., and Kelman, I. (eds) *Routledge Handbook of Hazards and Disaster Risk Reduction*, London: Routledge, pp. 629–640.

O'Donnell, I. and Smart, K. (2009) Responding to urban disasters: Learning from previous relief and recovery operations. ALNAP lessons. Provention Consortium. Available from http://www.alnap.org/pool/fi les/alnap-provention-lessons-urban.pdf.

Olowoporoku, A.O., Longhurst, J.W.S., Barnes, J.H., and Edokpayi, C.A. (2011) Towards a new framework for air quality management in Nigeria. In: Brebbia, C., Longhurst, J.W.S., and Popov, V. (eds) *Air Pollution XIX*, Southampton: WIT Press.

O'Riordan, T. and Ward, R. (1997) Building trust in shoreline management: creating participatory consultation in shoreline management plans. *Land Use Policy*, 14, pp. 257–276.

Ouimet, W.B. (2010) Landslides associated with the May 12, 2008 Wenchuan earthquake: Implications for the erosion and tectonic evolution of the Longmen Shan. *Tectonophysics*, 491, pp. 244–252.

Pacione, M. (2006) 'City profile Mumbai', *Cities*, v 23, no. 3, p.p. 229–238.

Parnell, S., Simon, D., and Vogel, C. (2007) Global environmental change: conceptualising the growing challenge for cities in poor countries, *Area*, 39(3), pp. 357–369.

Pauchard, A., Aguayo, M., Pena, E., and Urrutia, R. (2006) Multiple effects of urbanization on the biodiversity of developing countries: The case of a fast-growing metropolitan area (Concepcion, Chile). *Biological Conservation*, 127, pp. 272–281.

Pelling, M. (2003) *The Vulnerability of Cities: Natural Disasters and Social Resilience*, London: Earthscan.

Pelling, M. (2012) Hazards, risk and urbanization. In: Wisner, B., Gaillard, J.C., and Kelman, I. (eds) *Routledge Handbook of Hazards and Disaster Risk Reduction*, London: Routledge.

Pelling, M. and High, C. (2005) Understanding adaptation: What can social capital offer assessments of adaptive capacity. *Global environmental change*, 15, pp. 308–319.

Pelling, M. and Wisner, B. (2009) *Disaster Risk Reduction: cases from urban Africa*, London: Earthscan.

Preston, B.L., Yuen, E.J. and Westaway, R.M. (2011) Putting vulnerability to climate change on the map: a review of approaches, benefits, and risks. *Sustainability Science*, 6, 177–202.

Pretty, J. and Ward, H. (2001) Social capital and the environment. *World development*, 29, 209–227.

Puppim de Oliveira, J.A. (2009) The implementation of climate change related policies at the subnational level: An analysis of three countries, *Habitat International*, 33(3), pp. 253–259.

Rakodi, C. (2006) Relationships of power and place: The social construction of African cities. *Geoforum*, 37, pp. 312–317.

Ranger, N., Milner, A., Dietz, S., Fankhauser, S., Lopez, A., and Ruta, G. (2010) *Adaptation in the UK: a decision making process*. Policy Brief, Grantham Research Institute on Climate Change and the Environment: London.

Ranger, N., Hallegatte, S., Bhattacharya, S., Bachu, M., Priya, S., Dhore, K., Rafique, F., Mathur, P., Naville, N., Henriet, F., Herweijer, C., Pohit, S., and Corfee-Morlot, J. (2011) An assessment of the potential impact of climate change on flood risk in Mumbai. *Climatic Change*, 104, pp. 139–167.

Ravindra, K., Wauters, E., Tyagi, S.K., Mor, S., and Grieken, R.V. (2006) Assessment Of Air Quality After The Implementation of Compressed Natural Gas (Cng) As Fuel In Public Transport In Delhi, India. *Environmental Monitoring and Assessment*, 115, pp. 405–417.

Redclift, M., Manuel-Navarrete, D. and Pelling, M. (2011) *Climate Change and Human Security: the challenge to local governance under rapid coastal urbanization*, London: Edward Elgar.

Red Cross (2004) *World Disasters Report*, Geneva: Red Cross.

Renn, O. and Schweizer, P.-J. (2009) Inclusive risk governance: concepts and application to environmental policy making. *Environmental Policy and Governance*, 19, pp. 174–185.

Renwick, M.E. and Green, R.D. (2000) Do Residential Water Demand Side Management Policies Measure Up? An Analysis of Eight California Water Agencies. *Journal of Environmental Economics and Management*, 40(1), pp. 37–55.

Roberts, D. (2010) Prioritizing climate change adaptation and local level resilience in Durban, South Africa. *Environment and Urbanization*, 22, pp. 397–413.

Rodolfo, K.S. and Siringan, F.P. (2006) Global sea-level rise is recognised, but flooding from anthropogenic land subsidence is ignored around northern Manila Bay, Philippines. *Disasters*, 30(1), pp. 118–139.

Roehr, D. and Kong, Y.W. (2010) Runoff reduction effects of green roofs in Vancouver, BC, Kelowna, BC, and Shanghai, PR China. *Canadian Water Resources Journal*, 35, pp. 53–67.

Romero Lankao, P. (2007) How do Local Governments in Mexico City Manage Global Warming? *Local Environment: The International Journal of Justice and Sustainability*, 12(5), pp. 519–535.

Rutland, T. and Aylett, A. (2008) The work of policy: actor networks, governmentality, and local action on climate change in Portland, Oregon. *Environment and Planning D: Society and Space*, 26(4), pp. 627–646.

Salagnac, J.-L. (2012) Global political initiatives and overtones. In: Booth, C.A., Hammond, F.N., Lamond, J.E., and Proverbs, D.G. (eds) *Climate change challenges for the built environment*, London: Blackwell.

Satterthwaite, D. (2008) *Climate change and urbanization: Effects and implications for urban governance*, United Nations, New York.

Satterthwaite, D. (2011) Why is community action needed for disaster risk reduction and climate change? *Environment and Urbanization*, 23, pp. 339–349.

Satterthwaite, D. (2011) How urban societies can adapt to resource shortage and climate change. *Phil Trans R Soc A*, 369, pp. 1762–1783.

Scholz, M. and Hedmark, A. (2010) Constructed wetlands treating runoff contaminated with nutrients. *Water Air and Soil Pollution*, 205, pp. 323–332.

Schreurs, M.A. (2008) From the Bottom Up: Local and Subnational Climate Change Politics, *The Journal of Environment Development*, 17(4), pp. 343–355.

Scott, M., Gupta, S., Jáuregui, E., Nwafor, J., Satterthwaite, D., Wanasinghe, Y.A.D.S., Wilbanks, T., Yoshino, M., Kelkar, U., Mortsch, L., and Skea J. (2001) Human settlements, energy, and industry. Climate change 2001: impacts, adaptation, and vulnerability. In: McCarthy, J.J., Canziani, O.F., Leary, N.A., Dokken, D.J., and White, K.S. (eds) *Contribution of working group II to the third assessment report of the intergovernmental panel on climate change*, Cambridge: Cambridge University Press.

Sekovski, I., Newton, A., and and Dennison, W.C. (2012) Megacities in the coastal zone: Using a driver-pressure-state-impact-response framework to address complex environmental problems. *Estuarine, Coastal and Shelf Science*, 96, pp. 48–59.

Setzer, J. (2009) Subnational and transnational climate change Governance: evidence from the state and city of São Paulo, Brazil, Fifth Urban Research Symposium Cities and Climate Change: Responding to an Urgent Agenda, Marseille.

Sharma, A.K., Grant, A.L., Grant, T., Pamminger, F., and Opray, L. (2009) Environmental and Economic Assessment of Urban Water Services for a Greenfield Development. *Environmental Engineering Science*, 26(5), pp. 921–934.

Sharma, A., Burn, S., Gardner, T., and Gregory, A. (2010) Role of decentralised systems in the transition of urban water systems. *Water Science & Technology: Water Supply*, 10(4), pp. 577–583.

Shetye, S. R. and Shenoi S. S. C. (1988) 'Seasonal cycle of surface circulation in the coastal north Indian ocean', *Proceedings of the Indian Academy of Sciences (Earth and Planetary Sciences)*, pp. 9753–62

Shi, C., Hutchinson, S.M., Yu, L., and Xu, S. (2001) Towards a sustainable coast: an integrated coastal zone management framework for Shanghai, People's Republic of China. *Ocean & Coastal Management*, 44, pp. 411–427.

Shih, S.C.W. and Nicholls, R.J. (2007) Urban managed realignment: Application to the Thames Estuary, London. *Journal of Coastal Research*, 23, pp. 1525–1534.

Shuto, N. (1987) The effectiveness and limit of tsunami control forests. *Coastal Engineering in Japan*, 30(1), pp. 143–153.

Silva, A., Nascimento, N., Seidl, M., and Vieira, L. (2010) SWITCH in Belo Horizonte, Brazil: infiltration and detention systems for more sustainable stormwater control in Belo Horizonte. *Reviews in Environmental Science and Biotechnology*, 9, pp. 7–13.

Smith, V.H. (2003) Eutrophication of freshwater and coastal marine ecosystems: a global problem. *Environmental Science and Pollution Research*, 10, pp. 126–139.

SouthSouthNorth (2011) Project Portfolio And Reports http://www.southsouthnorth.org/ (accessed March 2011)

Spranger, M. (2008) Natural Catastrophe Risk Insurance Mechanisms for Asia and the Pacific. *Special Nature of Disaster Risk in Megacities*. Tokyo, Japan.

Srinivasan, V., Gorelick, S.M., and Goulder, L. (2010) Sustainable urban water supply in south India: Desalination, efficiency improvement, or rainwater harvesting? *Water Resources Research*, 46(W10504).

Stein, R.S., Aykut, A.B., and Dietrich, J. H. (1997) Progressive failure on the North Anatolian fault since 1939 by earthquake stress triggering. *Geophysical Journal International*, 128, pp. 594–604.

Stein, R.S., Toda, S., Parsons, T., and Grunewald, E. (2006) A new probabilistic seismic hazard assessment for greater Tokyo, *Proceedings of the Royal Society of London A*, 364, pp. 1965–1988.

Stern, N. (2008) The economics of climate change. *American Economic Review: Papers and Proceedings*, 98(2), pp. 1–37.

Swyngedouw, E. (2004) *Social Power and the Urbanisation of Water: Flows of Power*. New York: Oxford University Press.

Takizawa, S. (2008) *Groundwater Management in Asian Cities*, Tokyo: Springer Verlag.

Tal, A. (2011) The Desalination Debate – Lessons Learned Thus Far. *Environment: Science and Policy for Sustainable Development*, 53(5), pp. 34–48.

Tanaka, N., Nandasena, N.A.K., Jinadasa, K.B.S.N., Sasaki, Y., Tanimoto, K., and Mowjood, M.I.M. (2009) Developing effective vegetation bioshield for tsunami protection. *Civil Engineering and Environmental Systems*, 26, pp. 163–180.

Taubenböck, H., Esch, H. T., Felbier, A., Wiesner, M., Roth, A. and Dech, S. (2012) Monitoring urbanization in mega cities from space, *Remote Sensing of Environment*, 117 pp. 162–176.

Taubenböck, H., Wurm, M., Netzband, M., Zwenzner, H., Roth, A., Rahman, A., Dech, S., (2010) Flood risks in urbanized areas – multi-sensoral approaches using remotely sensed data for risk assessment. *Natural Hazards and Earth System Science*, 11, pp. 431–444.

Tribbia, J. and Moser, S.C. (2008) More than information: what coastal managers need to plan for climate change. *Environmental Science & Policy*, 11, pp. 315–328.

Tompkins, E.L., Few, R., and Brown, K. (2008) Scenario-based stakeholder engagement: Incorporating stakeholders preferences into coastal planning for climate change. *Journal of Environmental Management*, 88, pp. 1580–1592.

Twigg, J. (2004) *Disaster Risk reduction: mitigation and preparedness in development and emergency programming*. Humanitarian Practice network, #9, London: ODI.

UNESCO (undated) Marine Spatial Planning Initiative, www.unesco-ioc-marinesp.be/ (last accessed 9 April 2013).

UNESCO (2000) *Reducing megacity impacts on the coastal environment. Alternative livelihoods and waste management in Jakarta and the Seribu Islands*, Jakarta: UNESCO.

UNHABITAT (2007) *Global report on human settlements, 2007: Enhancing urban safety and security*, London: Earthscan.

UNHABITAT (2008) *State of the World's Cities 2008/2009: Harmonious Cities*. London and Sterling, VA: Earthscan.

UNHABITAT (2011) *Global report on human settlements, cities and climate change: policy directions*, United Nations Human Settlements programme report, London: Earthscan.

Van Seters, T., Rocha, L., Smith, D., and MacMillan, G. (2009) Evaluation of green roofs for runoff retention, runoff quality, and leachability. *Water Quality Research Journal of Canada*, 44, pp. 33–47.

Varis, O. and Somlyódy, L. (1997) Global urbanization and urban water: can sustainability be afforded? *Water Science and Technology*, 35(9), pp. 21–32.

Webster, C.J. (2005) The New Institutional Economics and the evolution of modern urban planning: Insights, issues and lessons. *Town Planning Review*, 76(4), pp. 455–484.

WHO/UNICEF Joint Monitoring Programme (2011) *World Health Statistics*, Geneva: WHO.

Wilbers, G., Zwolsman, G., Klaver, G., and Hendriks, A. (2009) Effects of a drought period on physico-chemical surface water quality in a regional catchment area. *Journal of Environmental Monitoring*, 11(6), pp. 1298–1302.

Wiser, R., Yang, Z., Hand, M., Hohmeyer, O., Infield, D., Jensen, P.H., Nikolaev, V., O'Malley, M., Sinden, G., and Zervos, A. (2011) *Wind Energy*. In: Edenhofer, O., Pichs Madruga, R., Sokona, Y., Seyboth, K., Matschoss, P., Kadner, S., Zwickel, T., Eickemeier, P., Hansen, G., Schlömer, G., and von Stechow, C. (eds) *IPCC Special Report on Renewable Energy Sources and Climate Change Mitigation*, Cambridge: Cambridge University Press.

Wolters, M., Bakker, J.P., Bertness, M.D., Jefferies, R.L., and Moller, I. (2005) Saltmarsh erosion and restoration in south-east England: squeezing the evidence requires realignment. *Journal of Applied Ecology*, 42, pp. 844–851.

Wong, N.H., Tay, S.F., Wong, R., Ong, C.L., Sia, A. (2003) Life cycle cost analysis of rooftop gardens in Singapore. *Building and Environment*, 38, pp. 499–509.

World Bank and United Nations (2010) *Natural hazards, unnatural disasters*, Washington, DC, the International Bank for Reconstruction and Development.

Wright, M. (2004) Coastal defence – current UK practice. *Proceedings of Institution of Civil Engineers: Municipal Engineer*, 157, pp. 147–155.

Yang, J., Yu, Q., and Gong, P. (2008) Quantifying air pollution removal by green roofs in Chicago. *Atmospheric Environment*, 42, pp. 7266–7273.

Zaisheng, H. (1998) Groundwater for urban water supplies in northern China – An overview. *Hydrogeology Journal*, 6(3), pp. 416–420.

Zamith, L.R. and Scarano, F.R. (2006) Restoration of a resting sandy coastal plain in Brazil: Survival and growth of planted woody species. *Restoration Ecology*, 14, pp. 87–94.

Zhixin, W., Chuanwen, J., Qian, A., and Chengmin, W. (2009) The key technology of offshore wind farm and its new development in China. *Renewable and Sustainable Energy Reviews*, 13(1), pp. 216–222.

Zoleta Nantes, D. (2000) Flood Hazards in Metro Manila: Recognizing Commonalities, Differences, and Courses of Action. *Social Science Diliman*, 1(1).

Chapter 6

# Urban development, climate change and disaster risk reduction: interaction and integration

Lead authors: *Joern Birkmann (United Nations University, Germany), Matthias Garschagen (United Nations University, Germany), Adriana Lopez (Universidad del Pacifico, Colombia), Mark Pelling (King's College London, UK), Nina Qaiem Maqami (King's College London, UK) and Qingnian Yu (Hohai University, China)*

Contributing authors: *Samuel Carpenter (Humanitarian Futures, King's College London), Adriana Lopez (United Nations University, Germany), Albert Adu-Gyamfi (Hohai University, China), Guoqing Shi (Hohai University, China), Matthias Garschagen (United Nations University, Germany)*

## 1 Introduction

Coastal regions are particularly vulnerable to the impacts of global environmental change (GEC), such as coastal erosion, sea-level rise and an increase in extreme hazard events (Nicholls *et al*., 2007). A failure to contain or mitigate the environmental consequences of intensive consumption and production in coastal regions adds additional pressure to environmental systems providing ecosystem services from clean air to coastal defence, while the adaptive capacity of most urban centres remains limited (McGranahan, 2007). This needs not be the case (Romero-Lankao and Dodman, 2011). Cities are the products of myriad social processes, and concentrate both risks and opportunities. Urban poverty, social exclusion and environmental health hazards are expressions of heightened vulnerabilities in light of both GEC and broader socio-economic shifts. Yet, cities are also hubs of innovation and provide a range of livelihood opportunities. The trade-offs between managing vulnerability to environmental risk and the benefits afforded by urban living have been the focus of the preceding chapters. For megacities connected to the global economy demands for a safe and attractive environment can be reinforcing, though often resources are caught in competition between the interests of poverty alleviation and global competitiveness.

It is clear that ongoing urbanisation in large cities and megacities poses new challenges to urban development, risk management and climate change adaptation and mitigation strategies (Solecki *et al*., 2011). Changing patterns in the frequency, distribution and magnitude of natural hazards resulting from local and global environmental change, and in the demographic and social dynamism of cities, impact on infrastructure systems that are often slow to evolve with built-in inertia (from the checks and balances of decision-making systems to the physical robustness of built infrastructure). This allows a gap to build over time between existing capacity and the evolving hazard and vulnerability context. This is a context driven by human as well as environmental process, local and global as we have seen in earlier chapters. The challenge is how to narrow this gap while also meeting the goals of sustainable development (social and economic equity and ecological integrity) (O'Brien *et al.*, 2012; De Sherbinin *et al.*, 2007). Established risk reduction measures seeking to reduce exposure to coastal hazard risk may no longer be adequate as policy benchmarks shift and uncertainty about the future is recognised, as described in Chapter 3. This is of particular concern for megacities in medium- and low-income countries, which often lack the infrastructure and financial, technical and administrative capacity to address existing risks, placing large populations at risk (Wenzel *et al.*, 2007), but also for richer countries where political stability and economic competitiveness rests on a safe and comfortable environment.

Responding to this challenge is likely to include the need for joined-up thinking that can integrate, and where possible make virtuous and cost-saving linkages between everyday development resources and practice, risk management and climate change adaptation and mitigation. This requires more than protecting existing development from the external threat of climate change – and rather calls for a recognition that risk has multiple sources and under such uncertainties is potentially best managed by building generic capacity at sites of potential vulnerability. Capacity that can be inspired by a specific hazard to help build policy momentum and popular support, but should be able to build capacity that can be applied to multiple risk scenarios. This is an approach to risk management that restates many existing development goals – a demand to reduce inequality in access to basic services, to reduce income poverty and enhance social cohesion and the responsiveness of government and promote ecological integrity and considering interactions between urban areas and their hinterland. The role of information and science is likely to increase in this approach. Both professional science that can outline uncertainty and local knowledge that can help to contextualise hazard and vulnerability as well as capacity to resolve the interdependencies of development, risk management and global environmental change.

The changing risk landscape requires risk reduction strategies capable of addressing both direct and indirect risks, the extremes and incremental shifts of climate change (Field *et al.*, 2012). Given that the impact of extreme events and incremental environmental stresses affect different places in different ways, management responses would do well to be aligned to local contexts. However, much of current adaptation action and broader socioeconomic development interventions are not based on a sufficiently in-depth understanding of place-specific capacities, challenges and constraints. Further, though the respective fields of disaster risk reduction (DRR), climate change mitigation (CCM) and adaptation (CCA) and broader urban development share conceptual and practical overlaps, their focus and scope differs in a number of important ways. Current efforts to address these linkages and conflicts systematically are only recently emerging and these are the primary focus of this chapter.

We first summarize the nature and impact of emerging environmental risk in coastal megacities based on the previous chapters, focusing on future uncertainties and the framing of vulnerability. We then examine synergies and constraints between disaster risk management and climate change adaptation in coastal megacities and highlight the potential for more integrated approaches in both research and practice.

## 2 The nature of environmental risk in coastal megacities

Chapter 3 describes the wide range of environmental hazards coastal megacities are exposed to, stemming from oceanographic, geophysical, meteorological and anthropogenic processes in the context of GEC. Coupled with urbanisation and globalisation, they impact on the infrastructure and the socioeconomic fabric of cities and take place within distinct political frameworks that shape the urban space. Conceptualising cities as complex adaptive systems helps in understanding these manifold dynamic and uneven impacts (Young *et al.*, 2006; Grove, 2009). The ecosystem approach introduced in Chapter 2 further highlights how much large urban systems operate at different scales and, increasingly, remotely: they sustain themselves on the 'carrying capacity' of global commons elsewhere (Rees, 2003). The illustrative ecological footprint analysis enhances the scope of so far often spatially confined approaches to environmental risks in places like Istanbul or Mumbai. As attested earlier, there are significant knowledge gaps on process interactions across temporal and spatial scales, which are so crucial for an informed understanding of coastal ecosystem functions and decline (Nicholls *et al.*, 2007).

For example, urbanisation is a reflexive process – it feeds back onto itself, generating risks. These feedback processes are often challenging to describe, as research is often limited to individual steps in cycles of risk production, such as new insights into local ecosystem decline but may, however, not inform on the trade-offs involved. While it is clear that urban development decisions create benefits and losses across societal groups, little is known, for example, about the longer-term outcomes of flood plain conversion into urban housing projects from a cross-sector perspective. Some examples of urbanisation outcomes include Sanli *et al.* (2008), who correlate deforestation in the Istanbul area with a 33 per cent population growth, resulting in the loss of 3–4 per cent of the surrounding forest area over a period of just 13 years. Likewise, in Rio de Janeiro's Guanabara Bay 90 per cent of mangroves and salt marshes were removed to make way for beach development for tourism purposes (de Sherbinin *et al.*, 2007). In Manila Bay, out of around 54,000 hectares of mangrove forest in 1890, only 2,000 hectares were left in 1990 due to conversion into aquaculture and salt beds, urban expansion and industrial development (Jacinto *et al.*, 2006).

Urbanisation is thus one of the key drivers of environmental processes, which can translate into real pressures on the environment as a result of excessive production and consumption patterns (Jago-on *et al.*, 2009; Sekovski *et al.*, 2011). The removal of vegetative cover has a direct impact on risk exposure: deforestation increases surface run-off and erosion, conducive for flooding and landslides, whereas mangrove destruction removes a key natural barrier against storm surge and tsunami waves. Similar relationships exist between urban water demand, subsidence and flood risk (Rodolfo and Siringan, 2006; Shalizi and Lecocq, 2010) or GHG emissions, air quality and health – each expression of the complexity of urban change and its double-edged effects.

Urban and maritime transport systems equally constitute examples of dynamic drivers of heightened environmental change due to their effect on water and air quality. Rapid mega-harbour development, particularly in Asia, puts pressures on the surrounding habitat and impacts on water quality and wider hazard risks such as oil spills. Further, industrial activity and energy production and consumption impact on the environment as well, with different types of energy exerting different types of pressures, depending on consumption patterns and industry. While oil is the dominant energy source in Manila, Shanghai and Kolkata use coal, Mumbai relies on natural gas and Los Angeles on hydroelectric power (Sekovski *et al.*, 2011; Decker *et al.*, 2002).

Table 6.1 describes the multiple pathways in pressure and states that connect individual drivers with observed impacts in the coastal urban zone, based on Ruth and Coelho (2007). Integrated urban assessments such as this seek to capture how change in one system element (e.g. excessive groundwater use) may trigger changes elsewhere, thus creating all kinds of effects often with time-lagged and non-linear relationships.

More recent assessments seek to account for both infrastructures and the actual social dynamics that shape urban vulnerability to GEC; however, larger system analysis has difficulty in capturing individual components (the particular migrant community, economic sectors, etc.) limiting its practical application (UNEP, 2011; Holman *et al.*, 2005 on the UK). This may be a particular problem for coastal megacities, where much economic activity lies beyond the control of individual municipal governments and hard infrastructure is often privately managed or owned (such as most cargo terminals in the Asia region) (Nicholls *et al.*, 2008). Precisely how future vulnerability to sea-level rise and storm surges is framed differently across the public–private sphere is certainly shaped as much by temporal aspects as by issues of accountability and the different constituencies involved.

## 3 Future megacity vulnerability and risk management

In addition to environmental pressures and coastal hazards, several features make up the contemporary urban sphere and determine vulnerability to GEC impacts (Ruth and Coelho, 2007; Kraas, 2007; Pelling, 2003). These are characterised by a:

- convolution and simultaneousness of processes, of 'tipping points' and feedback-loops, such as the urbanisation features described above;
- high socioeconomic vulnerability, manifested in high population density, migrant populations and economic and political dominance that make specific social groups more vulnerable;
- high concentration of people and economic assets, critical infrastructures, capital and decision structures;
- limited steering capacities and a partial loss of governance, expressed through deficit administration, planning and budgets;
- decentralisation trends, leading to municipal competition and limited planning capacities.

However, these are by no means features unique to megacities alone and, crucially, we find considerable variations for each of these characteristics across the globe. That is, the adaptive capacity of communities within and across cities differs: both Tokyo and Karachi share some of the highest population densities in the world, yet, their respective levels of vulnerability to extreme weather events differs considerably and,

*Table 6.1* Cause and consequence of environmental stress on cities in the coastal zone

| *Drivers* | *Pressures* | *States* | *Impact* |
|---|---|---|---|
| Urbanisation | Waste generation | Degraded air quality | Coastal erosion |
| Energy | Gas emissions into atmosphere | Degraded water quality | Decline in biodiversity |
| Transport | Alterations of hydrological and sediment flux | Degraded water and sanitation services | Changes in ecosystem functions |
| Agriculture | Pressures on groundwater recharge and supply | Decrease in coastal sanitation | Human health impacts |
| Industry | Habitat loss | Land subsidence | Socioeconomic impacts |
| Water consumption | Disrupted ecosystems | Degraded aquatic and riparian ecology | |

Based on: Sekovski *et al.* (2011), De Sherbinin *et al.* (2007) and Grimm *et al.* (2008)

crucially, is far from being static. Thus, increasingly, assessments of future risk combine environmental hazards with an analysis of place-specific features that include socio-economic, demographic and cultural variables, broader development patterns, and governance structures as key determinants of adaptive capacity and the likely impact of actual adaptation actions. Further, despite the challenges with assessment methodologies noted above, the question if and how far these respond to social equity issues is becoming important, i.e. which groups are served by what type of adaptation responses (UN-Habitat, 2011; European Commission, 2012; Schipper, 2012).

It is clear that future vulnerability in the urban coastal zone will be an even stronger expression of the manifold dynamics of the coupled socio-ecological systems described in Chapters 1 and 2 (Holling *et al.*, 2002; Young *et al.*, 2006). Coastal megacities are prime expressions of human-dominated ecosystems – biophysical systems routinely interact with social and economic systems – yet, our conventional understanding of risk, stability and control may be ill-equipped to grasp the surprise dynamics that consolidate urban vulnerabilities or enhance resilience. Similarly, as connectedness and complexity represent core attributes of the contemporary urban sphere (Eriksen and O'Brien, 2007), notions of linear cause-consequence are of limited value for more ambitious vulnerability assessments. While these seek to enhance predictability and causal narratives, a different reading of system dynamics highlights uncertainty, non-linearity and surprise as key characteristics of the contemporary urban form.

Further, recognising the embeddedness of urban regions into wider globalisation processes, understood as the 'widening, deepening, and speeding up of global interconnectedness' (Held *et al.*, 1999) draws attention to the myriad of dynamic processes at broader scales. They clearly influence the resilience, adaptive capacity and vulnerability of megacities but so far little is known about the cascading effects across the local and global level.

Acknowledging the complexity of urban systems and incorporating insights and frameworks from social–ecological systems analysis carries several challenges. In view of increasing uncertainty and multiple development trajectories, it is difficult to align mismatching views: policy makers need precise information on which to base decisions, whereas integrated assessments are based upon thinking through a range of contrasting scenarios. As Ruth and Coelho (2007: 333) argue: 'the challenge will be to identify strategies that are robust for a wide range of possible scenarios'. This is arguably a challenge for many hard infrastructure investment decisions. Long lifetimes may in fact hinder adaptation – once an underground sewer network is in place, adjusting it to sunken groundwater levels is virtually impossible.

Where infrastructure has built-in redundancy this may offer some scope for adaptation in form or function in the future, but with costs in the present. This is an example of the new efficiency that resilience points us towards, one where the potential future costs of failed infrastructure are given more prominence in current decisions. The considerable infrastructure investment made in megacities places them at the forefront of this conversation of development decision making for resilience.

Addressing vulnerabilities takes many different forms; adaptation to increased flood risks can be planned or top down, autonomous, reactive or anticipatory. Multiple adaptations unfold simultaneously, will often be disconnected, unaware of each other and potentially contradictory, especially across scale but also over time. Contradictions that feed into existing social difference – for example across economic class or ethnic divisions, may be recognised and will further reify ongoing narratives of difference in the city. Contestation over adaptation choices may in this way be worked out through the languages and logics of culture and identity as much as economic rationality and cost. In this way risk reduction and climate change actions are as embedded in social and political processes as everyday development and a reflection of respective relations among social groups, individuals and networks. It is unhelpful to withdraw disaster risk and climate change behind the screen of risk management, tempting though this may be to technologically trained urban planners or humanitarian and development NGOs. In short, urban risk management needs to better recognise that it takes place in the context of distributional conflicts, inequalities and persistent socio-ecological systems – in short, adaptation actions necessarily produce winners and losers that impact life chances going beyond hazard mitigation (Adger *et al.*, 2011; O'Brien *et al.*, 2004; Cannon and Müeller-Mahn, 2010). So far, most types of adaptation interventions are relatively low cost, enjoy broad public support and co-benefits, but often this may be less because of the particular characteristics of the adapting system (slum dwellers, urban transport authorities) but relate to the level where vulnerability strategies are developed and implemented, as well as by whom. Therefore, they always reflect corresponding interests (Adger *et al.*, 2011; Reid, 2009).

Adaptation measures initiated by external donors involving large technology transfers or securing land tenure rights for slum dwellers are the outcome of political negotiations different to those one may find at local initiatives, and they frame action accordingly. Similarly with regard to temporal scales: creating mechanisms for long-term planning proves extraordinarily difficult not least due to the uncertainty typical for future GEC impacts. Therefore, the concept of strategic and adaptive planning gain more importance (see Birkmann *et al.* 2010). Most adaptation measures have fiscal implications through public expenditures and it is likely that these influence policy options. Yet, the macroeconomic modelling required to adequately display the complex cost-benefit calculations involved is in its infancy. Even within

OECD countries calls for multi-annual budgeting remains largely unheard even without any reference to climate change (UNEP, 2011; Lamhauge *et al.*, 2012). And while progress tracking has emerged as a central issue in the international climate mitigation regime, how to account for uncertainty, establish baselines, attribute cause and effect and identify actual performance indicators or, more specific to adaptation, determine adaptive capacity constitutes a major challenge. It also sits at odds with the strong focus on results and evaluation of intervention outcomes, undoubtedly shaping urban risk discourse: the space for experimentation and social learning, identified as crucial (O'Brien *et al.*, 2012) and a key characteristic of resilient systems, may in fact remain very small. No regrets options such as investments in public health and sanitation are sensible under practically any climate scenario. Yet, the value of other policy options really depends on climate uncertainties as outcomes, such as in river basin adjustments, differ quite profoundly. Both deterministic (e.g. cost-benefit analysis) and probabilistic approaches usually seeking to address these types of risk uncertainties may not be the most adequate risk management strategy in the long term (Setiadi *et al.*, 2010; Cannon, 2008).

The challenge of navigating between the often short term and private interests of development, set against the collective concerns for managing risk, will never be easy. Government and planning capacities are central to any forward movement (Pelling and Wisner, 2007). Particularly the dynamics of megacities in low- and middle-income countries may not be conducive for reducing hazard risk (Sattherthwaite *et al.*, 2007). Yet, the heterogeneity characteristic for many large urban regions may also be a driver for transformative change, as described in Chapter 1.

### *3.1 New risks due to climate change*

Climate change is not necessarily the most pressing issue faced by coastal megacities today and often, other environmental trends such as subsidence caused by excessive groundwater extraction, the public health concerns generated by lack of safe housing and access to basic sanitation and drinking water, or functioning public transport systems, are perceived as more urgent problems. Yet, the scope, uncertainty and complexity posed by the projected increase in extreme weather events mean sea-level rise, and more gradual changes in precipitation and temperature, undeniably challenges the coastal megacities – and their underlying logic and trajectories of location (UNEP, 2011). These risks sit alongside those shared with other large cities, such as heat waves, urban drought and food security, susceptibility to pandemics and existing geophysical hazards, which themselves can have increasing consequences as cities expand into areas of higher hazard exposure (for example up-slope) or generate risk (for example through land reclamation and associated liquefaction risk).

Chapter 3 describes the possible consequences of more intense atmospheric and marine hazards for coastal urban environments. Those most vulnerable to climate-change related impacts are the urban poor, residents of informal settlements and vulnerable groups such as children or minority communities (World Bank, 2011; Shaw, 2009). The key features of urban informal settlements – where nearly one-third of the global urban population lives – that limit local adaptive capacity are outlined in Table 6.2

As argued above, managing these risks calls for innovative, integrated and tailored response strategies whose formulation depends on urban governance structures, political leadership and municipal risk management capacity.

*Table 6.2* Trends driving vulnerability in informal urban settlements

| *Global and national trends* | *Local vulnerability trends* | *Risk impacts* |
|---|---|---|
| *Inequalities in economic development* | | |
| • Weak land administration and regulation | • Deterrence of public/private investment and service provision, due to insecure tenure or illegality of settlements | • Exacerbated flooding because of substandard drains blocked with uncollected garbage |
| *Rural–urban migration* | | |
| • Creation of informal settlements in high-risk areas (flood plains, steep hillsides) characterised by:<br>– Lack of access to land and secure land tenure<br>– Lack of access to basic services (for example, water, sanitation, electricity) | • Lack of economic assets (e.g. property or credit), political influence and education needed to adapt effectively | • Mortality and property loss, both localised or remote |
| *Lack of affordable urban housing* | | |
| • Overcrowding and stress on existing infrastructure<br>• Social fragmentation, conflict | • Concentrations of vulnerable communities and economic assets exposed to extreme hazards (for example, storm surges or heavy rains)<br>• Structural weaknesses in housing and infrastructures (for example, blocked drainage)<br>• Environmental degradation (for example, no waste disposal, untreated waste-water, reduction of protective ecosystem functions)<br>• Public health concerns that weaken the adaptive capacity of a community (for example, water-borne diseases or infant mortality) | • Faster spread of communicable diseases from stagnant flood waters and compromised immune systems<br>• Exacerbated poverty and barriers to economic development<br>• Increased stress on city services (for example, emergency response)<br>• Further social fragmentation |

Sources: Moser and Sattherthwaite (2011), World Bank (2010) and World Bank (2011)

### *3.2 Managing information*

Over the next decades, decision makers will have a significantly increased research output to guide their risk and vulnerability assessment, preparedness and response efforts. However, the way in which knowledge is mediated through policy processes in the context of scientific uncertainty constitutes an ongoing challenge for investment in effective disaster risk reduction or proactive climate change adaptation. Aid trends over the past decade show little increase in expenditures. In 2009, for every US$100 spent on response to humanitarian need, only 0.9 per cent of official development assistance (equalling US$346.3m) was spent on prevention or risk reduction measures (and 75 per cent of this going to only four countries) (Kellet and Sparks, 2012).

While the narrative of climate change lends new weight to the case for DRR, data for vulnerability, hazard and impact at the sub-national levels are still difficult to obtain. Megacities, with human capacity advantages and often significant academic institutions have a key advantage and offer scope for systematic data collection, with potential for common frameworks to be developed that might allow the root causes of vulnerability and resilience to be compared through comparative analysis. Improved observational data collection systems will likely benefit from tighter interlocking between research and policy communities to make sure the right questions are being asked and the correct data collected, coupled with better efforts to embed affected communities as actors in this process through participatory mechanisms as a means to bolster legitimacy for public policy decisions and support local risk reduction activities, and to build a broader consensus with local stakeholders for adaptation and risk reduction strategies in the context of various uncertainties linked to future climatic and socio-economic conditions. Here, the need to find broadly accessible tools to visualise risk factors is paramount. The routine call in both academia and practice for a greater convergence of DRR, CCA and development agendas seems only to reinforce the argument for investment in knowledge dissemination and information management (Thomalla *et al.*, 2006, Mitchell and van Aalst, 2008).

The multiplicity of competing knowledge providers and information formats in the hazards and risk, climate mitigation and adaptation field is broad and fast expanding. Effective oversight through coordination of information systems and disaster management procedures to facilitate decision making is likely to only grow in importance in the coming years, especially for megacities with multiple layers of government responsibility where intervention options are complex and involve the combination of several policy tools. For example, with regard to the urban heat island effect, are city-wide initiatives or focused investments the best solution? Can government encourage developers to use cooling pavement covers through tax breaks and other incentives? How many municipalities have to collaborate in this for it to be effective? Or, would it be more efficient to concentrate only in areas where high-risk residents such as the elderly live? Similarly, with opportunities for policy transfer from one city to another, how valid are current response strategies to urban heat islands in US coastal cities (Hoverter, 2012) for Asian megacities? Heat-wave risk management is not alone in presenting a number of contrasting strategies with attendant policy directions. It may not be possible to know without uncertainty what the most effective strategy will be and in this knowledge gap political bargaining can be decisive. Those wishing to understand risk management decisions need also to understand political process. The data resources and knowledge of private sector actors that play a key role in providing technology and institutional capacity to megacities (from insurers to engineering and development consultancies), and the role played by this sector in maintaining dominant development trajectories, even through risk management initiatives, has received minimal academic attention despite its considerable resource and influence. Analytical and management tools to help make such spaces of deliberation more transparent will only help build trust and legitimacy in the difficult decisions ahead. But this requires a new openness in the governance of megacities and one that holds an inquiring academic lens to private sector interests and especially those associated with state agendas. In addition, a closer cooperation and governance system is needed between megacity and its hinterland including peri-urban areas in order to address the challenges outlined before.

A distinctive feature of much of the knowledge on GEC and climate change is that it is generated outside those communities experiencing its impacts. This poses particular challenges for local stakeholder engagement. Current efforts to complement scientific expertise with locally generated knowledge, and to lend confidence to people's own capacity to weigh the risks they face, are important elements of a better understanding of the political – local as well as city and national – of resilient development, that which takes risk reduction seriously (Hardoy and Romero Lankao, 2011).

Developing and applying building standards, risk assessment tools and the adherence to international frameworks such as the Hyogo Framework for Action remain important instruments structuring the decision space and revealing the influence of politics and vested interest in risk and development decision making. Yet beyond this, arguably, the need for tailored information is paramount. Innovative use of social media and crowd-sourcing technologies can potentially provide a means to streamline information management systems in early warning systems and bridge science–end-user gaps, such as by climate forecasting services for urban disaster preparedness planning (RCCC, 2012). Participatory mapping tools such as OpenStreetMap can aid disaster relief efforts and a myriad of user-generated cycling maps in Chinese cities alone promote more sustainable transport. Other emerging initiatives involve community participation in urban planning processes in novel and cost-effective ways (e.g. MetroQuest, SurveyLA in Los Angeles and PlaNYC2030 in New York City). The dynamic features of social media technology may thus challenge current standard methodologies of public consultation processes for urban infrastructure projects and question conventional notions of information sovereignty by public authorities. In the face of the uncertainties surrounding modelling and projections of future urban–coastal risks, the flexibility associated with such platforms may also prove useful in terms of building adaptive capacity into information management systems.

## 4 Synergies and conflicts between climate change mitigation, disaster risk management, climate change adaptation and socioeconomic development in coastal megacities

In order to understand the complex processes and policy-making agendas that influence response strategies to GEC impacts and climate change related vulnerabilities, four domains of research and policy making have to be disentangled. These are climate change mitigation, disaster risk reduction (DRR) and climate change adaptation

(CCA), and socioeconomic development. Their potential synergies and trade-offs are examined below.

Mitigation actions refer to the reduction of GHG emissions – these have increased by 70 per cent between 1970 and 2004 (UNFCCC, 2012). Reducing emissions has been the primary aim of climate change policy in the past two decades. Given their economic output and intensive energy and resource consumption patterns, coastal megacities are an important component in achieving mitigation targets. A wide range of policies and instruments is available to create incentives for action amongst high consumption businesses and citizens (e.g., taxes and charges, tradable permits, regulations and standards, or voluntary agreements). Urban responses largely concentrate on four sectors: urban planning and the built environment, energy, transport and, to a much lesser extent, carbon sequestration (UN-Habitat, 2011). Yet, while some municipalities in OECD countries invest progressively in mitigation efforts, by and large it remains a marginal issue characterised by policy fragmentation, and so far no major coastal city has adopted a comprehensive framework that stretches beyond single municipalities to gain the benefits of scale that megacities and their extended urban regions offer.

DRR is the 'systematic development and application of policies, strategies and practices to minimize vulnerabilities, hazards and the unfolding of disaster impacts throughout a society, in the broad context of sustainable development' (UNISDR, 2004: 3). The increase in hazard events calls for a more prominent role of DRR in urban adaptation efforts. Similarly, 'for DRR to be successful, it needs to take account of the shifting risks associated with climate change and ensure that measures do not increase vulnerability to climate change in the medium to long-term' (Mitchell and van Aalst, 2008: 1). There exist important overlaps and differences between DRR and CCA in terms of institutional responsibilities, funding sources and methodologies used to assess risks and vulnerabilities. The conceptual commonalities and differences between the two concepts have been the subject of several studies (Mitchell and van Aalst, 2008; Thomalla *et al.*, 2006; Birkmann *et al.*, 2009; Mercer, 2010). The consensus identifies a distinction between CCA, which is aimed at facilitating long-term adjustment to changing average climate conditions, and DRR, which entails engaging with a wide variety of potential extreme events. DDR addresses hazards, ranging from earthquakes, landslides to floods whereas CCA deals with climate related hazards associated with changes in the average mean climate conditions. These include the influence of drought-induced salinisation on critical infrastructures, or gradually mounting pressure on food production and distribution systems. Furthermore, while climate change adaptation is concerned with marshalling resources including social and political power to better face future uncertainty, coping describes the actions of individuals and collectives exposed to hazard now and their efforts to avoid, carry, share or transfer impact (Pelling, 2011; Lavell *et al.*, 2012).

In response to the increase in hazard and risk intensity the field of DRR is starting to converge with adaptation efforts at some levels, particularly in the context of the discussion of UNFCCC's programme to address loss and damage. Fuelled by concern over the recognition on the limits of mitigation policies, several municipalities in megacities – within Manila, greater Shanghai, Istanbul, New York, Dhaka or London, to name a few – have developed adaptation plans, policies and actions, conceptualising the issue in numerous ways ('who adapts to what and how to what extent'). Yet, the need for more coordinated policy responses across sectors and institutions has widely been recognised (Mitchell *et al.*, 2010), and organisations in disaster management circles have started to incorporate actions to deal with gradual and incremental stresses and are redefining their understanding of extreme events. Forward-looking strategies are being developed, ranging from, for example, legislative changes to investments in early-warning systems and scenario mapping on the consequences of climate-related stressors (World Bank, 2011; Setiadi *et al.*, 2010; Tucker *et al.*, 2010; Moench *et al.*, 2011).

The key benefits for better integration can be summarised as:

(a) reduced climate related losses through comprehensive DRR measures;
(b) better use of resources, i.e. with regard to aid effectiveness and
(c) enhanced sustainability of CCA and DRR approaches

(Tearfund, 2008; Birkmann *et al.*, 2009)

Given the fact that, in reality, people themselves do not differentiate between hazards that can be attributed to climate change or other causes and have long been adapting to changes in their environment, integration is inevitable and can overcome the exceptionalism of disaster risk management that association with catastrophic events brings.

As noted earlier, urban communities may face a future different to that experienced so far, and the changing risk landscape calls for new types of coping and adaptation strategies. Table 6.3 provides an overview of key differences and current efforts to integrate the DRR and CCA fields.

Efforts to address both vulnerability and poverty, sometimes called sustainable adaptation measures, seek to broaden current development policy and climate change policy paradigms (Eriksen and O'Brien, 2007; Schipper and Pelling, 2006). Targeting the specific climate risks that the urban poor may face includes investment in water and social infrastructure and livelihood diversification to strengthen coping capacity, lobbying for legislative changes to align disaster preparedness within the broader municipal development planning process or facilitating the trade-offs between short-term versus long-term gains. So far, however, most analyses have focused on strengthening the poverty–adaptation nexus within a rural context and the applicability of emerging tools for the megacity context could be limited.

### 4.1 Barriers to integration

Despite clear signs of convergence, current efforts aiming to bridge the gap between DRR, CCA and wider socioeconomic development in both academia and practice often remain abstract, making it hard to translate suggestions into practice (Birkmann *et al.*, 2009; Davies *et al.*, 2008; Tearfund, 2008; Ahmed, 2010). While many publications underscore the need to address the root causes of vulnerability to disasters, either through structural or non-structural measures (see e.g. Birkmann, 2006; Bankoff *et al.*, 2004; Wisner *et al.*, 2004; Flynn, 2007; Benson and Clay, 2004; Kelkar *et al.*, 2006), practical tools and legal measures that determine urban development and adaptive capacities (e.g. land-use plans, property rights, environmental impact assessments) have not been sufficiently revisited or amended.

The two main barriers to mainstreaming both DRR and CCA across policy sectors relate to the challenge of adopting long-term planning approaches and to temporal and spatial trade-offs. Further, a number of crucial differences can be identified between

*Table 6.3* Conceptual and practical differences between DRR and CCA

| *Differences* | | *Signs of convergence* |
|---|---|---|
| *Disaster risk reduction* | *Climate change adaptation* | |
| Relevant to all hazard types | Relevant to climate and weather-related hazards | DRR programmes react to the impact of climate change on hazard frequency/ magnitude and vulnerabilities |
| DRR part of disaster cycle and humanitarian assistance | Part of international climate change policy processes, rooted in scientific theory | Joint programming, mainstreaming into development sectors |
| Targets present/near-future risks based on historical records/local knowledge | Targets short-, medium- and long-term future – i.e. addressing uncertainty and new risks from CC impacts | 'No regrets' options, DRR increasingly forward looking, CCA takes present data as entry point for action |
| Traditional knowledge as basis for community-based DRR | Traditional knowledge needs to be coupled with outside expertise as new risks emerge | Interventions that combine local knowledge with scientific expertise |
| Risk as a function of hazard, vulnerability, exposure and capacity | Vulnerability mainly understood as physical exposure | Increasing focus on CCA in local vulnerability assessments and improvement of social resilience. IPCC special report on extreme events (IPCC, 2011; EC, 2011) |
| Full range of established tools (legislation and institutional arrangements, early warning systems, insurance, building design codes, siting, ecosystem protection) | Range of tools under development, expanding rapidly | Fast progress on tool integration |
| Moderate – low political interest | High political interest, new agenda | Increasing recognition that both need to be addressed together |
| Funding streams often unpredictable and low | Funding streams increasing though implementation difficult | DRR community engaging in CCA funding mechanisms |

Based on: Mitchell *et al.* (2010) and World Bank (2011)

the concepts of CCA and socioeconomic development with regard to timeframes, objectives, institutional setup and scientific as well as normative backgrounds (Adger, 2003; Brooks *et al.*, 2011). Particularly the distinction between their respective objectives is often not sufficiently acknowledged: put broadly, development interventions seek to improve living conditions whereas 'adaptation does not necessarily make life better, but simply renders it possible – under the condition of anticipated changes' (Cannon and Mueller-Mahn, 2010: 629).

Brooks *et al.* (2011) point out that many climate change response frameworks are based on the assumption that adaptation can 'neutralise' climate change impacts so that originally set development targets can be met. Yet, 'such frameworks underestimate the potential need for transformative change' (Brooks *et al.*, 2011: 6). Whereas this has been well recognised in the scientific literature, much of adaptation policy and practice remains primarily concerned with making investments robust and to preserve the status quo, such as the protection strategies for water and sanitation infrastructure or coastal rehabilitation with no real strategies that extend beyond the near future (Pelling, 2011). The preceding chapters outlined, however, how GEC impacts can change the very conditions and context in which development takes place. This assumption that existing systems, such as the dominant pattern of socioeconomic development, can continue with adjustments fits to emerging resilience concepts (Pelling, 2011). It is also linked to economic growth strategies, but how sustainable these paradigms are, even if they provide much needed development benefits, is not clear and a good example for the aforementioned trade-offs along timescales and constituencies. For example, large-scale investments in coastal tourism can create employment, and property sales bring revenue to local governments, however, typical features of the tourism sector can increase vulnerability not only to climate-related hazards but short-sighted construction booms often cause massive groundwater and subsidence problems. Yet, there are also positive trade-offs, such as in the case of pollution. As it negatively affects attractiveness, several cities have identified the potential for positive gains when investing in mitigation policies (Kamal-Chaoui and Robert, 2009). The conditions under which decision makers weigh these trade-offs likely relate to factors such as the respective competitiveness of a given location, citizen pressure, timing, etc., and certainly deserve more research attention.

## 5 Solutions

### 5.1 Considering the significance of climate change uncertainty in urban governance

As the additional uncertainties of climate change present a challenge to informal and formal institutions, how to link different levels of governance to support communities at risk is one of several forward-pressing questions (Brooks *et al.*, 2011). While there are now extensive reviews of the infrastructural options available (e.g. Hallegatte, 2009), there is as yet only limited experience when it comes to related governance mechanisms. Attributes such as 'decentralization and autonomy, accountability and transparency, responsiveness and flexibility, participation and inclusion, experience and support' (UN-Habitat, 2011: 39; Tanner *et al.*, 2009) are routinely presented as key features of management capacities that it is argued can help to address urban climate change impacts, reminiscent of wider discourse on good governance (e.g. Grindle, 2007; Rodrik, 2004; Kaufmann, 2003). Yet, the multitude of local governance arrangements and the proliferation of networks of social capital in the absence of effective state services in urban areas of many megacities need much more acknowledgement (Bicknell *et al.*, 2009). In addition, decentralization strategies might also hamper effective governance systems and regimes between megacities and its peri-urban areas in the hinterland. Calls on cities to make better use of their regulatory authority are illusory where there is limited authority in the first place. Instead of assessing how existing arrangements measure up to a normative standard or mirror the experience of cities in OECD countries, the complexity of local arrangements needs to be disentangled – to see what works and to identify alternative means that increase constructive state–society engagements across the whole spectrum of public action (Mitlin, 2012).

Overall, there is an urgent need to improve the linkages and synergies between governmental formal adaptation programmes and urban development strategies on the one hand, and autonomous and spontaneous adaptation processes and risk reduction efforts on the other (Birkmann *et al.*, 2010; Schipper, 2012). Further, the reality of multilevel governance needs to be acknowledged more broadly. Municipal authorities work with national government or collaborate with other cities, as well as the private sector and civil society to leverage authority, technical expertise and funding (Bulkeley and Betsill, 2003). This not only involves vertical coordination among different levels of government, but horizontal coordination across sectors and agencies (Kamal-Chaoui and Robert, 2009). In how far municipal governance in the megacity context actually differs from other urban settings in its autonomy from national policies or regional arrangements needs to be tested empirically and certainly is likely to vary greatly as political contexts change. It may also show the limits of transferring rural response strategies to the coastal megacity context. For example, economic growth dynamics may conflict with municipal leverage to champion adjustments in urban planning schemes seeking to account for future heat waves, and the privatisation of public infrastructure and land property often narrows the space for municipal action.

## 5.2 *Improving the synergies between risk management, CCA, CCM and development*

Some examples of solutions to multiple problems already exist. In richer cities where consumption levels are very high a priority will be to link risk reduction to climate change mitigation – for example urban green infrastructure provides cooler microclimates and reduces surface water runoff and thus contributes to adaptation, while also providing potential corridors to protect ecological diversity and enhance the utility and economic value of urban locations (see Chapter 3). There are few megacities that could not learn from these examples but for those poorer cities with large informal populations at risk these priorities will also need to recognise the demands for basic needs including security from these groups.

The literature points to a number of principles for integrating DRR and CCA into sustainable urban planning, as outlined below:

- Define a legal framework: establishing hierarchies and responsibilities of the municipalities in areas that have high or medium threat to any catastrophic event.
- Define areas of environmental protection and ecological importance that should be developed in a sustainable form.
- Clearly establish the areas that involve hazards and vulnerabilities, potential areas for mitigation actions by defining short-, medium- and long-term procedures.
- Develop a system to identify and monitor the most vulnerable areas to be prioritised at the municipal level.
- Identify the type of actions and interventions required to ensure an integrated management.
- Define a hazard zoning clearly identifying the different regions, their potentials and constraints for future development.

- Establish policies and guidelines for land use planning that take into account the comprehensive management of urban risk.
- Involve the community in risk management decisions through processes of citizen participation.

Effective communication of future climate and urbanisation risks is a new challenge for local governments, and one where partnerships with private sector professionals or universities has proven successful. This can be a powerful driver for partnerships among local people, local universities and urban officials to build up knowledge and awareness. Such participatory programmes for city-level information gathering and disseminations have been implemented in cities in Thailand, India and Vietnam (Brown *et al.*, 2012).

### *5.3 Managing adaptation in a radically changing environment*

Increased awareness on changing risks as part of development has the potential to raise levels of strategic planning processes in cities that have thus far considered disasters as external, exceptional events (Birkmann and von Teichman, 2010). The variability of climate change and GEC impacts requires a different response, with an enhanced emphasis on adaptable and learning-based management systems (Berkhout *et al.*, 2006; Ziervogel *et al.*, 2006). To date this approach has been applied most widely in rural and natural resource management contexts (O'Brien *et al.*, 2012), but it has huge potential for urban systems. The inherent dynamism of urban centres open to shifting demography, exposed to global market fluctuations and political and cultural events, makes adaptive learning an attractive proposition even without climate change. Although limits to flexibility are imposed by the costs and lifetime of physical infrastructure, work is needed to think through and pilot the application of adaptive management approaches in the urban context.

Coastal megacities have some advantage in this regard. They can draw on planning models and existing relationships built from integrated coastal zone management, where flexibility and communication between sectors is a basis for planning. A declaration of the International Workshop on Coastal Megacities in Hangzhou, China recognised this, suggesting that improved linkages between risk reduction, urban development and climate change adaptation might be enhanced with integrated coastal zone management. This also echoes our call for enhanced information and learning capacity through proposals for:

(1) the importance of education, training and public awareness;
(2) a network of coastal megacities to provide a global source of knowledge exchange on coastal urban planning management, science and technology; and
(3) case studies with possible indicators for sustainability and resilience.

(see also Brown *et al.*, 2012, World Bank, 2008 and Twigg, 2007).

The shift from viewing environmental hazard, including climate change, as an external to an internal problem for development (Pelling, 2011) brings with it consequences for the scope of urban planning. An external viewpoint allowed risk management to be delivered principally through local hazard mitigation. This can avoid the deeper transformations in everyday behaviour required for climate change mitigation or vulnerability reduction. But a recognition that disaster risk, including that which can be associated with climate change, is a product of the development practices inherent in contemporary urbanisation make these practices and underlying values the focus of attention (Pelling, 2011). Calls for a reflection and possible action on everyday practices and on dominant development pathways is challenging, especially given the narrow remit of urban planning, coastal zone management and related professionals in the city. But, this is also a grand opportunity if urban development can move towards a more joined-up outlook and move beyond the narrow agendas of what today constitutes acceptable change, closing the space for innovation.

There is, though, a need to highlight the possibility of positive gains if climate change mitigation and resolving the root causes of vulnerability are to gain ground amidst competing, often conflicting demands for political space and budget lines. Some progress has been made in wealthier cities, where both agendas have become connected with wider agendas for 'liveable and healthy' cities. This places environmental management, urban renewal and poverty alleviation as part of the responsibilities of urban managers and planners in making cities competitive. It is argued that high-quality living environments are needed to attract the high-skill populations and inward capital investments these cities rely upon. There are fewer examples from poorer cities where the emphasis on development is less about the city being competitive and more about risk management. The case needs to be made that risk management as part of pro-poor development can be a driver for socio-economic development and competitiveness.

## 6 Conclusions

### *6.1 Reconceptualising megacities and the coast as integrated systems*

Megacities and coastal zones are arguably the two most dynamic human and natural environments on the planet. It is little wonder that both have attracted so much study. It is, though, surprising that only little work has prioritised the integration of knowledge and experiences from these two realms. In short to move beyond a disciplinary orientation to a systems viewpoint that can draw out the feedbacks, trade-offs and compounding processes in the urban–coast space is long overdue.

The task of bringing together the urban planning and coastal management communities is complicated by the multiple levels at which they interact. At the forefront of a good deal of contact are marginalised populations – low-income housing settlements on the coastal fringes or artisanal and informal sector fishers. These are not easy populations to integrate into planning processes, nor have planners and urban decision-making institutions more broadly often proven their legitimacy to these groups. More often planning has as its de facto priority economic growth and this may not coincide with the interests of the urban poor. Redevelopment of low-income settlements post-disaster or as part of DRR or CCA that does not provide enhanced access to basic needs and working opportunities is all too common. The dynamic processes that lie behind the coevolution of political, economic and physical systems are also ill captured by standard static development data.

There is a need for realism in such an integrative agenda of research and policy. It is tempting to romanticise the capacities of local actors at risk – often but not always the poor. Building the entitlements of these groups is likely to be a core element of any approach to resilience or sustainable development. But as recent events have shown, for example in Bangkok flood events, or Hurricane Katrina, risk reduction also requires strategic and substantial investment in infrastructure planning and in shaping the right institutions and organisational capacities from the local up to the international. Katrina is a prime example here with science having for some time identified risk, and even as the hurricane moved towards and eventually hit New Orleans an array of organisational failures revealed themselves. Vulnerability here was a product of poverty, some would argue of structural racism too but most fundamentally of failed governance. For coastal megacities the challenge is not only to recognise this but to then follow the lines of weakness out and away from the city to national or international organisations or standards of behaviour and norms that have allowed vulnerability to accumulate or look set to fail should an event unfold. Integration, in other words, also includes bringing together bottom-up and top-down strategies focused on social, infrastructural and institutional targets with the knowledge that risk comes from the weakest link.

A lack of perspective can also be seen in the identification of key drivers for risk in megacities. It is tempting, given the scale of such urban areas, to downscale global climate models and have this as the starting point for analysis for risk. The IPCC is clear in the significance of climate change for some hazard (urban heat wave for example) but for many hazard types changes that can be associated with climate change are overwhelmed by shifts in exposure driven by demography or local hazard conditions related to subsidence, coastal erosion and pollution of the air and water. The trick is to integrate the medium (five years plus) and long-range (ten years plus) strategic planning tasks that must incorporate climate change with short-term planning decisions. The short-term will set the opportunity space for medium- and long-term planning and needs to be framed as such, in this sense even where climate change does not appear to be a critical driver it needs to be considered as a boundary condition for decision making. Distorted perspectives also follow disaster events. All too often the priority for risk reduction in disaster reconstruction is framed by the disaster past, not oriented to the range of risks that a city and its neighbourhoods and external critical service provision hubs face going forward. Following the Indian Ocean Tsunami investment in tsunami early warning and land-use planning, despite their importance, displaced resources for more likely risk scenarios – flooding, landslides and windstorms. Political and public priorities have distorted science-based knowledge. Better understanding the coproduction of policy in these contexts is a first step to a smarter engagement between science and politics.

The existing literature and underlying practice on climate change and large-scale urbanisation is unhelpful in its polarisation (Romero-Lankao and Dodman, 2011). There are many examples of climate change mitigation projects (normally driven by technological innovation) in richer cities, and of vulnerability reduction (driven by social or economic or land-use policy) in poorer cities. This bifurcation in policy and vision is unhelpful insofar as it makes transfer of knowledge and expertise difficult. Vulnerability reduction is tainted as part of the policy discourse of a city at risk (with all the implications this has for poor governance, widespread poverty, etc). The result:

imaginative opportunities for learning from the global South are lost. Local governments are the central actors in urban risk management and should be central for any revisioning of the city and for shared learning going forward (UNISDR, 2010). Global and regional networks such as ICLEI or the ACCCRN in Asia offer some scope for moving beyond this and a focus on coastal cities as a 'hotspot' of the international scientific collaborative, LOICZ, can facilitate better flow of ideas and experience globally. Similarly, the high technology demonstrated through climate change mitigation schemes, and large-scale local hazard mitigation e.g. through flood defences, drainage, etc., can be alienating for poorer cities. Megacities have an advantage here, being centres of global wealth creation even in poorer countries so that opportunities for taking up high-tech and large-scale solutions can be considered. Tension lies in transferring a technology-based solution into a context where high levels of poverty both make the poor vulnerable to development – the poor have a small voice in strategic development of this scale, even when it is their livelihoods, homes and communities that might be the immediate 'losers'. Technology needs to be embedded in a social solution in such contexts: mitigation in vulnerability reduction and livelihood enhancement. Integration again is the watchword.

How might we reconceptualise megacities and large urban regions on the coast as integrated systems? A starting point suggested by this book is to consider the multiple and overlapping boundary spaces between the coast and interior – boundaries set by wind, temperature, water and chemical flows – not simply the high water mark. Work on social–ecological systems offers a coherent framework and one that is compatible with work on climate change adaptation and mitigation scholarship and with natural resource management. This offers exciting scope for cross-over research with fisheries and other coastal ecology and the human ecology of the city. Social–ecological theory presents us with a number of testing metaphors that could usefully push an integrated research agenda on coastal cities. We are asked to consider thresholds at which systems flip from one stable state to another and the extent to which changes in systems at one scale might infect or be suppressed, even provoked, by other scaled subsystems. Finally we are asked to consider the merits of resilience as stability seeking or transformation. This last question is perhaps the most profound and offers a mechanism for linking the two great challenges faced by the cities focused on in the book – mitigation and adaptation. Can there be pathways for transforming development so that it better meets social needs through risk reduction at the local and global scales?

Megacities are focal points of wealth, human capacity and political weight, they have already shown themselves willing to take on global leadership for climate change mitigation, often going beyond national governments. But these and other large and rapidly growing urban spaces – and the rural or global networks of resources that they depend upon – have yet to be conceptualised as coupled social–environmental systems with and precision. Chapter 1 in this volume presented both the risks and opportunities that megacities face. The argument put forward here is that it is not sufficient to compile urban planning and coastal management knowledge; a new way of seeing cities on the coast is required, one based on seeking ways of integrating across domains of practice and knowledge generation. This is a task for all developed and developing urban areas, the focus on megacities as icons of human endeavour is necessary but not sufficient.

### 6.2 Towards integration

Important to improving synergies are information and data deficiencies, administrative and governance constraints, funding limitations, as well as a need to define basic metrics of progress – for example, measures of increased resilience and adaptive capacity and decreased vulnerability (Tol and Yohe, 2007; da Silva *et al.*, 2012).

Projections indicate that policy making has to better reflect the number of interconnected drivers of urban vulnerability. These include urbanisation processes, infrastructural decay, cyber attacks and systems closures, and shifting spatial dynamics of poverty and inequality, all of which could place increased pressures on social cohesion and political stability. Given the range of location-specific geographic conditions, the nature of public infrastructure, resource accessibility and the character of local communities, local government will play a key role in designing adaptation strategies that fit an area's specific requirements (Laukkonen *et al.*, 2009). This impediment makes it difficult to come out with broad policies and programmes that can extend to reducing vulnerability in coastal megacities. The institutions responsible for implementing and enforcing the broad policies of climate change are at an early stage and poorly resourced (Kithiia, 2010). Moreover, the present planning architecture in many countries does not provide room for local governments to implement adaptation, even though the effective responses to climate change calls for an incorporated planning process in which local government should play a leading role (Okereke *et al.*, 2009).

Building risk management, climate change adaptation and mitigation into urban planning for the future can draw on many existing experiences, from integrated coastal zone management to multi-hazard risk analysis and the multi-agency approaches employed for water and transport systems management in many cities today. The required step change in political will and policy focus and the necessity for broader, inclusive approaches that involve the most vulnerable urban communities suggests the following priorities for furthering risk management in coastal megacities:

- High-level, proactive leadership to initiate and coordinate adaptive planning for the critical infrastructure of cities and surroundings.
- The incorporation of climate change risks into stakeholder agency and organisation operations, management and planning. This in turn requires long-term planning horizons and decision-making procedures that can expose and tolerate uncertainty. Most immediately this supports the identification of low-regrets policies.
- To promote a dynamic planning process among city government, public and private stakeholders, and experts to support a risk-management approach to incorporate climate change and to begin to implement flexible adaptation pathways.
- Develop linkages (including organisational forms or joint working groups on legislation) to enable cross-agency and cross-policy coordination in adaptation plans to cope with extreme and compound hazard events, ecological stresses and human impacts in the coastal zone.

Earlier chapters have highlighted the continually expanding knowledge base on the vulnerabilities of populations inhabiting megacities on the coast. This trend is likely to be compounded by continuing processes of globalisation, particularly the growing extensity, intensity and velocity of information flows (Held and McGrew, 2007). The ways in which information is mediated through governance processes in the context of scientific uncertainty will likely be a growing challenge in the future, particularly as policy makers seek to move from a focus on response to enhancing resilience to disasters. Policy makers will need to ask the right kinds of questions of scientists in order to ensure that they have clarity on the level of certainty present in any research on which they may choose to ground their decisions (see HFP, 2011). Moreover, research, policy and practice communities will need to avoid working in isolation from each other. This will mean increasingly coming together under the banner of resilience to break down artificial boundaries such as those between disaster risk reduction (DRR) and climate change adaption (Thomalla *et al.*, 2006), or between humanitarian action and development (HERR, 2011).

In tandem with the expanding knowledge base on future vulnerabilities proposed above, there is likely to be a multiplicity of competing sources and formats of information on hazards, including research, monitoring, forecasts and early warning systems. Therefore actors and organisations with humanitarian responsibilities for urban regions on the coast will need to ensure that their information systems and disaster management procedures are effectively coordinated and codified in order to facilitate effective decision making. Lessons could be learnt here from the development of the Integrated Food Security Phase Classification (IPC), a standardised scale (integrating food security, nutrition and livelihood data) developed by a group of leading international humanitarian agencies in 2004 to help coordinate response in food security emergencies worldwide.

But while coordination tools such as the IPC or bodies such as the UNISDR will remain important in this respect, the innovative use of social media and crowdsourcing technologies can also provide a means to streamline information management systems in emergency response. Following the Haiti earthquake in January 2010, users of openstreetmap.org assisted rescue workers and humanitarian agencies by providing missing information on the layout of Port-au-Prince. The technology is now also being used to enhance preparedness efforts in coastal cities such as Padang, Indonesia, as well as in many other crisis-prone situations across the world. In the face of the uncertainties surrounding modelling and projections of future urban–coastal risks, the flexibility associated with such open source platforms may also prove useful in terms of building adaptive capacity into information management systems.

The future vulnerabilities and capacities within coastal cities will be affected by decision-making processes from the global to the community level. This highlights the increasing disjointedness of policy making and the need to foster increased coherence through participatory engagement between policy actors, both horizontally and vertically. This can be done by bringing together hazard researchers and policy makers (both within and across organisations and sectors), while also ensuring that their engagement extends directly to the relevant at-risk communities. For example, an ongoing pilot project led by the Humanitarian Futures Programme (HFP) is bringing together climate scientists, humanitarian and development policy makers, and communities in Kenya and Senegal with the aim of enabling climate science to better support specific decision-making contexts.

As this engagement with the scientific community attests, it is increasingly clear that future humanitarian action, from risk reduction to response, will not only involve the decisions of conventional authorities, from affected governments to humanitarian agencies, but is also likely to involve a number of 'non-traditional' humanitarian actors, including the scientific community and the private sector. Innovative modes of coordination need to be developed in order for these additional capacities to be brought to bear on the overwhelming needs the humanitarian system is tasked with addressing. One such platform for engagement is that of NetHope, a collaboration of 33 international humanitarian organisations working to improve coordination by partnering with leading foundations and corporations to ensure their members benefit from the efficiencies of the latest advances in communications technology (see www.nethope.org/, last accessed 10 April 2013).

Projecting from our current situation, it is possible that we may see increasing ambivalence from host country governments in disaster situations towards large-scale interventions by the traditional actors of the international humanitarian system – United Nations agencies, the Red Cross and Red Crescent movement, and international NGOs. Such ambivalence was evident, for example, in Myanmar in 2008 when Cyclone Nargis devastated coastal areas of the country (Harvey, 2009). This ambivalence also relates to the increasing political centrality of crises and resulting moves towards the reassertion of sovereignty on behalf of crisis-affected states (Kent and Burke, 2011), particularly in regions such as Southeast Asia. In this context, regional disaster management frameworks are becoming increasingly important, especially for urban regions on the coast, as seen in the case of the Association of Southeast Asian Nations' (ASEAN) Agreement on Disaster Management and Emergency Response, signed in 2009 (ASEAN, 2009). The development of such frameworks will be similarly important for other regional organisations, and some are already responding to this need. For example, the Economic Community of West African States (ECOWAS) is currently implementing a programme of action running to 2014 to operationalise its own DRR policy (ECOWAS, 2006).

Yet national- and local-level disaster management will also be critical to increasing the resilience of coastal megacities. According to the Hyogo Framework for Action, states have the primary responsibility for taking effective measures to reduce disasters: 'disaster risk reduction is a national and local priority' (UNISDR, 2005: 6). In addressing the vulnerabilities of coastal megacities, local actors are likely to prove particularly important. As noted in a recent UNISDR study, local governments have a vital role to play in DRR in terms of multi-stakeholder coordination, community engagement, utilising their own institutional capacities, and developing and implementing innovative practices (UNISDR, 2010). But in delivering on this mandate, local governments will have to address evolving and contested municipal and patronage structures as well as processes of decentralisation and government or economic reform. Further, in engaging with those at the community level, particularly in coastal slums, authorities will have to learn to utilise existing networks of slum leadership, whether considered legitimate or not (Hudson-Edwards and Raford, 2010). However, these are not challenges that policy makers must face alone. Initial work by UNISDR is showing the potential of city-to-city exchange and learning on DRR issues, bringing together mayors and other municipal officials from across Asia as part of its 'Making Cities Resilient' campaign.

Given the evidence and arguments developed in this book, the pressure on policy makers with risk management responsibilities at the global, regional, national and local level to meet the needs of vulnerable urban populations on the coast looks likely only to increase in the coming decades. Therefore dealing with this challenge through an integrated, collaborative and strategic approach that can consider short- and long-term needs – and that links formal and informal planning tools – is likely to prove essential to meeting the challenge of delivering safer and more resilient coastal megacities for their future inhabitants.

## References

Adger, W.N., Brown, K., Nelson, D.R., Berkes, F., Eakin, H., Folke, C., Galvin, K., Gunderson, L., Goulden, M., O'Brien, K., Ruitenbeek, J. and Tompkins, E.L. (2011) Resilience implications of policy responses to climate change. *Climate Change*, 2, 757–766.

Ahmed, A. (2010) Mapping of Climate Change Adaptation and Disaster Risk Management related Governance – Cambodia, Indonesia and the Philippines. *Strengthening Climate Resilience*. Brighton: Institute for Development Studies.

ASEAN (2009) Regional Disaster Management Agreement Enters into Force, www.asean.org/24136.htm (accessed 9 March 2012).

Bankoff, G., Frerks, G. and Hilhorst, D. (eds) (2004) *Mapping Vulnerability: Disasters, Development, and People*. London: Earthscan.

Benson, C. and Clay, E. (2004) *Understanding the Economic Financial Impacts of Natural Disasters*. Washington, DC: World Bank.

Berkhout, F., Hertin, J. and Gann, D. (2006) Learning to adapt: organizational adaptation to climate change impacts. *Climate Change*, 78, 135–156.

Bicknell, J., Dodman, D. and Satterthwaite, D. (eds) (2009) *Adapting Cities to Climate Change: Understanding and Addressing the Development Challenges*. London: Routledge.

Birkmann, J. (2006) *Measuring vulnerability to natural hazards – towards disaster resilient societies*. Tokyo: UNU-Press.

Birkmann, J. and von Teichman, K. (2010) Integrating disaster risk reduction and climate change adaptation: key challenges – scales, knowledge, and norms. *Sustainability Science* 5(2), 171–184.

Birkmann, J., Garschagen, M., Krauke, K. and Quang, N. (2010) Adaptive urban governance: new challenges for the second generation of urban adaptation strategies to climate change. *Sustainability Science*, 5, 185–206.

Birkmann, J. *et al.* (2009) Addressing the Challenge: Recommendations and Quality Criteria for Linking Disaster Risk Reduction and Adaptation to Climate Change. In: Birkmann, J., Tetzlaff, G. and Zentel, K. (eds) *DKKV Publication Series 38*, Bonn: DKKV.

Brooks, N., Anderson, S., Ayers, J., Burton, I. and Tellam, I. (2011) Tracking Adaptation and Measuring Development. *IIED Working Paper No. 1*. London: International Institute for Environment and Development.

Brown, A., Dayal, A. and Rumbaitis Del Rio, C. (2012) From practice to theory: emerging lessons from Asia for building urban climate change resilience. *Environment and Urbanization*, 24(2), 531–556.

Bulkeley, H. and Betsill, M. (2003) *Cities and Climate Change: urban sustainability and global environmental governance*. London: Routledge.

Cannon, T. (2008) Reducing people's vulnerability to natural hazards: Communities and resilience. *UNU-WIDER Research Paper*, No. 2008/34.

Cannon, T. and Müller-Mahn, D. (2010) Vulnerability, Resilience and Development Discourses in Context of Climate Change. *Natural Hazards*, 55, 621–635.

Da Silva, J., Kernaghan, S. and Luque, A. (2012) A systems approach to meeting the challenges of urban climate change. *International Journal of Urban Sustainable Development*, 1, 1–21.

Decker, E.H., Elliot, S. and Smith, F.A. (2002) Megacities and the environment. *The Scientific World Journal*, 2, 374–386.

De Sherbinin, A., Schiller, A. and Pulsipher, A. (2007) The vulnerability of global cities to climate hazards. *Environment and Urbanization*, 19(1), 39–64.

ECOWAS (2006) *ECOWAS Policy for Disaster Risk Reduction*. August. Humanitarian Affairs Department (DHA).

Eriksen, S.H. and O'Brien, K. (2007) Vulnerability, poverty and the need for sustainable adaptation measures. *Climate Policy*, 7(4), 337–352.

European Commission (2012) *Communication on the EU Adaptation Strategy 2012*. Brussels: EC.

Flynn, D.T. (2007) The impact of disasters on small business disaster planning: a case study. *Disasters*, 31, 508–515.

Grimm, N.B., Faeth, S.H., Golubiewski, N.E., Redman, C.L., Wu, J., Bai, X. and Briggs, J.M. (2008) Global Change and the Ecology of Cities. *Science*, 319(5864), 756–760.

Grindle, M.S. (2007) Good Enough Governance Revisited. *Development Policy Review*, 25(5), 533–574.

Hallegatte, S. (2009) Strategies to adapt to an uncertain climate change. *IOP Conference Series: Earth and Environmental Science*, 6(41), 4120–4123.

Hardoy, J. and Romero-Lankao, P. (2011) Latin American cities and climate change: challenges and options to mitigation and adaptation responses. *Current Opinion in Environmental Sustainability*, 3(3), 158–163.

Held, D. and McGrew, A. (2007) *Globalization/Anti-Globalization: Beyond the Great Divide*. Cambridge: Polity.

Held, D., McGrew, A. Goldblatt, D. and Perraton, J. (1999) *Global Transformations*. Stanford, CA: Stanford University Press.

Humanitarian Futures Programme (HFP) (2011) *Making Space for Science–Humanitarian Policy Dialogue: Unlocking the Potential for Effective Crisis Prevention, Preparedness, Response and Early Recovery, Working Draft*. London: Humanitarian Futures Programme, King's College London.

Holling, C.S., Gunderson, L.H. and Ludwig, D. (2002) In quest of a theory of adaptive change. In: Gunderson, L.H. and Holling, C.S. (eds) *Panarchy: understanding transformations in human and natural systems*. Washington, DC: Island Press.

Holman, I.P., Rounsevell, M.D.A., Shackley, S., Harrison, P.A., Nicholls, R.J., Berry, P.M. and Audsley, E. (2005) A regional, multi-sectoral and integrated assessment of the impacts of climate and socio-economic change in the UK. Part I. Methodology. *Climatic Change*, 71, 9–41.

Hoverter, S. (2012) *Adapting to Urban Heat: A tool kit for local governments*. Harrison Institute for Public Law. Georgetown Climate Center. www.georgetownclimate.org/adaptation-tool-kit-urban-heat (last accessed 26 August 2012).

Humanitarian Emergency Response Review (HERR) (2011) Chaired by Lord (Paddy) Ashdown, 28 March 2011.

Jacinto, G.S., Azanza, R.V., Velaxquez I.B. and Siringan, F.P. (2006) Manila Bay: Environmental Challenges and Opportunities. In: Wolanski, E. (ed.) *The Environment in Asia Pacific Harbours*. Berlin/Heidelberg: Springer-Verlag.

Jago-On, K.A.B., Kaneko, S., Fujikura, R., Fujiwara, A. and Imai, T. (2009) Urbanization and subsurface environmental issues: An attempt at DPSIR model application in Asian cities. *Science of the Total Environment*, 407, 3089–3104.

Kamal-Chaoui, L. and Robert, A. (eds) (2009) Competitive Cities and Climate Change. *OECD Regional Development Working Papers N° 2*. Paris: OECD.

Kelkar, U., James, C.R. and Kumar, R. (2006) The Indian insurance industry and climate change: exposure, opportunities, and strategies ahead. *Climate Policy*, 6, 658–671.

Kellett, J. and Sparks, D. (2012) *Disaster Risk Reduction. Spending where it should count.* Development Initiatives: Global Humanitarian Assistance. www.globalhumanitarianassistance.org/reports (last accessed 26 August 2012).

Kithiia, J. (2010) Old notion – new relevance: setting the stage for the use of social capital resources in adapting East African coastal cities to climate change. *International Journal of Urban Sustainable Development*, (1), 17–32.

Lamhauge, L., Lanzi, E. and Agrawala, S. (2012) *Monitoring and evaluation for adaptation: lessons from development co-operation agencies.* OECD Environment Working Papers. Paris: OECD.

Lavell, A., Oppenheimer, M., Diop, C., Hess, J., Lempert, R., Li, J., Muir-Wood, R. and Myeong, S. (2012) Climate change: new dimensions in disaster risk, exposure, vulnerability, and resilience. In: Field, C.B., Barros, V., Stocker, T.F., Qin, D., Dokken, D.J., Ebi, K.L., Mastrandrea, M.D., Mach, K.J., Plattner, G.-K., Allen, S.K., Tignor, M. and Midgley, P.M. (eds) *Managing the Risks of Extreme Events and Disasters to Advance Climate Change Adaptation.* A Special Report of Working Groups I and II of the Intergovernmental Panel on Climate Change (IPCC). Cambridge: Cambridge University Press.

McGranahan, G. (2007) Urban Transitions and the Spatial Displacement of Environmental Burdens. In: Marcotullio, P.J. and McGranahan, G. (eds) *Scaling Urban Environmental Challenges. From Local to Global and Back.* London: Earthscan.

Mercer, J. (2010) Disaster Risk Reduction or Climate Change Adaptation: Are we reinventing the wheel? *Journal of International Development*, 22, 247–264.

Mitchell, T. and van Aalst, M. (2008) *Convergence of Disaster Risk Reduction and Climate Change Adaptation*, Review for DFID, UK. www.preventionweb.net/files/7853_ConvergenceofDRRandCCA1.pdf (accessed 26 August 2012).

Mitchell, T., Van Aalst, M. and Silva Villanueva, P. (2010) Assessing Progress on Integrating Disaster Risk Reduction and Climate Change Adaptation in Development Processes. *Strengthening Climate Resilience Discussion Paper* 2. Brighton: University of Sussex, Institute of Development Studies.

Mitlin, D. (2012) Lessons from the urban poor: collective action and the rethinking of development. In: Pelling M., Manuel-Navarrete, D. and M. Redclifft (eds) *Climate change and the crisis of capitalism.* London: Routledge.

Moench, M., Tylor, S. and Lange, J. (2011) Catalyzing urban climate resilience: applying resilience concepts to planning practice in the ACCCRN Program (2009–2011). Bangkok: ISET- Boulder.

Moser, C. and Sattherthwaite, D. (2011) Toward pro-poor adaptation to climate change in the urban centers of low and middle income countries. In: Mearns, R. and Norton, A. (eds.) *Social Dimensions of Climate Change Equity and Vulnerability in a Warming World.* Washington, DC: World Bank.

Nicholls, R.J., Wong, P.P., Burkett, V.R., Codignotto, J.O., Hay, J.E., McLean, R.F., Ragoonaden, S. and Woodroffe, C.D. (2007) Coastal systems and low-lying areas. Climate Change 2007: Impacts, Adaptation and Vulnerability. Contribution of Working Group II to the Fourth Assessment Report of the Intergovernmental Panel on Climate Change. M.L. Parry, M.L., O.F. Canziani, O.F., J.P. Palutikof, J.P., P.J. vna der Linden, P.J. and C.E. Hanson, C.E. (eds) Cambridge: Cambridge University Press.

O'Brien, K., Leichenko, E., Kelkar, U., Venema, H., Aandahl, G., Tompkins, H., Javed, A., Bhadwal, S., Barg, S., Nygaard, L. and West, J. (2004) Mapping vulnerability to multiple stressors: climate change and globalization in India. *Global Environmental Change*, 14(4), 303–313.

O'Brien, K., Pelling, M., Patwardhan, A., Hallegatte, S., Maskrey, A., Oki, T., Oswald-Spring, U., Wilbanks, T. and Yanda, P.Z. (2012) Toward a sustainable and resilient future. In: Field, C.B., Barros, V., Stocker, T.F., Qin, D., Dokken, D.J., Ebi, K.L, Mastrandrea, M.D., Mach, K.J., Plattner, G.-K., Allen, S.K., Tignor, M. and Midgley, P.M. (eds.) *Managing the Risks of Extreme Events and Disasters to Advance Climate Change Adaptation: A Special Report of Working Groups I and II of the Intergovernmental Panel on Climate Change (IPCC)*. Cambridge: Cambridge University Press.

Okereke, C., Bulkeley, H. and Schroeder, H. (2009) Conceptualizing Climate Governance Beyond the International Regime. *Global Environmental Politics*, 9, 58–78.

Pelling, M. (2011) *Adaptation to climate change: from resilience to transformation*. London: Routledge.

Pelling, M. and Wisner, B. (eds) (2008) *Disaster Risk Reduction: cases from urban Africa*, London: Earthscan.

Red Cross/Red Crescent Climate Centre (RCCC) (2012) *Forecasting tools. Forecast-based humanitarian decisions: Designing tools and processes to link knowledge with action*. www.climatecentre.org/site/forecast-based-decisions (accessed 26 August 2012).

Rees, W. (2003) Understanding Urban Ecosystems: An Ecological Economics Perspective. In: Berkowitz, A. *et al.* (eds) *Understanding Urban Ecosystems*. New York: Springer-Verlag.

Reid, H. (2009) Community-based Adaptation to Climate Change: An Overview. Community-based Adaptation to Climate Change. *Participatory Learning and Action (PLA) Series*, 60, 11–33.

Rodolfo, K.S. and Siringan, F.P. (2006) Global sea-level rise is recognised, but flooding from anthropogenic land subsidence is ignored around northern Manila Bay, Philippines. *Disasters*, 30, 118–139.

Rodrik, D. (2004) *Getting Institutions Right*, CESifo DICE Report, February.

Romero-Lankao, P. and Dodman, D. (2011) Cities in transition: transforming urban centres from hotbeds of GHG emissions and vulnerability to seedbeds of sustainability and resilience. *Current Opinion in Environmental Sustainability*, 3(3), 113–120.

Ruth, M. and Coelho, D. (2007) Understanding and managing the complexity of urban systems under climate change. *Climate Policy*, 7(4), 317–336.

Satterthwaite, D. (2008) Cities' contribution to global warming: notes on the allocation of greenhouse gas emissions. *Environment and Urbanisation*, 20(2), 539–549.

Satterthwaite, D., Huq, S., Reid, H., Pelling, M. and Romero Lankao, P. (2007) *Adaptation to climate change in urban areas: the possibilities and constraints in low- and middle-income nations*. London: International Institute for Environment and Development.

Schipper, L. (2012) *Transformative adaptation*. http://tinyurl.com/97aszk9 (accessed 16 September 2012).

Schipper, L. and Pelling, M. (2006) Disaster risk, climate change and international development: scope for, and challenges to, integration. *Disasters*, 30(1), 19–38.

Shaw, R. (2009) Climate Disaster Resilience: Focus on coastal urban cities in Asia. *Asian Journal of Environment and Disaster Management*, 1.

Sekovski, I., Newton, A. and Dennison, W.C. (2011) Megacities in the coastal zone: Using a driver-pressure-state-impact-response framework to address complex environmental problems. *Estuarine, Coastal and Shelf Science*, 1–12.

Setiadi, N., Birkmann, J. and Buckle, P. (eds) (2010) Disaster Risk Reduction and Climate Change Adaptation: Case Studies from South and Southeast Asia. *SOURCE Publication Series No. 14*. Bonn: United Nations University EHS.

Shalizi, Z. and Lecocq, F. (2010) To mitigate or to adapt: is that the question? Observations on an appropriate response to the climate change challenge to development strategies. *World Bank Research Observer*, 25(2), 295–321.

Solecki, W., Leichenko, R. and O'Brien, K. (2011) Climate change adaptation strategies and disaster risk reduction in cities: connections, contentions, and synergies. *Current Opinion in Environmental Sustainability*, 3(3), 135–141.

Tanner, T., Mitchell, T., Polack, E. and Guenther, B. (2009) Urban Governance for Adaptation: Assessing Climate Change Resilience in Ten Asian Cities. *IDS Research Summary 315*. Brighton: Institute for Development Studies.

Tearfund (2008) *Linking Climate Change Adaptation and Disaster Risk Reduction*. http://tinyurl.com/bpwnxdz (accessed 26 August 2012).

Thomalla, F., Downing, T., Spanger-Siegfried, E., Han, G. and Rockström, J. (2006) Reducing hazard vulnerability: towards a common approach between disaster risk reduction and climate adaptation. *Disasters*, 30(1), 39–48.

Tol, R.S.J. and Yohe, G.W. (2007) The weakest link hypothesis for adaptive capacity: An empirical test. *Global Environmental Change*, 17(2), 218–227.

Tucker, C.M., Eakin, H. and E.J. Castellanos (2010) Perceptions of risk and adaptation: Coffee producers, market shocks, and extreme weather in Central America and Mexico. *Global Environmental Change*, 20(1), 23–32.

Twigg, J. (2007) *Characteristics of a disaster-resilient community: a guidance note*. London: DFID Disaster Risk Reduction Interagency Coordination Group.

UNFCCC (2012) *Fact sheet: Climate change science – the status of climate change science today*. http://unfccc.int/press/fact_sheets/items/4987.php (accessed 16 September 2012).

UNISDR (2004) *Terminology: Basic Terms of Disaster Risk Reduction*. Geneva: UNISDR.

UNISDR (2005) *Hyogo Framework for Action 2005–2015*. Geneva: UNISDR.

UNISDR (2010) *Local Governments and Disaster Risk Reduction: Good Practices and Lessons Learned*. A contribution to the "Making Cities Resilient" Campaign. Geneva: UNISDR.

UNEP (2011) *Mainstreaming Climate Change Adaptation into Development Planning. A Guide for Practitioners*. UNDP-UNEP Poverty Environment Initiative. (online) www.unpei.org (accessed 28 August 2012).

UN-Habitat (2011) *Cities and Climate Change: policy directions. Global report on human settlements 2011*. London: Earthscan.

Wenzel, F., Bendimerad, F. and Sinha, R. (2007) Megacities – megarisks. *Natural Hazards*, 42(3), 481–491.

Wisner, B., Blaikie, P., Cannon, T. and Davis, I. (2004) *At Risk: Natural Hazards. People's Vulnerability and Disasters*. London: Routledge.

World Bank (2010) *Climate Risks and Adaptation in Coastal Asian Megacities. A Synthesis Report*. Washington, DC: World Bank Group.

World Bank (2011) *Guide to Climate Change Adaptation in Cities*. Washington, DC: World Bank Group.

Young, O.R., Berkhout, F., Gallopin, G., Janssen, M.A., Ostrom, E. and van der Leeuw, S. (2006) The globalization of socio-ecological systems: An agenda for scientific research. *Global Environmental Change*, 16(3), 304–316.

Ziervogel, G., Bharwani, S. and Downing, T.E. (2006) Adapting to climate variability: pumpkins, people, and policy. *Natural Resources Forum*, 30, 294–305.

Chapter 7

# Case studies

## Governing social and environmental transformation in coastal megacities

Lead authors: *Mark Pelling (King's College London, UK), Sophie Blackburn (King's College London, UK)*

Contributing authors: *Adeniyi Asiyanbi (King's College London, UK), Ayansina Ayanlade (King's College London, UK), Antje Bruns (Humboldt University Berlin, Germany), Kirsten Butcher (King's College London, UK), José Bolanos (King's College London, UK), Robert Coates (King's College London, UK), Ayden Cumming (King's College London, UK) Aliendheasja Fawilia (King's College London, UK), Steven Forrest, (King's College London, UK) Thomas van der Klugt (King's College London, UK), Davina Layne (King's College London, UK), Eleanor Lewis (King's College London, UK), Irwin Gabriel Lopez (King's College London, UK), Felix Mallin (King's College London, UK), Sara J. Casallas Ramirez (King's College London, UK), Matt Richmond (King's College London, UK), William Solecki (CUNY, USA)*

Megacities on the coast are dynamic and complex systems. To this point our book has provided specific viewpoints from environmental and social positions on the coupled processes that determine the security, liveability and sustainability of cities and the surrounding coastal environment. This chapter serves to provide a more integrated view through the analysis of seven case study cities. Each city has its unique properties determined by development legacy, political, economic and environmental context. The aim in presenting these studies is to illustrate the interactions between themes already introduced in preceding chapters and draw out the importance of context in shaping present and future sustainability, and through this reveal the extent to which transformation has taken place and may be necessary to reach more sustainable paths for urban development.

The case studies are organised around a standard structure to facilitate comparison, and are presented in alphabetical order:

1. Jakarta
2. Lagos
3. Manila
4. Mumbai
5. New York
6. Rio de Janiero
7. Shanghai.

## JAKARTA, INDONESIA

***Thomas van der Klugt, Felix Mallin and Aliendheasja Fawilia***

Jakarta's local environmental impacts, and the challenges it faces from climate change, are not separate issues, but interact closely with one another, generating a negative symbiotic relationship. It is this complex pattern of interaction and feedback that makes responses at all levels a challenge. Since the 1950s, Jakarta's rapid population growth has transformed it into one of the world's megacities (Silver, 2008). After 2000, the population of the Special Capital Territory of Jakarta (known as DKI Jakarta) grew at an annual average rate of 1.41 per cent to a total of 9.6 million by 2010. Occupying an absolute land area of 664 kilometres square, the city has an average population density of 14,469 inhabitants per square kilometre (BPS, 2012). This process of urbanisation has produced a wide range of environmental impacts. DKI Jakarta's residents generate 25,687 metres cubed of garbage per day (Pasang *et al.*, 2007), while residents of Jakarta Metropolitan Area – the conurbation area known locally as Jabodetabek – generate more than 35,000 metres cubed of garbage per day (UNESCO, 2000). Approximately 20 per cent of garbage generated in Jakarta each day is disposed of into rivers and drains or simply burnt by households (Pasang *et al.*, 2007); around 1,000 metres cubed of waste reaches Jakarta Bay, leaving the coastline highly polluted (Munawir, 2005; UNESCO, 2000).

Jakarta's 72-kilometre coastline used to feature a rich coral reef complex, but this has changed dramatically as Jakarta has urbanised (Arifin, 2008; van der Meij *et al.*, 2010); Tomascik *et al.* (1997) have found that several species of coral have disappeared since the late 1920s. The decline in the number and variety of coral reefs has precipitated a decline in the fish population of Jakarta Bay over the last 40 years (Djamali and Parino, 2008; Tomascik *et al.*, 1997). Meanwhile, the development of highways, docks and residential and recreational areas along Jakarta's coastline has also reduced the coastal mangrove swamp area by almost 83 per cent over the last 40 years (Pramudji, 2008).

Within Jakarta, a massive use of vehicle transport, compounded by a lack of public transport infrastructure – or a vision for it – has caused serious air pollution (Resosudarmo and Napitupulu, 2004; Steinberg, 2007). Jakarta desperately needs a firm policy to overcome the air pollution problem: one author has suggested that the rate of increase in health problems due to air pollution exceeds the growth rate of Indonesia's economy (Resosudarmo and Napitupulu, 2004; see also Yusuf and Resosudarmo, 2009).

As well as suffering from the effects of its local environmental impacts, DKI Jakarta has been identified as one of the regions in Southeast Asia most vulnerable to climate change. The population density of Central Jakarta – at over 20,000 inhabitants per square kilometre it is one of the most densely populated urban areas in Southeast Asia – leaves it especially exposed to climate change risks: it has been ranked as the single most vulnerable urban region in Southeast Asia by Yusuf and Francisco (2009).

To date, flooding has produced the most disastrous effects on Jakarta, and thus poses the primary environmental threat to the city. A set of environmental factors such as sea-level rise, heavy rain and tropical storms, combined with internal issues – most significantly, rapid land subsidence – produce a complex hazard environment combined with growing vulnerable populations and increasing asset investments to present a progression of urbanisation and local and global environmental change (Firman *et al.*, 2011) and risk.

All five municipalities of DKI Jakarta have been highlighted as highly prone to the effects of flood-related disasters – and as much as 40 per cent of the entire metropolitan area lies below sea level. The districts carrying the highest risk, however, are those located closest to the coast – North and Central Jakarta – where exposure to a mix of tidal flooding and inundation caused by overflowing rivers and canals is greatest (World Bank, 2011). Continuous heavy flooding occurred between 2002 and 2007, affecting approximately a quarter of Jakarta's land area. The most disastrous tidal flood to date took place in 2007, inundating between 40–70 per cent of DKI Jakarta and leaving 422,300 people displaced (Brinkmann and Hartman, 2008; World Bank, 2011).

Continuously increasing population pressure and out-dated, ineffective drainage systems reinforce this problem (Fuchs, 2010). Steinberg (2007: 360) suggests that three main causes for the flooding events are a 'lack of carrying capacity of flood control infrastructure . . . the reduction of capacity of existing systems, due to uncontrolled garbage dumping . . . [and the] reduction of rainwater absorption due to urbanisation and deforestation'.

Funding issues and other problems have beleaguered the city's flood canal system (James, 2008); the East Flood Canal remains uncompleted, while the West Flood Canal lacks enough capacity. Little is done to clear garbage from Jakarta's numerous other waterways – instead, the problem tends to be blamed on informal settlements – although in recent years, attempts to think more strategically about mitigating flooding, such as instituting upstream river basin rehabilitation schemes, have emerged (Steinberg, 2007).

DKI Jakarta's administrative division into several different cities means that information on rising water levels is not always shared between municipalities, and the opening of sluice gates is poorly coordinated. Jakarta's water management systems are outdated with rivers, canals and sewage pipes suffering from severe pollution and a legacy of poor maintenance (Fuchs, 2010; James, 2008).

Jakarta's urbanisation has also led to large-scale groundwater extraction (Onodera *et al.*, 2009), exacerbated by the inadequacy of the city's piped water supply. Recent improvements in the piped water system have often mirrored the colonial tendency to favour middle- and upper-class income households, and even here, the supply remains unreliable (Bakker, 2007; Colbran, 2009). Consequently, homes and businesses that lack a reliable water supply extract groundwater to compensate. This causes rapid land subsidence – in North Jakarta, an annual subsidence rate of up to 12 centimetres was measured – which has significantly increased exposure to inundation risks (Brinkmann and Hartman, 2008). The falling groundwater level also leads to seawater intrusion and the salinisation of fresh water supplies (Colbran, 2009). The impact of a recent increase in groundwater extraction fees is not yet clear (Fuchs, 2010).

In dealing with all these issues, the government of Jakarta does face a challenging historical legacy. The city's canals were initially a product of Dutch colonial planning, and present both problems and opportunities in terms of tackling flooding. Much more recently, the 1997–8 Asian economic crisis drastically reduced the resources available to the city's management agencies, and the capacity has never been rebuilt (Steinberg, 2007). As Jakarta has expanded since the 1950s, a succession of city planning documents – in particular, the 1987 New Master Plan for DKI Jakarta – have been paid scant attention, leading to unregulated urbanisation (Steinberg, 2007). Government agencies often do not ensure that the development of open spaces set aside for the mitigation of flooding impacts takes place according to official environmental requirements (Fuchs, 2010). Peri-urban areas, such as the sprawl between JMA and the neighbouring Bandung Metropolitan Area create urban areas of uncertain administrative status (Firman, 2009).

The government of DKI Jakarta has faced severe criticism over its lack of strategies to tackle the city's numerous environmental problems, and specifically for its inadequate handling of the devastating 2002 and 2007 floods. Research suggests that Jakarta's policies have not been pro-poor and have failed to take into account the needs and desires of stakeholders and the general population; direct elections for the governor of DKI Jakarta were only instituted in 2007 (Steinberg, 2007). Those people living in informal settlements – an estimated 5 per cent of the total population of DKI Jakarta – are amongst those most vulnerable to external shocks. As in many other megacities, the majority of the city's slums and squatter areas are located in highly disaster-affected areas (World Bank, 2011).

Despite this, there has been a growth in civil society and community-based responses, which have also sought to connect with external actors. The UNDP, for instance, provided funding and technical assistance to the Indonesian Society for Disaster Preparedness, which played a key role in pressuring the Indonesian government to produce national plans for disaster reduction (UNDP, 2005). In 2006, Jakarta organised a Disaster Awareness Day and participated in the International Day of Disaster Reduction, which brought together communities and international NGOs (James, 2008). James (2008) has argued that the 2004 tsunami precipitated something of a transformation in Indonesian disaster management strategy, from an approach that had traditionally been reactionary, to one more focused on preparation and mitigation. Certainly, the tsunami – combined with anxieties concerning environmental and climate change issues – has prompted a growing regional dialogue examining environmental impacts and risks common to many Asian coastal cities and possible responses to them (Fuchs, 2010; James, 2008).

However, in Jakarta, as in many other megacities, a key problem is the difficulty of separating impacts, risks and responses. Policy responses – or a lack of them – on one issue can exacerbate the development of others. Jakarta's response to the interrelated problems of water management and supply, pollution, waste disposal and flood risks – and to the possible impact of climate change on this nexus of issues – arguably remains piecemeal.

## References

Arifin, Z. (2008) Kajian kecenderungan Perubahan Kontaminan Logam Berat di Perairan Teluk Jakarta. In: Ruyitno, A. Syahailatua, M. Muchtar, P., Sulistijo and Susana, T. (eds), *Kajian Perubahan Ekologis Perairan Teluk Jakarta*. Jakarta: LIPI Press – Pusat Penelitian Oseanografi.

Bakker, K. (2007) Trickle Down? Private sector participation and the pro-poor water supply debate in Jakarta, Indonesia. *Geoforum*, 38(5), 855–868.

BPS (2012) Statistik Indonesia – Statistical Yearbook of Indonesia 2011. *Badan Pusat Statistik*. http://dds.bps.go.id/eng/flip/flip11/index3.php (last accessed 11 March 2012).

Brinkmann, J. and Hartman, M. (2008) Jakarta Flood Hazard Mapping Framework. *Jakarta Flood Management Project*. www.hkv.nl/documenten/Jakarta_Flood_Hazard_Mapping_Framework_MH.pdf (last accessed 10 March 2012).

Colbran, N. (2009) Will Jakarta be the next Atlantis? Excessive groundwater use resulting from a failing piped water network. *Law Environment and Development Journal*, 5(1), 20–36.

Djamali, A. and Parino (2008) Jenis-jenis ikan dari Teluk Jakarta. In Ruyitno, A., Syahailatua, M., Muchtar, P., Sulistijo and Susana, T. (eds) *Kajian perubahan ekologis perairan Teluk Jakarta*. Jakarta: LIPI Press – Pusat Penelitian Oseanografi.

Firman, T. (2009) The continuity and change in mega-urbanization in Indonesia: A survey of Jakarta–Bandung Region (JBR) development. *Habitat International*, 33(4), 327–339.

Firman, T., Surbakti, I.M., Idroes, I.C. and Simarmata, H. (2011) Potential climate-change related vulnerabilities in Jakarta: Challenges and current status. *Habitat International*, 35(2), 372–378.

Fuchs, R.J. (2010) Cities at Risk: Asia's Coastal Cities in an Age of Climate Change. *Analysis from the East-West Center*, 96, Asia Pacific Issues series, pp. 1–12.

James, E. (2008) Getting ahead of the next disaster: recent preparedness efforts in Indonesia. *Development in Practice*, 18(3), 424–429.

Munawir, K. (2005) Pemantauan kadar pestisida organoklorin di beberapa muara sungai di perairan teluk Jakarta. *Oseanologi dan Lomnologi di Indonesia*, (37), 15–25.

Onodera, S., Saito, M., Sawano, M., Hosono, T., Taniguchi, M., Shimada, J., Umezawa, Y., Lubis, R.F., Buapeng, S. and Delinom, R. (2009) Erratum to "Effects of intensive urbanization on the intrusion of shallow groundwater into deep groundwater: Examples from Bangkok and Jakarta." *Science of the Total Environment*, 407(9), 3209–3217.

Pasang, H., Moore, G.A. and Sitorus, G. (2007) Neighbourhood-based waste management: a solution for solid waste problems in Jakarta, Indonesia. *Waste management (New York, N.Y.)*, 27(12), 1924–1938.

Pramudji (2008) Informasi hutan mangrove di kawasan pesisir Teluk Jakarta. In: Ruyitno, A., Syahailatua, M., Muchtar, P., Sulistijo, and Susana, T. (eds) *Kajian perubahan ekologis perairan Teluk Jakarta*. Jakarta: LIPI Press – Pusat Penelitian Oseanografi.

Resosudarmo, B.P. and Napitupulu, L. (2004) Health and Economic Impact of Air Pollution in Jakarta. *The Economic Record*, 80(1), 65–75.

Steinberg, F. (2007) Jakarta: Environmental problems and sustainability. *Habitat International*, 31(3-4), 354–365.

Tomascik, T., Janice Mah, A., Nontji, A. and Moosa, M. (1997) *The ecology of Indonesian Seas: Part two*. Singapore: Periplus Editions (HK) Ltd.

UNDP (2005) *Report on UN Activities for the 1st Month After the Earthquake and Tsunami Disaster*. Banda Aceh: Office of the Humanitarian Coordinator.

UNESCO (2000) *Reducing megacity impacts on the Coastal Environment. Alternative livelihoods and waste management in Jakarta and the Seribu Islands*. Paris: UNESCO.

World Bank (2011) Jakarta Case Study Overview. Climate Change, Disaster Risk and the Urban Poor: Cities Building Resilience for a Changing World. http://siteresources.worldbank.org/INTURBANDEVELOPMENT/Resources/336387-1306291319853/CS_Jakarta.pdf (last accessed 2 March 2012).

Yusuf, A.A. and Francisco, H. (2009) *Climate Change Vulnerability Mapping for Southeast Asia*. Economy and Environmental Program for Southeast Asia edition. Singapore: EEPSEA.

Yusuf, A.A. and Resosudarmo, B.P. (2009) Does clean air matter in developing countries' megacities? A hedonic price analysis of the Jakarta housing market, Indonesia. *Ecological Economics*, 68(5), 1398–1407.
van der Meij, S., Suharsono, S. and Hoeksema, B. (2010) Long-term changes in coral assemblages under natural and anthropogenic stress in Jakarta Bay (1920–2005). *Marine Pollution Bulletin*, 60(9), 1442–1454.

## LAGOS CASE STUDY

***Adeniyi Asiyanbi, Antje Bruns, Ayansina Ayanlade and Eleanor Lewis***

### Introduction

Lagos, Nigeria's economic hub, is the second-largest and one of the fastest-growing megacities in Africa (UN-HABITAT, 2010). The most recent census (NPC, 2006) puts the city's population at eight million, a figure that is deeply contested (Potts, 2012). In 2009, UN-DESA estimated Lagos to have a population of 10.2 million, making it the twenty-first largest city in the world, with a projected rise to eleventh by 2025. With a growth rate of 3.1 per cent, Lagos adds an estimated 248,000 people to its population every year (Potts, 2012). Such a high level and rate of urbanisation without corresponding economic and infrastructural growth has been described as over-urbanisation, leaving the city with a massive infrastructure deficit and about 70 per cent of the city's population living in informal or 'slum' settlements (UN-HABITAT, 2003: 106; Adelekan, 2010). In addition, the geographical features of Lagos, particularly its coastal location and low-lying landmass portend a severe threat to this teeming megacity. These factors contribute to heightened concerns regarding city–coast risks in Lagos.

### Urbanisation impacts on the environment

Researchers have noted the immense pressure of the rapid growth of Lagos on the environment (Adekanmbi and Ogundipe, 2009; Adeoye *et al.*, 2009; Breitmeier *et al.*, 2009). First, mangrove, swamp and other coastal wetland vegetations around Lagos have been under threat from increasing urbanisation and consequent high demand for land (Adekanmbi and Ogundipe, 2009; Braimoh and Onishi, 2007). Vegetated areas in Lagos have declined from 180,384 hectares in 1984 to 140,568 hectares in 2000 (Osei *et al.*, 2006). Mangroves are important not only for biodiversity but also for the natural barrier they provide against storms, reducing flooding, which often has the strongest impact on the poorest inhabitants due to the location of their shelter on hazardous land (Douglas *et al.*, 2008; Breitmeie *et al.*, 2009). Whilst outcomes are uncertain, climate change is expected to increase the severity of storms, further increasing the vulnerability of the inhabitants of Lagos (DFID, 2009).

Second, in the absence of adequate water systems, over pumping of water has degraded aquifers through accelerated saltwater intrusion. It is estimated that about ten million gallons of water are withdrawn from Lagos' multilayer aquifers for domestic and industrial usage diurnally (Adepelumi *et al.*, 2009). Most of the drilled boreholes in the coastal area of Lagos have had to be abandoned due to severe saltwater intrusion and pollution from the surface into the aquifer (Oyedele *et al.*, 2009).

Another major environmental impact of urbanisation in Lagos is air pollution. Accounting for more than 60 per cent of Nigeria's industrial and commercial activities in an area of less than 3,568.61 square kilometres, Lagos has a very high industrial and population density according to the Lagos State Government (Adelekan, 2010). The associated industrial and transport emissions mean Lagos' emission levels are extremely high. Lagos' annualised total carbon monoxide emission (1.44 Tg/yr) is higher than those of Beijing (1.23Tg/yr), Tokyo (0.46 Tg/yr) and London (0.13 teragram per year) (Hopkins *et al.*, 2009). With over six million vehicles on the roads of Lagos every day, the transport sector contributes hugely to air pollution. Fumes from generators, widely used to supplement the inadequate supply of electricity, further contribute to air pollution in Lagos. Air pollution is already impacting human health in Lagos and respiratory ailments are becoming common. Uran food production impacts have also been detected; vehicular fumes absorbed by leaf surface concentrate potentially toxic heavy metals such as lead (Pb) and cadmium (Cd) in vegetables grown in Lagos (Bakare *et al.*, 2004). Okoye (1991) traced disproportionately high Pb levels in fish from the Lagos Lagoon to vehicular fumes containing Pb from leaded petrol.

## City–coast risks and vulnerability

While rapid urbanisation in Lagos continues to have serious impacts on the environment, perhaps more profound are the risks that the geographical features of Lagos pose to its population and assets. The coastal location of Lagos and its low-lying topography predispose it to a particularly severe threat from climate change. This threat is exacerbated by the social vulnerability of its population and the physical weakness of its infrastructure. Most of Lagos has an elevation ranging from just above sea level to 15 metres above sea level (Iwugo *et al.*, 2003). In light of this, even IPCC's conservative prediction of average global sea-level rise of 18–59 centimetres by 2100 would be highly significant for many parts of Lagos. Much higher and faster global sea-level rise estimates of up to six metres by 2100 (Jevrejeva *et al.*, 2006) would herald a much greater threat for Lagos. This is of particular concern for those areas of informal settlement, since many slum dwellers have built their settlement on stilts over swamps and other coastal wetlands near the lagoons, which act as flood basins. This ensures that the areas inhabited by the poorest are also most vulnerable to regular flooding and its consequences (Douglas *et al.*, 2008) and lie at the heart of an integrated management approach.

Beyond flooding, sea-level rise will result in increased salinisation of surface and ground water, killing freshwater plants and animals, adversely affecting coastal agriculture and contaminating water supplies. DFID (2009) reports that a 1°C rise in surface temperature in the Atlantic will precipitate the conditions required to create hurricanes off the coast of Nigeria. These possibilities combined with the grossly inadequate infrastructure and the water table rising 0.5 metres per year (Oyedele *et al.*, 2009) compound the risk situation for Lagos. A study evaluating the vulnerability of 136 coastal cities to climate extremes projected that Lagos will have the fifth most exposed population by 2070 (Nicholls *et al.*, 2007).

Ocean surges are already becoming more severe and frequent in Lagos (Tsakporhore and Archibong, 2011). In combination with the clearing of mangroves, ocean surges render the coastline extremely susceptible to shoreline retreat (Sunday *et al.*, 2006). The most recent occurrence in August 2011 displaced 1,000 residents from Lagos' Alpha Beach, wiping out properties and washing away roads and other infrastructure. Government officials claimed shoreline erosion at Alpha Beach encroached ten meters inland between June to August 2011, a rate of shoreline retreat higher than the oft cited 25–30 metres per year observed by Ibe and Quelennac (1989). Questions exist as to whether reclamation efforts by the government are able to effectively deal with emerging coastal threats. Many blamed reclamation at the Bar Beach site of the proposed Eko Atlantic City (EAC) for the severity of the August 2011 surge. Although the government rebuffed speculation, a study by Van Bentum (2012) shows that the building of the 6.5-kilometre-long wall to protect the EAC site will require excavation of huge amounts of sediment, creating a severe sediment deficit that could foster coastal erosion elsewhere. Whilst time will reveal how well the EAC itself will weather the whip of the Atlantic, it is clear that a significant sea-level rise and higher temperatures resulting from climate change will devastate the coastal length of Lagos, especially if serious efforts are not put into mitigating their likely impacts.

Prior to the August 2011 surges, there was a severe flashflood in July 2011. The heavy downpour lasted for 17 hours, precipitating the total amount of rainfall expected for the month of July. The flood led to the collapse of buildings, bridges and other public infrastructure. The Nigerian Red Cross (IFRC, 2011) estimated that about 5,393 households were affected, and immediate media reports indicated 20 deaths. The impacts of such a widespread disaster in Lagos can be very difficult to ascertain, due to the population density, poor administrative capacity and very vulnerable living conditions. Adelekan (2010) has shown that though rainy days are getting fewer in Lagos, rainfall is more intensive. Faced with the task of providing adequate urban infrastructure and combating the many critical environmental challenges that confront Lagos, how the city responds could further distance development from sustainability or be an opportunity for transformation, and regional leadership from Africa's second-most populous city.

## Responses

Lagos State Government (LASG) is undertaking several large-scale infrastructure projects in partnership with a range of actors to address some of the city's major problems, in the light of predicted environmental risks.

Flooding is the single most important infrastructure challenge Lagos faces (World Bank, 2006; Ibem, 2011). The Lagos State Ministry of Environment (LASME) has undertaken an extensive programme of drain rehabilitation, construction and dredging. Over 676,603 metric tonnes of silt have been removed from the cities' drains and canals (LASME, 2011b), however, it is not only silt that blocks the drains and causes flooding. Lack of a citywide solid waste management, and especially refuse collection has often led to rubbish being dumped in the canals. Lagos Waste Management Authority (LAWMA), in conjunction with a range of private actors, is overhauling the waste management infrastructure. In addition to investment in new refuse collectors and landfill rehabilitation, new recycling facilities have been created along with a 'waste-to-wealth' project that turns 45 per cent of Lagos' 9,000 tonnes of daily refuse into compost for use in the cities' 'greening' projects (LAWMA, 2011). It is, however, unclear how far these increased services serve the informal settlements of Lagos, as they could.

The EAC has been heavily promoted by the government as protecting the erosion of the Victoria Island coastline. To protect EAC against storm surges and associated erosion, a 6.5-kilometre long wall, called the Great Wall of Lagos, is currently being built. However, little study has been done to determine the potential impact of reclaiming 1,000,000 square metres of land on the rest of the coast. Van Bentum (2012), using numerical modelling, shows that the changing of the coastline will increase the speed of coastal erosion further down the shore, placing communities at risk. The EAC is a highly exclusive venture, privately financed and designed to only accommodate Lagos' commercial district. While attempts to increase land availability are valid in such a densely populated city, LASG needs to ensure that projects such as EAC and the dredging of 100,000 hectares of wetland (LASME, 2011a) do not create problems in the future. The consequences of climate change adaptation measures need to be considered well into the future to ensure that maladaptation does not take place (Kasperson *et al.*, 2001).

The infrastructure upgrade undertaken by LASG is positive; however, it does not seem to be translating quickly into better living conditions for the enormous portion of the population living in informal settlements. Urban governance is still largely exclusionary, denying the poorest households access to basic public services (UN-HABITAT, 2010; Gandy, 2006). Municipal governments remain relatively weak with low public confidence in government-led service provision leading to a continuing expansion of the informal social and economic sector. LASG has much more to do to demonstrate that it can manage and provide for its rapidly growing population and position the city for a sustainable future.

## References

Adekanmbi, O.H. and Ogundipe, O. (2009) Mangrove Biodiversity in the Restoration and Sustainability of the Nigerian Natural Environment. *Journal of Ecology and Natural Environment*, 1(3), 64–72.

Adelekan, I.O. (2010) Vulnerability of poor urban coastal communities to flooding in Lagos, Nigeria. *Environment and Urbanisation*, 22, 443–450.

Adeoye, N.O., Ayanlade, A. and Babatimehin, O. (2009) Climate Change and Menace of Floods in Nigerian Cities: Socio-economic Implications. *Advances in Natural and Applied Sciences*, 3(3), 369–377.

Adepelumi, A.A., Ako, B.D., Ajayi, T.R., Afolabi, O. and Omotoso, E.J. (2009) Delineation of saltwater intrusion into the freshwater aquifer of Lekki Peninsula, Lagos, Nigeria. *Environmental Geology*, 56, 927–933.

Bakare, S., Denloye, A.A. and Olaniyan, F.O. (2004) Cadmium, Lead and Mercury in Fresh and Boiled Leafy Vegetables Grown in Lagos, Nigeria. *Environmental Technology*, 25(12), 1367–1370.

Braimoh, A.K. and Onishi, T. (2007) Spatial determinants of urban land-use change in Lagos, Nigeria. *Land Use Policy*, 24, 502–515.

Breitmeier, H., Schwindenhammer, S. and Kuhn, J. (2009) Analyzing Urban Adaptation Strategies to Climate Change: A Comparison of the Coastal Cities of Dhaka, Lagos and

Hamburg, DVPW-Kongress 2009 "Politik im Klimawandel – Keine Macht für gerechte Lösungen?", 21–25, September, Kiel.

DFID (2009) *Impact of Climate change on Nigeria's Economy*. Prepared by Environmental Resources Management Limited for the Department for International Development, London. www.scribd.com/doc/47086875/27-Impact-of-Climate-Change-on-Nigerias-Economy (accessed 4 March 2012).

Douglas, I., Alam, K., Maghenda, M., McDonnell, Y., McLean, L. and Campbell, J. (2008) Unjust waters: climate change, flooding and the urban poor in Africa. *Environment and Urbanization*, 20, 187–205.

Gandy, M. (2006) Planning, Anti-Planning and the Infrastructure Crisis Facing Metropolitan Lagos, *Urban Studies,* 43(20), 371–396.

Hopkins, J.R., Evans, M.J., Lee, J.D., Lewis, A.C., Marsham, J.H, McQuaid, J.B., Parker, D.J., Stewart, D.J., Reeves, C.E. and Purvis, R.M (2009) Direct estimates of emissions from Lagos. *Journal of Atmospheric Chemistry and Physics*, 9, 8667–8682.

Ibe, A.C. and Quelennac, R.E. (1989) *Methodology for Assessment and Control of Coastal Erosion in West Africa and Central Africa*. UNEP Regional Sea Reports and Studies No. 107. New York: United Nations Environment Programme.

Ibem, E.O. (2011) Challenges of disaster vulnerability reduction in Lagos Megacity Area, Nigeria. *Disaster Prevention and Management*, 20(1), 27–40.

IFRC (2011) *Nigeria: Flash Floods*. A release of the Disaster Relief Emergency Fund of the International Federation of Red Cross and Red Crescent 25 July. www.ifrc.org/docs/appeals/11/MDRNG010.pdf (accessed 2 March 2012).

Iwugo, K.O., D'Arcy, B. and Andoh, R. (2003) Aspects of land-based pollution of an African coastal megacity of Lagos Conference Poster Papers, Diffuse Pollution Conference, Dublin. www.ucc.ie/dipcon/docs/theme14/theme14_32.PDF (accessed 2 March 2012).

Jevrejeva, S.A. Grinsted, J., Moore, C. and Holgate, S. (2006) Nonlinear trends and multi-year cycle in sea level records. *Journal of Geophysical Research*, 111.

Kasperson, R.E., Kasperson, J.X., Dow, K., Ezcurra, E., Liverman, D.M., Mitchell, J.K., Samuel, J.R., O'Riordan, T. and Timmerman, P. (2001) Introduction: global environmental risk and society. In Kasperson, J.X. and Kasperson, R.E. (eds) *Global Environmental Risk*. Tokyo: United Nations University Press.

Lagos State Ministry of the Environment (LASME) (2011a) The Climate Change Challenge in Lagos State: Turning Liability into an Asset. Paper presented at Resilient Cities 2011, 2nd World Congress on Cities and Adaptation to Climate Change, Bonn, Germany, 3–5 June. http://resilient-cities.iclei.org/fileadmin/sites/resilient-cities/files/Resilient_Cities_2011/Presentations/C/C5_and_D5_Banire.pdf (last accessed 2 March 2012).

Lagos State Ministry of the Environment (LASME) (2011b) Adaption Strategy on Stormwater Drainage Management and Flood Control. Paper presented at Resilient Cities 2011, 2nd World Congress on Cities and Adaptation to Climate Change, Bonn, Germany, 3–5 June. http://resilient-cities.iclei.org/fileadmin/sites/resilient-cities/files/Resilient_Cities_2011/Presentations/C/C5_and_D5_Ogunbambi.pdf (last accessed 2 March 2012).

Lagos Waste Management Authority (LAWMA) (2011) Cities and Adaptation to Climate Change – Solid Waste Management in Lagos State. Paper presented at Resilient Cities 2011, 2nd World Congress on Cities and Adaptation to Climate Change, Bonn, Germany, 3–5 June. http://resilient-cities.iclei.org/fileadmin/sites/resilient-cities/files/Resilient_Cities_2011/Presentations/C/C5_and_D5_Olubori.pdf (last accessed 2 March 2012).

National Population Commission (NPC) (2006) *Population Census and Household Data Survey for Nigeria*. Published by the Federal Government of Nigeria. www.population.gov.ng/images/stories/Priority%20Tables%20Volume%20I-update.pdf (last accessed 7 March 2012).

Nicholls, R.J., Hanson, S., Herweijer, C., Patmore, N., Hallegatte, S., Corfee-Morlot, J., Château, J. and Muir-Wood, R. (2007) *Ranking Port Cities with High Exposure and Vulnerability to Climate Extremes: Exposure Estimates*. OECD Environment Working Paper No 1. www.oecd.org/env/workingpapers (accessed 2 March 2012).
Okoye, B.C.O. (1991) Heavy metals and organisms in the Lagos Lagoon. *International Journal of Environmental Studies*, 37(4), 285–292.
Osei, A., Merem, E.C. and Twumasi, Y.A. (2006) Use of Remote Sensing Data to Detect Environmental Degradation in the Coastal Region of Southern Nigeria. In: *Proceedings of the ISPRS Commission VII Mid-term Symposium "Remote Sensing: From Pixels to Processes"*. Enschede.
Oyedele, K.F., Ayolabi, E. Adeoti, L. and Adegbola, R.B. (2009) Geophysical and hydrogeological evaluation of rising groundwater level in the coastal areas of Lagos, Nigeria. *Bulletin of Engineering Geology and the Environment*, 68, 137–143.
Potts, D. (2012) Challenging the Myths of Urban Dynamics in Sub-Saharan Africa: The Evidence from Nigeria. *World Development*, [Accessed 03/03/2012].
Silver, C. (2008) Planning the Megacity: Jakarta in the Twentieth Century, Routledge: Abingdon (Oxfordshire).
Sunday, A.O. and John, T.O. (2006) Lagos shoreline change pattern: 1986–2002, *American–Eurasian Journal of Scientific Research*, 1(1), 25–30.
Tsakporhore, O. and Archibong, E.O. (2011) *Ocean Surge: Dynamics, impacts, past and future changes in Nigerian coastline: A case of Bar Beach V.I. Lagos*. A paper presented at the LOICZ Open Science Conference 2011, 12–15 September, Yantai, China.
UN-DESA (2009) *Urban Agglomerations 2009*, New York: Department of Social and Economic Affairs, United Nations. http://esa.un.org/unpd/wup/index.htm (accessed 10 March 2012).
United Nations Human Settlements Programme (UN-HABITAT) (2003) *The Challenge of Slums. Global Report on Human Settlements 2003*. London: UN-HABITAT.
United Nations Human Settlements Programme (UN-HABITAT) (2010) *The State of African Cities: Governance, Inequality and Urban Land Markets*. Nairobi: UN-HABITAT.
Van Bentum, K.M. (2012) The Lagos Coast: Investigation of the long-term morphological impact of the Eko Atlantic City project. http://repository.tudelft.nl/view/ir/uuid%3A794318f9-9279-4a3c-9030-2f9fab7c6562/ (accessed 2 March 2012).
World Bank (2006) Lagos Metropolitan Development and Urban Governance Project, Report No: AB1396. Project Information Document. www-wds.worldbank.org/external/default/main?pagePK=64193027&piPK=64187937&theSitePK=523679&menuPK=64187510&searchMenuPK=51351213&theSitePK=40941&entityID=000104615_20060508153648&searchMenuPK=51351213&theSitePK=40941 (accessed 2 March 2012).

## METRO MANILA

### *Sara J. Casallas Ramirez, José Bolanos, Irwin Gabriel Lopez*

With a population of over 11.5 million people (WHO, 2012) and projected to grow in economic terms (IMF, 2011), Metro Manila (MM), the Philippines' National Capital Region, is a coastal megacity where rapid and fragmented growth has occurred at the expense of the environment. As such, the effects of anthropogenic climate change on the city and vice versa – for example increasing rainfall and floods – have highlighted the need for effective policies and strategies. However, lack of coordination between the various stakeholders, and the structural division of power, hinders all efforts. Starting with an image of MM as part of a global picture, following with a description of its political structure, and finalising with an overview of the challenges posed by climate change, this case study will provide a brief summary of these realities.

## Metro Manila: development and characteristics

Covering only 0.2 per cent of the country, MM is responsible for 80 per cent of the country's exports, almost two-thirds of the jobs and 30 per cent of the GDP (Sovacool and Brown, 2010). Manila has historically attracted migrants from all over the country, resulting in an expanding city where the urban fringe keeps moving outwards (Sovacool and Brown, 2010; Malaque and Yokohari, 2007). Furthermore, over 7 per cent of the population live below the poverty line (Albert and Collado, 2004) and informal settlements (Shatkin, 2004) and slum housing are increasing at a rate of 8 per cent per year (Ballesteros, 2010).

Metro Manila is highly dependent on the global economy and its politics and policies are based on market-led approaches to development (Lambino, 2010; Shatkin, 2000). In parallel, but driven by other logics, are the city's local and community-based disaster preparedness (CBDP) groups and activities. Finding common and supportive ground between popular politics and community movements on one hand and speculative, market-driven urban development on the other is not easy; both are powerful and clear forces in Metro Manila (Allen, 2006; Porio, 2011). The resulting lack of cohesive and comprehensive approaches to risk and development is manifest in high socioeconomic losses in the event of common natural hazards such as typhoons and floods (Tharoor, 2009; Hranjski and Gomez, 2011). Losses are not only direct but also felt when critical infrastructure systems are damaged. During Typhoon Ondoy in 2009, for example, the city had to shut down causing economic losses of up to 2.7 per cent of the country's GDP (PDNA, 2009). With stakes that high, it is imperative to understand the local political context, and the challenges it poses, in order to understand the lack of cohesive action that results in such massive losses.

## Political structure and its effect on mitigation and adaptation initiatives

While other Philippine regions are made up of provinces, Metro Manila is comprised of 16 cities, one municipality and 1,705 barangays (neighbourhood committees) and is managed by the Metropolitan Manila Development Authority (MMDA), a regional administrative agency (NSCB, 2011). According to Section 2 of the Republic Act (RA) 7924 (Congress of the Philippines, 1995), the MMDA is responsible for planning, monitoring, regulating and supervising the delivery of various metro-wide services such as solid waste management, flood control and public safety, which are directly or indirectly linked to climate change adaptation and mitigation. In the *MMDA* vs. *Bel-Air Village Association* case (Supreme Court of the Philippines, 2000), however, the Supreme Court held that the MMDA does not have the power to enact laws, a move that significantly reduced its power but reinforced the Metro Manila Council's role as the region's policy-making body.

At the local level, Section 12(c) of the RA 10121 (Congress of the Philippines, 2009a) requires cities and municipalities to create their own disaster risk reduction and management (DRR/M) plans. In addition, laws like the RA 9729 (Congress of the Philippines, 2009a, Section 14) – Climate Change Law of 2009 – and the RA 7160 (Congress,

1991, Section 20-c) require Local Government Units (LGUs) to prepare and implement other development-related plans. DILG (2008) highlighted the need to streamline these plans to avoid weak implementation and fragmented actions. Likewise, Ramos (2011) acknowledged the difficulties faced by LGUs in updating their disaster plans and in harmonizing them with other local plans or initiatives.

Lastly, Metro Manila's barangays play a crucial role in community-based disaster preparedness especially in disaster response. In Makati City for instance, 16 out of its 33 barangays have their own rescue teams (Lopez, 2011). In 2010 however, the RA 10121 (Congress of the Philippines, 2009a, Section 11) increased the barangays' role by abolishing all Barangay Disaster Coordinating Councils (BDCC) and transferring their powers to the Barangay Development Councils (BDC), which are multi-sectoral bodies in charge of development programmes. In doing so, barangays are no longer just involved in disaster response, but in a much wider context of disaster preparedness linked with development as well.

Working at the community scale, many of the 60,000 registered NGOs in the Philippines (WANGO, 2012) are directly involved in development and resilience. Existing DRRM policies recognise NGOs as key stakeholders, but weak linkages between communities and city or municipal governments are constant obstacles for them (Ramos, 2011). Overall, awareness on disaster preparedness and response remains low, and resource constraints stand in the way of capacity-building programmes (Ramos, 2011).

## Global environmental change and the city

As a coastal megacity located in a semi-alluvial plain that forms a large drainage basin of major river systems – Pasig, Pampanga, Talisay and Meycauayan river basins among other – MM's urbanisation has a direct effect on the environment (PEMSEA, 2012; Porio, 2011). While the city contributes to global and regional anthropogenic impacts on the environment, the most visible effects of urbanisation are seen at the local level.

## City–coast environmental change and risk

In the available literature, environmental change associated with the coastal nature of the city and global environmental change acting at different scales seem to be divided into high and low risk, with the latter receiving the most attention. For the purpose of this paper, low risk refers to those less visible longer-onset changes, usually associated with coastal–urban ecosystem services such as the loss of biodiversity and coastal erosion, the reduction in the infiltration ability of the soil (Mirza, 2011), depletion of fisheries and deforestation of mangroves and coastal vegetation (Gedan *et al.*, 2011). High risk refers to those city-level land-use changes that have the potential to directly affect business as usual. These are local conditions open to influence from the dominant climate systems affecting the country (Table 7.1) and that in turn produce the precipitation regime and together with landscape features the flood regime for the city and its region (World Bank, 2011).

Changes in these processes are intensified by sea-level rise (SLR) and subsidence. Increase hazard can be attributed to increasing precipitation, temperature and sea surface temperature (Adnan and Atkinson, 2011). Currently, increased rainfall or drought during the monsoon is expected and exaggerated during El Niño and La Niña

*Table 7.1* Major climate processes that drive climate in the Philippines

| *Climate process* | *Impact on climate* |
|---|---|
| Southwest Monsoon | Rainfall from June to August |
| Northeast Monsoon | Rainfall from November to February |
| El Niño/La Niña (ENSO) events | Droughts or increased rainfall |
| Low pressure over the Pacific or South China Sea | Tropical cyclones or typhoons |

Adapted from World Bank (2011)

(ENSO) events respectively (Hilario *et al.*, 2009; PAGASA, 2011). Longer wet and dry seasons are also predicted while worse floods, with increased magnitude due to SLR, in conjunction with subsidence, are already observed (World Bank, 2011). Additionally, increases in the magnitude of tropical storms have also been observed, resulting in increased rainfall and flooding as well as damage due to storm surges and the stronger wind speed of typhoons (Dasgupta and Laplante, 2011).

The socioeconomic consequences of the development approaches and resulting environmental changes described above can be devastating. For example, increased rainfall and floods combined with poor waste management has led to an increase in the incidence of dengue, flu and other viral diseases (Sia Su, 2008; Porio, 2011), affecting particularly vulnerable, poor populations who live in highly exposed areas such as floodplains (PDNA, 2009). When the effects of climate change and poor CBDP strategies come together the result is alarming. This was the case of Typhoon Ondoy in 2009, which was followed by Typhoon Pepeng within a week. These two typhoons left major economic losses of up to US$4.42 billion, 9.3 million people severely affected (most of them living in Metro Manila) and increased the number of poor in the city (PDNA, 2009).

## Responses

The impacts of current and future climate change on MM cannot be overlooked. Initiatives such as the CBDP frameworks noted above, interest on lowering industrial impacts (Corpuz, 1999) and better resource management (Ebarvia, 1997) reflect an interest on acting. But it is difficult to be optimistic when natural hazards continue to halt the city on a regular basis, poverty continues to reduce the capacity for resilience (Porio, 2011) and CBDP frameworks 'place greater responsibility on . . . local people without necessarily proportionately increasing their capacity to formulate initiatives' (Allen, 2006: 97). Duhaylungsod and Mendoza (2009) even suggest that mitigation of climate change is no longer an option and more effort should be put into adaptation. Leaping from one climate change agenda to the other misses the grand opportunity that coastal megacities more than any other form have of combining these agendas and energies in creative and impactful ways. Here lies a global opportunity for MM, but one that is yet to take hold in popular or political imagination. However, what the analysis provided here has shown is the integration and synchronising strategies between the different scales and an understanding of the impacts of the city on the environment are prerequisite for any action towards climate change and DRM.

## References

Adnan, N.A. and Atkinson, P.M. (2011) Exploring the impact of climate and land use changes on streamflow trends in a monsoon catchment. *International Journal of Climatology*, 31(6), 815–831.

Albert, J.R.G. and Collado, P.M. (2004) *Profile and Determinants of Poverty in the Philippines*. Paper presented at the 9th National Convention on Statistics (NCS), EDSA Shangri-La Hotel, October 4–5, 2004, Manila. http://nscb.gov.ph/ncs/9thncs/papers/poverty_Profile.pdf (last accessed 25 March 2012).

Allen, K.M. (2006) Community-based disaster preparedness and climate adaptation: local capacity building in the Philippines. *Disasters*, 30(1), 81, 101.

Ballesteros, M.M. (2010) Linking Poverty and the Environment: Evidence from Slums in Philippine Cities. *Discussion paper series No. 2010-33*. Makati City: Philippine Institute for Development Studies.

Bernardo, E. C. (2008) Solid-Waste Management Practices of Households in Manila, Philippines. *Annals of the New York Academy of Sciences*, 1140(1), 420–424.

Congress of the Philippines (1991) *LAWNUM: RA07160, The local government code of 1991. S. No. 155, H. No. 31046* Manila: Congress of the Philippines. Available from: http://www.congress.gov.ph/download/ra_08/Ra07160.pdf (accessed 27 March 2012).

Congress of the Philippines (1995) *Republic Act No 7924, H. No. 14170*. Manila: Congress of the Philippines. www.congress.gov.ph/download/ra_09/Ra07924.pdf (last accessed 27 March 2012).

Congress of the Philippines (2009a) *Republic Act No. 9729, S. No. 2583, H. No. 5982*. Manila: Congress of the Philippines. www.congress.gov.ph/download/ra_14/RA09729.pdf (last accessed 27 March 2012).

Congress of the Philippines (2009b) *Republic Act No 10121, S. No. 3086, H. No. 6985*. Manila: Congress of the Philippines. Available from: http://www.congress.gov.ph/download/ra_14/RA10121.pdf (accessed 27 March 2012).

Corpuz, C.F.J. (1999) *Pollution tax for controlling emissions from the manufacturing and power generation sectors: Metro Manila*. Singapore: Economy and Environment Program for Southeast Asia (EEPSEA).

Dasgupta, S. and Laplante, B. (2011) Exposure of developing countries to sea-level rise and storm surges. *Climatic Change*, 106(4), 567–579.

DILG (2008) *Rationalized Planning System of the Philippines – A Source Book*. Manila: Department of Interior and Local Government.

DILG (2012) *Metro Manila Quick Facts* [Online]. Manila: Department of Interior and Local Government (DILG). Available from: http://ncr.dilg.gov.ph/ (accessed 25 March 2012).

Duhaylungsod, N.C. and Mendoza, R.B. (2009) *Philippines National Policy on Climate Change (SUD-Net Cities in Climate Change Initiative)*. Nairobi: UN-HABITAT.

Ebarvia, M.C.M. (1997) *Pricing for groundwater use of industries in Metro Manila, Philippines*. Singapore: Economy and Environment Program for Southeast Asia (EEPSEA).

Gedan, K.B., Kirwan, M.L., Wolanski, E., Barbier, E.B. and Siliman, B.R. (2011) The present and future role of coastal wetland vegetation in protecting shorelines: answering recent challenges to the paradigm. *Climatic Change*, 106(1), 7–29.

Hilario, F., Guzman, R., Ortega, D., Hayman, P. and Alexander, B. (2009) El Niño Southern Oscillation in the Philippines: Impacts, Forecasts and Risk Management. *Philippine Journal of Development*, 66(36), 9–34.

Hranjski, H. and Gomez, J. (2011) Manila: Typhoon Nesat Causes Huge Floods. *The Huffington Post*, 27 November. www.huffingtonpost.com/2011/09/27/manila-typhoonnesat_n_982769.html (last accessed 12 March 2012).

IMF (2011) *World Economic Outlook, April 2011: Tensions from the Two-Speed Recovery Unemployment, Commodities, and Capital Flows*. Washington, DC: International Monetary Fund.

Lambino, J.P. (2010) The Economic Role of Metro Manila in the Philippines: A Study of Uneven Regional Development under Globalization. *The Kyoto Economic Review*, 79(2), 156–195.

Lopez, I. (2011) *Scenario-Based Earthquake Response Capacity Assessment and Its Implications to Planning: The Case of Makati City*. Masters Thesis. Diliman: University of the Philippines.

Malaque, I.R. and Yokohari, M. (2007) Urbanization process and the changing agricultural landscape pattern in the urban fringe of Metro Manila, Philippines. *Environment and Urbanization*, 19(1), 191–206.

Mcleod, E., Hinkel, J., Vafeidis, A. T., Nicholls, R. J., Harvey, N., Salm, R. (2010) Sea-level rise vulnerability in the countries of the Coral Triangle. *Sustainability Science*, 5(2), 207–222.

Mirza, M.M.Q. (2011) Climate change, flooding in South Asian and implications. *Regional Environmental Change*, 11(1), S95–S107.

NSCB (2012) *Philippine Standard Geographic Code*. Makati: National Statistics and Coordinating Board. www.nscb.gov.ph/activestats/psgc/listreg.asp (last accessed on 25 March 2012).

PAGASA (2011) *Climate Change in the Philippines*. Quezon City: PAGASA.

PEMSEA (2012) *Manila Bay*. Manila: Partnerships in Environmental Management for the Seas of East Asia. http://beta.pemsea.org/profile/pollution-hotspots/manila-bay (last accessed 20 March 2012).

PDNA (2009) *Typhoons Ondoy and Pepeng: Post Disaster Needs Assessment*. Manila: *Post-Disaster Needs Assessment*, The World Bank.

PIDS (2010) Linking Poverty and the Environment: Evidence from Slums in Philippine Cities. *Discussion paper series No. 2010-33*. Makati City: Philippine Institute for Development Studies (PIDS).

Porio, E. (2011) Vulnerability, Adaptation, and Resilience to Floods and Climate Change-Related Risks among Marginal, Riverine Communities in Metro Manila. *Asian Journal of Social Science*, 39(4), 425–445.

Ramos, B. (2011) *Philippines – National Progress report on the Implementation of the Hyogo Framework For Action (2009–2011) – interim*. www.preventionweb.net/files/18619_phl_NationalHFAprogress_2009-11.pdf (last accessed 25 March 2012).

Sovacool, B.K. and Brown, M.A. (2010) Twelve metropolitan carbon footprints: A preliminary comparative global assessment. *Energy Policy*, 38(9), 4856–4869.

Shatkin, G. (2000) Obstacles to empowerment: Local politics and civil society in metropolitan Manila, the Philippines. *Urban Studies*, 37(12), 2357–2375.

Shatkin, G. (2004) Planning to Forget: Informal Settlements as 'Forgotten Places' in Globalising Metro Manila. *Urban Studies*, 41(12), 2469–2484.

Sia Su, G.L. (2008) Correlation of Climatic Factors and Dengue Incidence in Metro Manila, Philippines. *Ambio*, 37(4), 292–294.

Supreme Court of the Philippines (2000) *Metropolitan Manila Development Authority, petitioner, vs. Bel-Air Village Association, Inc., respondent (G.R. No. 135962, March 27, 2000)* Manila: Supreme Court of the Philippines. http://sc.judiciary.gov.ph/jurisprudence/2000/mar2000/135962.html (last accessed 27 March 2012).

Tharoor, I. (2009) The Manila Floods: Why Wasn't the City Prepared? *Time*, 29 September.

World Bank (2011) *Vulnerability, Risk Reduction, and Adaptation to Climate Change*. Washington, DC: The World Bank Group.

Umezawa, Y., Hosono, T., Onodera, S., Siringan, F., Buapeng, S., Dlinom, R., Yoshimizu, C., Tayasu, I., Nagata, T., Taniguchi, M. (2008) Sources of nitrate and ammonium contamination in groundwater under developing Asian megacities. *Science of the Total Environment*, 404(2-3), 361–376.

WANGO (2012) *About the Philippines*. New York: World Association of Non-Governmental Organizations. www.wango.org/congress/about.aspx (last accessed 24 March 2012).

WHO (2012) *WHO: Urgent steps must be taken to make Metro Manila healthier*. Manila: World Health Organization Western Pacific Region (WHO). www.wpro.who.int/media centre/releases/2010/pr20100704/en/index.html (last accessed 25 March 2012).

## MUMBAI, INDIA

### *Steven Forrest and Kirsten Butcher*

Mumbai is an economic powerhouse of the Indian subcontinent, generating around 5 per cent of India's total GDP (Ranger *et al.*, 2011). However, approximately two-thirds of the city's population live in informal settlements, with no legal right to the land they occupy (Murthy *et al.*, 2001). The rapid urbanisation of Mumbai has led to social, environmental and climatic pressures on the city. Coastal mangrove degradation, water pollution and vehicle congestion are threatening both human health and marine ecosystems.

Mumbai faces risks including flooding, which, combined with a network of inadequate drainage, led to the 2005 floods disaster. City authorities are beginning to address these problems with disaster risk reduction and management schemes. However, inward migration from local–rural populations adversely affected by climate change may hamper attempts to manage these problems.

## Urban context

Mumbai has seen an annual population growth rate of 3 per cent since 1975 when the population stood at 7.3 million constituting a new super agglomeration (UN Population Division, 2001). By 2000 it was 16 million and the UN Population Division (2001) predict that Mumbai will become the second largest city in the world by 2015 with a forecasted population of 22.5 million. 'Explosive' rates of urban sprawl were evident around the more peripheral zones in the period from the mid-1970s to 1991 as space in the city centre became more constrained (Taubenböck, 2008: 79).

Mumbai's approach to informal settlements since 1985 consists of slum clearances alongside same-site replacement with government-subsidised housing (Mukhija, 2001). 'People-managed resettlement', which integrates slum community input, was implemented (Patel *et al.*, 2002: 1), but these efforts have recently become more commercially driven at the expense of residents (Arputham and Patel, 2010). Given their location on low-lying land, many vulnerable dwellings are exposed to hydrometerological hazards compounded by the regional tropical climate (Sherbinin *et al.*, 2007). The combination of a high population density with poor drainage systems (a key factor in the 2005 flash floods) is further worsened by sea-level rise and the hazard this presents, including an increase in fatalities from vector-borne diseases and the economic costs of displacement due to the increasing fragility of the natural environment (Kumar *et al.*, 2008).

Daytime temperatures in Mumbai range from 16°C to 33°C with an annual rainfall of 2,700 millimetres, of which 2,500 millimetres falls during the monsoon period of mid-June to September (Sherbinin *et al.*, 2007). The Canadian Climate Centre's 'business-as-usual' emissions (A2) Scenario, a division of the Climate Research Branch of Environment Canada that focuses on research analysis, predicts that the region will experience a decrease in average annual precipitation of 2 per cent by the year 2100 (Patankar, 2010). This water deficit, if a continuing trend, will lead to droughts outside of the monsoon season and a migration pattern away from rural areas, placing a further strain on the city's resources and potentially undermining the population growth predictions (Sherbinin *et al.*, 2007).

## Urbanisation impacts on the environment

Adverse environmental impacts of mega-urbanisation include mangrove deforestation in addition to water and air pollution at the local and global scales. The coastal city was originally formed from land reclamation and large-scale mangrove clearing (Kathiresan, no date). Mangrove deforestation has continued in the modern age with a 39.32 per cent decrease in the area of mangroves in the Mumbai suburban region observed from 1991–2001; most of this land is now used for informal settlements (Vijay *et al.*, 2005).

Mangroves prevent shoreline erosion and were found to protect the coastline villages from sea surges in the 2004 Asian tsunami (Murthy *et al.*, 2001; Kathiresan and Rajendran, 2005). Coastal mangroves such as those in Mahul Creek provide a habitat for marine life and a large number of migratory birds (Vijay *et al.*, 2005; Verma *et al.*, 2002).

Unregulated waste treatment has led to solid waste disposal in open dumping grounds that encourage the proliferation of pests (Pacione, 2006). It has also exposed the Thane and Mahim Creeks to large volume discharges of partially treated and untreated industrial effluents and municipal sewage (Murthy *et al.*, 2001). These effluents contain toxic heavy metals such as lead and have been detected over nine kilometres away in mangrove roots (Shete *et al.*, 2007). This pollution has decreased concentrations of dissolved oxygen reducing the amount of marine life present (Murthy *et al.*, 2001). It is also having a significant impact on the coastal aquaculture with a heightened risk of undesirable coastal blooms (Zingde, 1999). The World Bank and local authorities are establishing effluent treatment plants and discharging waste further from the coastline to reduce their impact on management systems (Murthy *et al.*, 2001). In addition, indiscriminate and unregulated sand mining is having a severe impact on coastal ecosystems (Murthy *et al.*, 2001).

The urbanisation of Mumbai has led to an increase in the observed mean temperature due to the large number of fossil fuel-burning vehicles and high-rise buildings absorbing more solar radiation (Dhorde *et al.*, 2009). The large number and poor maintenance of vehicles also contributes to the high levels of carbon monoxide, nitrous oxides, Volatile Organic Compounds (VOCs) and suspended particulates that negatively affect human health (Murthy *et al.*, 2001; Srivastava, 2004; Pacione, 2006). City regulations seeking to reduce air pollution have been introduced to reduce congestion by relocating large offices and industries to the outskirts of Mumbai (Gupta, 2007; Sherbinin *et al.*, 2007).

## City-coast risks

Mumbai also functions as the country's industrial centre with approximately 9,000 factories producing anything from fertilizers to thermal power (Murthy *et al.*, 2001).

This concentration of industry in a coastal region susceptible to environmental hazards is a real risk to the economic growth of India.

Large sections of the city are seated on reclaimed land situated one to five metres above sea level (Dodamn, 2009) and in certain areas below the high-tide mark (Hallegatte *et al.*, 2010). The combination of these two factors impede upon the city's natural drainage system, which increases the exposure of the shoreline to hazardous events, such as sea surges. This explains why the region is prone to flooding (Munich Re, 2005) and has suffered annual flooding between 2004 and 2007, which can also be partially attributed to land reclamation and building on landfill areas with insufficient attention paid to man-made drainage systems (Ranger *et al.*, 2011).

In 2011, there were 2.787 million people exposed to the risk of flooding in Mumbai and it is predicted that this will rise to 11.418 million by 2070 (Hanson *et al.*, 2011). The value of assets exposed to coastal flooding was also expected to increase from US$46.20 billion to US$1,598.05 billion in 2010 (Hanson *et al.*, 2011). Some of the worst floods to hit Mumbai occurred on 26 July 2005, when 994 millimetres of rain fell in a single 24-hour period (Jenamani *et al.*, 2006). Rapid urbanisation and land use change resulted in the inadequate drainage systems backing up and overflowing, temporarily paralysing the city (Gupta, 2007). The ripple effects of the flood caused the Northwest region of the subcontinent to suffer approximately US$3–5 billion in economic damages affecting 20 million individuals (Ranger *et al.*, 2011). It is also estimated that climate change will triple total financial losses associated with a one-in-100 year event, so these costs are likely to be dwarfed by the impact of future occurrences (Ranger *et al.*, 2011).

The World Bank (2011) estimates that 75 per cent of sewage in Mumbai is untreated and currently discharged directly into local water courses and coastal waters. This water pollution in and around Mumbai presents an increased risk of epidemics of water-borne diseases such as leptospirosis on a city already challenged by the overcrowded conditions associated with population expansion (Lau *et al.*, 2010).

## Responses

India has installed a centralised disaster management agency, the National Disaster Management Authority, to manage the destruction caused by hazards in the form of national disaster management policies and plans. In the late 1990s the Maharashtra region developed an urban disaster management plan for Mumbai; the first of its kind in India (Revi, 2008).

In response to the 2005 floods the government of Maharashtra appointed a fact-finding committee to identify the causes of the flood and recommend future management practices (Gupta, 2007). The Mithi River Development Authority was created from their findings to reduce land encroachment that was narrowing the river channel (Gupta, 2007). The plan included improving the conditions of informal settlements through land-use policies, increased preparedness and more effective monitoring (Ranger *et al.*, 2011). Automatic weather stations distributed around the city were installed to monitor real-time rainfall volumes and allow authorities to issue 'risk warnings' (Gupta, 2007). Mumbai has also implemented compulsory rainwater harvesting in new developments to reduce surface run-off from high rainfall events (Gupta, 2007).

The Earthquakes and Megacities Initiative (EMI) is a member of the UN Global Platform for Disaster Risk Reduction and has implemented schemes within Mumbai. These include capacity building activities, institutionalising Disaster Risk Management practice within the Municipal Corporation of Greater Mumbai (MCGM) and encouraging the collaboration of different stakeholders (EMI, 2012). Geographic Information Systems (GIS) were used by the EMI to conduct seismic risk assessments and by the MCGM to map hazard spots (floods and landslides) (EMI, 2010).

To aid in future disaster events, the Mumbai Emergency Control Centre was upgraded and the MCGM created 'Task Force One', which is composed of 26 highly trained disaster search and rescue specialists (Gupta, 2007). However, government officials are not always building flexibility to future changes into their flood management schemes (Hallegatte *et al.*, 2010; Ranger *et al.*, 2011). This 'maladaptation' may succeed in the short-term but fails to consider the long-term impacts of climate change (Pelling, 2011).

## References


Arputham, J. and Patel, S. (2010) Recent developments in plans for Dharavi and for the airport slums in Mumbai. *Environment and Urbanisation*, 22(2), 502–504.

Dhorde, A., Dhorde, A. and Gadgil, A.S. (2009) Long-term Temperature Trends at Four Largest Cities of India during the Twentieth Century. *Journal of the Indian Geophysical Union*, 13(2), 85–97.

EMI (2010) *Earthquakes and Megacities Initiative: Geographic Information System GeoDatabase for Disaster Risk Reduction in Greater Mumbai.* www.emi-megacities.org/home/projects/111-mumbai/743-geographic-information-system-geodatabase-for-disaster-risk-reduction-in-greater-mumbai.html (last accessed 5 March 2012).

EMI (2012) *Earthquakes and Megacities Initiative: Disaster Risk Reduction in Greater Mumbai.* www.emi-megacities.org/home/projects/drr-in-greater-mumbai.html (last accessed 5 March 2012).

Gupta, K. (2007) Urban flood resilience planning and management and lessons for the future: a case study of Mumbai, India. *Urban Water Journal*, 4(3), 183–194.

Hallegatte, S., Ranger, N., Bhattacharya, S., Bachu, M., Priya, S., Dhore, K., Rafique, F., Mathur, P., Naville, N., Henriet, F., Patwardhan, A., Narayanan, K., Ghosh, S., Karmakar, S., Patnaik, U., Abhayankar, A., Pohit, S., Corfee-Morlot, J. and Herweijer, C. (2010) Flood Risks, Climate Change Impacts and Adaptation Benefits in Mumbai: An Initial Assessment of Socio-Economic Consequences of Present and Climate Change Induced Flood Risks and of Possible Adaptation Options. *OECD Environment Working Papers*, No. 27, OECD Publishing.

Hanson, S., Nicholls, R., Ranger, N., Hallegatte, S., Corfee-Morlot, J., Herweijer, C. and Chateau, J. (2011) A global ranking of port cities with high exposure to climate extremes. *Climate Change*, 104, 89–111.

Jenamani, R.K., Bhan, S.C. and Kalsi, S.R. (2006) Observational/forecasting aspects of the meteorological event that caused a record highest rainfall in Mumbai. *Current Science*, 90(10), 1344–1362.

Kathiresan, K. (no date) Threats to Mangroves: Degradation and Destruction of Mangroves. UNU-INWEH Course 1 – Training Course on Mangroves and Biodiversity. http://ocw.unu.edu/international-network-on-water-environment-and-health/unu-inweh-course-1-mangroves/Degradation-and-destruction-of-mangroves.pdf (accessed 2 March 2012).

Kathiresan, K. and Rajendran, M. (2005) Coastal mangrove forests mitigated tsunami. *Estuarine, Coastal and Shelf Science*, 65, 601–606.

Kumar, R., Jawale, P. and Tandon, S. (2008) Economic impact of climate change on Mumbai, India. *Regional Health Forum*, 12(1), 38–42.

Lau, C.L., Smythe, L.D., Scott, B.C. and Weinstein, P. (2010) Climate change, flooding, urbanisation and leptospirosis: fuelling the fire? *Transactions of the Royal Society of Tropical Medicine and Hygiene*, 104, 631–638.

Mukhija, V. (2001) Enabling slum redevelopment in Mumbai: Policy paradox in practice. *Housing Studies*, 16(6), 791–806.

Munich Re Group (2006) *Topics Geo: Significant Natural Catastrophes in 2005*. Knowledge Series.

Murthy, R.C., Rao, Y.R. and Inamdar, A.B. (2001) Integrated coastal management of Mumbai metropolitan region. *Ocean & Coastal Management*, 44(5-6), 355–369.

Pacione, P. (2006) Mumbai. *Cities*, 23(3), 229–238.

Patankar, A., Patwardhan, A., Andharia, J. and Lakhani, V. (2010) *Mumbai City Report*. Presented at International Workshop on Climate Change Vulnerability Assessment and Urban Development Planning for Asian Coastal Cities, Bangkok. http://startcc.iwlearn.org/doc/Doc_eng_16.pdf (last accessed 2 March 2012).

Pelling, M. (2011) *Adaptation to climate change: From resilience to transformation*. Routledge: Oxon.

Ranger, N., Hallegatte, S., Bhattacharya, S., Bachu, M., Priya, S., Dhore, K., Rafique, F., Mathur, P., Naville, N., Henriet, F., Herweijer, C., Pohit, S. and Corfee-Morlot, J. (2011) An assessment of the potential impact of climate change on flood risk in Mumbai. *Climate Change*, 104, 139–167.

Revi, A. (2008) Climate change risk: an adaptation and mitigation agenda for Indian cities. *Environment and Urbanization*, 20, 207–229.

Sherbinin, A.D., Schiller, A. and Pulsipher, A. (2007) The vulnerability of global cities to climate hazards. *Environment and Urbanization*, 19(1), 39–64.

Shete, A., Gunale, V.R. and Pandit, G.G. (2007) Bioaccumulation of Zn and Pb in *Avicennia marina* (Forsk.) Vierh. and *Sonneratia apetala* Buch. Ham. from Urban Areas of Mumbai (Bombay), India. *Journal of Applied Science and Environmental Management*, 11(3), 109–112.

Srivastava, A. (2004) Source apportionment of ambient VOCS in Mumbai city. *Atmospheric Environment*, 38(39), 6829–6843.

Taubenböck, H., Wegmann, M., Berger, C., Breunig, M., Roth, A. and Mehl, H. (2008) Spatiotemporal Analysis of Indian Mega Cities. *The International Archives of the Photogrammetry, Remote Sensing and Spatial Information Sciences*, XXXVII(B2), 75–82.

UN Population Division (2001) *World Urbanization Prospects: The 2001 Revision*. www.un.org/esa/population/publications/wup2001/WUP2001_CH6.pdf (last accessed 26 February 2012).

Verma, A., Chaturvedi, N., Balachandran, S. and Kehimkar, I. (2002) *Avian diversity in and around mangroves of Mahul Creek, Mumbai, India*. Proceedings of the National Seminar on Creeks, Estuaries and Mangroves – Pollution and Conservation, 28–30 November, Thane. http://md1.csa.com/partners/viewrecord.php?requester=gs&collection=ENV&recid=5743019&q=&uid=791541484&setcookie=yes (last accessed 2 March 2012).

Vijay, V., Biradar, R.S., Inamdar, A.B., Deshmukhe, G., Baji, S. and Pickle, M. (2005) Mangrove mapping and change detection around Mumbai (Bombay) using remotely sensed data. *Indian Journal of Marine Sciences*, 34(3), 310–315.

World Bank (2011) *From rationing to full service: water and sanitation challenge for Bombay*. http://lnweb90.worldbank.org/oed/oeddoclib.nsf/DocUNIDViewForJavaSearch/9A7152C54E38F2EE852567F5005D8DD5 (last accessed 28 February 2012).

Zingde, M.D. (1999) Marine Pollution – What are we heading for? *Ocean Science: Trends and Future Directions*, 229–245.

## NEW YORK CITY

*William Solecki*

New York City (NYC) is one of the most populous cities in the world. The City of New York has a population of 8,244,910 (US Census Bureau, 2011), and the extended metropolitan region has another approximately 14,000,000 residents. While there are significant income disparities, almost all the residents live in formal housing and have uninterrupted access to fresh drinking water, sanitation services and electricity. The city and metropolitan region are located in the Atlantic Ocean coastal zone and has a temperate continental climate. Four of the five boroughs (US counties) that make up NYC are islands or on islands, and as such have extensive coastline. New York City maintains some of the densest and most complex infrastructures in the world much of which is rapidly aging (Solecki *et al.*, 2012). Some of the infrastructure and materials may not be able to withstand the projected strains and stresses from a changing climate. The city's infrastructure can be considered within four main systems: energy, transportation, water and waste, and communications. Much of the infrastructure is near or on the coast or estuarine environments. Most of the more recent infrastructure of the city was built during the mid-twentieth century when most of the available sites were at the water's edge either on or near remnant wetlands or derelict industrial sites.

Although no single weather-related event can be attributed to climate change, New York has experienced climate extremes in its recent history that have brought attention to the potential risks posed by climate change to the city's critical infrastructure. Recent extreme climate-related events include Hurricane Irene in August of 2011, which caused the city government for the first time to implement its storm surge evacuation plan and associated risk reduction planning activities on a broad scale (e.g., shutting down the public transit system, including the subways, and authorising a mandatory evacuation of areas to be inundated under a Category 1 hurricane storm surge). While the storm surge flooding was not as much as expected, city agencies were able to test their emergency planning protocols.

Other recent weather extremes include the summers of 2010 and 2011, which were exceedingly hot and stormy. The summer of 2011 was particularly intense – July was one of the hottest months on record for New York, while August was one of the wettest. These events and others, which resulted in large social and economic costs, provide valuable insights into the impact that climate change could have in the future. They also highlight the need, even without climate change, to improve the city's resilience to environmental stressors, of which climate extremes are one of the most important. In many cases, linking adaptation efforts to the climate risks faced by the city today is an effective adaptation strategy and is now the centrepiece of the city's climate action.

New York's coastal location with almost 1,500 kilometres of shoreline (including tidal wetlands) make future climate change and associated sea-level rise and increased probability of storm-surge-associated coastal flooding widely seen as the most significant challenge to the city (Solecki *et al.*, 2012; NPCC, 2010). The sea-level rise projections derived from the IPCC-based methods from AR4, using the central range, or middle 67 per cent of the model distribution, illustrate mean annual sea-level rise in New York City as 5.1 to 12.7 centimetres by the 2020s; 17.8 to 30.5 centimetres by the 2050s; and 30.5 to 58.4 centimetres by the 2080s (for a full description of methods of sea-level rise projection development see Horton and Rosenzweig, 2010; Horton *et al.*, 2010). In response to discussion within the scientific community that the IPCC (2007) approach to sea-level rise may underestimate the range of possible increases, in large part because it does not fully consider the potential for land-based ice sheets to melt due to dynamical processes (e.g., Hansen *et al.*, 2007, Horton *et al.*, 2008), an alternative method that incorporates observed and longer-term historical ice-melt rates also were developed. The 'rapid ice-melt' approach suggests the sea level could rise by approximately 0.9 to 1.5 metres (with a central range of one to 1.4 meters) by the 2080s. The range in the rapid ice-melt scenario represents a combination of GCM model results and paleoclimatic uncertainties related to timing of deglaciation.

The shift in baseline elevation of the sea level has important implications for storm surge inundation and flood frequency in the city.

## Climate change adaptation in New York City

Throughout the period from 1800, New York City has experienced several environmental crises and transitions (Solecki *et al.*, 2012). The latest environmental challenge for New York that requires long-term strategic planning is climate change. It is projected to have wide impacts on the city's critical infrastructure through higher temperatures, more intense flooding events and sea-level rise. Because of its early recognition of the risks posed by climate change and its current commitment to mitigation of greenhouse gas (GHG) emissions as well as to adaptation, New York has become a national and international leader in responding to climate change (NPCC, 2010).

Current climate change adaptation efforts in the city build on previous assessments and studies. Within the metropolitan region, leading scientists, agency representatives, and nongovernmental organisation members have been studying issues related to climate extremes and climate change for more than a decade. In 2004, a climate change adaptation initiative was launched by the NYC Department of Environmental Protection. The major product of the NYC DEP Task Force was the Climate Change Assessment and Action Plan for the agency (DEP, 2008; Rosenzweig *et al.*, 2007). Since many climate change adaptations identified through this process help to increase the robustness of current systems managed by the agency, the NYC DEP Task Force had an immediate benefit by improving responses to present-day climate variability, such as managing episodes of intense precipitation in the upstate reservoirs. This work became the benchmark and exemplar of work soon to be carried by other New York City agencies.

New York's Mayor Michael Bloomberg created the Office of Long-Term Planning and Sustainability in 2006, with the goal of developing a comprehensive plan to create a greener, more sustainable city. Mitigating climate change was a central goal of the city's comprehensive sustainability plan, PlaNYC 2030, released in 2007. The PlaNYC work was expanded to include climate change adaptation in response to the importance of doing both climate change mitigation and adaptation simultaneously to protect the citizens and infrastructure of the city.

An immediate goal of PlaNYC was the creation of an interagency Climate Change Adaptation Task Force to protect the city's vital infrastructure in the face of a changing climate. The charge of the task force created in 2008 was to identify climate change risks and opportunities for the city's critical infrastructure and to develop coordinated adaptation strategies to address these risks. The task force consisted of approximately 40 city, state and federal agencies, regional public authorities, and private companies that operate, maintain or regulate critical infrastructure in the region related to energy, transportation, water and waste, natural resources, and communications. To support the task force, the city also convened a group of climate change and impact scientists, and legal, insurance and risk management experts as the New York City Panel on Climate Change (NPCC) to advise the city on climate change science, potential impacts and adaptation pathways specific to the city's critical infrastructure (Plate 16).

The NPCC consists of climate change and impacts scientists, and legal, insurance and risk-management experts, serves as the technical advisory body and was designed to function in an objective manner similar to the role that the Intergovernmental Panel on Climate Change (IPCC) plays on an international stage for nation-states. The work of the NPCC is to ensure that the city's adaptation efforts are based on sound science and a thorough understanding of climate change, its potential impacts and adaptation. To assist the city, the NPCC has analysed climate change hazards, studied impacts on the critical infrastructure of New York, and developed a risk management framework for adaptation planning, which, in turn, contributed to the development of the city's climate change adaptation planning framework.

A critical component of the NPCC's work was to define 'Climate Protection Levels' to address the issue of climate change impacts on the effectiveness of current regulations and design standards related to sea-level rise and storm surge, heat waves and inland flooding. Most important for the city is that in order to maintain a similar level of current risk it will be necessary to adjust the current building codes. This is another way in which climate change becomes integrated into the urbanisation process by influencing a set of climate risk-related construction guidelines – e.g. how to build for increased frequency and intensity of precipitation and flooding events, heat waves and extreme wind events.

The Adaptation Assessment Guidebook (AAG), another product of the NPCC, describes a detailed process designed to help stakeholders create an inventory of their at-risk infrastructure and to develop adaptation strategies to address those risks. The AAG includes three tools developed to aid the stakeholders in their adaptation planning process including an infrastructure questionnaire, risk matrix and prioritisation framework. The adaptation process was defined as a dynamic cycle of analysis and action followed by evaluation, further analysis and refinement (i.e., learn, then act, then learn some more). The steps outlined in the AAG are intended to become integral parts of ongoing risk management, maintenance and operation, and capital planning processes of the agencies and organisations that manage and operate critical infrastructure.

The adaptation approach developed by the NPCC fosters a Flexible Adaptation Pathways approach – originally developed by the London TE2100 – that can evolve over time as understanding of climate change improves and that concurrently reflect local, national and global economic and social conditions. Flexible Adaptation Pathways is a concept that encourages building climate change adaptation strategies that can be adjusted and modified over time to reflect the dynamic and ongoing climate change understanding.

## Recent activities

Since the release of the NPCC report, the climate change adaptation programming of the city has gone through a process of evolution in which the importance of climate resilience as a broad integrative approach with includes elements of increased preparedness for existing climate risks, potential future risk and enhanced sustainability has come to the fore. This ambition was codified with release of the PlaNYC update in 2011. Climate resilience efforts within the plan are focused on several elements:

- assess vulnerabilities and risks from climate change;
- increase the resilience of the built and natural environments;
- protect public health from the effects of climate change;
- increase the city's preparedness for extreme climate events;
- create resilient communities through public information and outreach.

(Source: PlaNYC, 2011)

A series policy initiatives and activities associated with these elements have been developed. A partial list of actions includes:

1. Elevation mapping: developing a high-resolution elevation model for the city – e.g. Lidar – to better demarcate the shoreline and areas of low elevation and exposed to storm surge.
2. Risk mapping: encourage the updating of the US Federal Emergency Management Agency (FEMA) flood insurance risk (FIRM) maps and a redrafting of the area of the 1% (a one-in-100-year event) flood zone.
3. Hazard modelling: contract for advanced coastal storm surge modelling to be developed – i.e. storm surge for extreme events such as hurricanes and northeasterly storms; the city also has just convened a natural hazards technical review panel that will assess the hazard modelling as well as other new hazard information.
4. Storm water management: develop a comprehensive storm management programme including a variety of actions, some as simple as sending out rapid response crews to clean drains in advance of large precipitation events to as a complex and comprehensive as a massive green infrastructure programme (>1.5 billion USD) to promote experimentation and implementation of techniques to promote blue roofs, increased ground water infiltration, and rainfall detention and retention.
5. Waterfront redevelopment: develop a comprehensive waterfront development plan that includes elements to promote coastal resilience to extreme flooding events and sea-level rise as well as promote recreation, resident development and commercial activity in the coastal zone.
6. Heat management: continue to develop and refine the city's heat response capacity including cooling centres, social network programmes to ensure that the highly vulnerable (e.g. elderly, young and health compromised) are looked after during heat waves, and demand-side management with local electricity utilities during high-demand heat events.

Arguably one of the most profound actions the city has taken has come through an extensive regulatory review of all existing building codes and standards to discern what sorts of adjustments, as demanded by climate change impacts and associated vulnerabilities, might need to be considered. This action by the city government illustrates the long-term impact of the initial New York Panel on Climate Change policy recommendations as they become woven into the everyday fabric of city function. Embodied within the NYC Green Codes taskforce, new initiatives include the creation and tying of building codes to flood maps with sea-level rise, the safeguarding of toxic materials in flood zones, ensuring toilets and sinks can operate during blackouts and enhancing building water supply, among other initiatives. In a small way, we see the actions of the NYC Green Codes taskforce as evidence of the urbanisation of climate change (Solecki, 2012), where climate change and a city's response become established elements of how a large city manages and identifies itself.

## Hurricane Sandy

Hurricane Sandy made landfall on the New Jersey shore just south of Atlantic City on 28 October 2012 and became one of the most damaging storms to strike the continental United States. While it is not possible to definitively attribute Sandy's unusual metrological character to climate change, its strength was certainly enhanced by the waters of the Atlantic Ocean near the coast being 3°C above normal, coupled with the region's coastline experiencing increasingly rapid sea-level rise.

The New York–New Jersey metropolitan area with millions of coastal zone residents had a dry run for a major hurricane event in 2011 when Hurricane Irene swept through the area, doing little damage to the coast but further alerting local and state officials about the need to be ready for a huge storm. The death toll from Sandy in the extended metropolitan region exceeded 100 with damage estimates including direct and indirect economic losses in the hundreds of billions of dollars. Many low-to-moderate income residents suffered damage or loss of their homes leaving tens of thousands displaced or homeless. As a specific subpopulation, the elderly and infirm were shown to be highly vulnerable, especially those living in the coastal evacuation zone and those on upper floors of apartment buildings left without elevator service. These individuals had limited adaptive capacity because they could not easily leave their residences. Personal losses were complicated by damage to public assets: critical infrastructure including roads, electrical cables and gas pipelines were severely damaged or totally destroyed. At its peak the storm cut electrical power for more than five million customers.

Sandy provides insight into the vulnerability of cities as complex socio-technological systems. This is illustrated through the consequences of failure at a Consolidated-Edison electricity distribution substation in lower Manhattan, which ceased operation at approximately 9:00 pm on Monday evening when its flood protection barrier (although designed to be 1.5 feet above the ten-foot storm surge of record from Hurricane Donna in 1960) was over topped by the 14-foot storm surge. As the substation stopped functioning, it immediately caused a system-wide loss of power for over 200,000 local customers. This event with other planned and unplanned outages resulted in much of Manhattan below 34th Street to be without electrical power for four or more days. The power outage revealed a diversity of vulnerabilities within the immediate affected area and the wider region. Residents in the numerous high-rise apartment buildings were left without heat and lights but also without elevator service and water (which must be pumped to upper floors). The outage shut businesses in office towers and in on-the-ground store fronts and stalled subway lines below the street, causing the local and regional economy great disruption. With public transportation out of service, people wanting to get to Manhattan soon resorted to using private cars, resulting in massive gridlock and traffic jams on the city's streets.

The effect of the storm would have been far worse if the climate resilience strategies in the region were not in place. The City of New York and the Metropolitan Transportation Authority, for instance, worked aggressively to protect life and property by stopping the operation of the city's subway before the storm hit and moving the train cars out of low-lying flood prone areas. At the height of the storm surge, all seven of the city's East River subway tunnels flooded. Catastrophic loss of life would have resulted if there were subway trains operating in the tunnels when the storm struck. The storm also fostered vigorous debate among local and state politicians, other decision makers and stakeholders about how best to prepare the region for future storms especially given the likelihood that this type of event will become more frequent with climate change, particularly in the context of accelerating sea-level rise. New York State Governor Cuomo and New York City Mayor Bloomberg both publically stated that the storm and its impact was evidence of climate change. Responding to the storm looks set to reshape the geography and economy of the city. The merits of a storm surge barrier and other hard infrastructure features as well as soft adaptation approaches such as policies to shift more of the risk burden to coastal property owners or to place restrictions on rebuilding have been discussed in the local media. Soon after the storm subsided, the corporate private sector began lobbying for a greater share of any reconstruction budget and subsequent ownership and management control over more critical infrastructure. Sandy has shown that adapting New York is more than even an economic and political as well as a social and engineering project.

## References

Hansen, J., Mki. Sato, P., Kharecha, G., Russell, D., Lea, W. and Siddall, M. (2007) Climate changes and trace gases. *Philosophical Transactions of The Royal Society*, 365, 1925–1954.

Horton, R. and Rosenzweig, C. (2010) Climate Risk Information. In: Rosenzweig, C. and Solecki, W. (eds) *New York City Panel on Climate Change. Climate Change Adaptation in New York City: Building a Risk Management Response*. Prepared for use by the New York City Climate Change Adaptation Task Force. New York: Annals of the New York Academy of Science. www.nyc.gov/html/planyc2030/downloads/pdf/nyc_climate_change_report.pdf

Horton, R., Herweijer, C., Rosenzweig, C., Liu, J., Gornitz, V. and Ruane, A.C. (2008) Sea level rise projections for the current generations CGCMs based on the semi-empirical method. *Geophysical Research Letters*, 35.

Horton R., Gornitz, V., Bowman, M. and Blake, R. (2010) Climate observations and projections. In: Rosenzweig, C. and Solecki, W. (eds) *Climate Change Adaptation in New York City: Building a Risk Management Response: New York City Panel on Climate Change 2010 Report*. New York: Annals of the New York Academy of Science.

Intergovernmental Panel on Climate Change (IPCC) (2007) Climate Change 2007: The Physical Science Basis. In: Solomon, S. *et al.* (eds) *Contribution of Working Group I to the Fourth Assessment Report of the Intergovernmental Panel on Climate Change.* Cambridge: Cambridge University Press.

New York City (2011) *plaNYC Update.* April. New York: City of New York.

New York City Department of Environmental Protection (DEP) (2008) *Assessment and action plan: A report based on the ongoing work of the DEP climate change Task Force.* New York: New York City Department of Environmental Protection Climate Change Program.

New York City Panel on Climate Change (NPCC) (2010) *Climate Change Adaptation in New York City: Building a Risk Management Response.* Rosenzweig, C. and Solecki, W. (eds) Prepared for use by the New York City Climate Change Adaptation Task Force. New York: Annals of the New York Academy of Science.

Rosenzweig, C., Major, D.C. and Demong, K. (2007) Managing climate change risks in New York City's water system: assessment and adaptation planning. *Mitigation and Adaptation Strategies for Global Change*, 12(8), 1391–1409.

Solecki, W.D. (2012) Moving Toward Urban Sustainability – Using Lessons and Legacies of the Past. In: Zeman, F. (ed.) *Handbook of Metropolitan Sustainability Understanding and Improving the Urban Environment*, Woodhead Publishing.

Solecki, W.D., Rosenzweig, C., Hammer, S. and Mehrotra, S. (2012) Urbanization of climate change: impacts, vulnerabilities and transformation pathways. In: Sclar, E., Volavka-Close, N. and Brown, P. (eds) *The Urban Transformation: Health, Shelter and Climate Change.* London: Earthscan.

US Census Bureau (2011) *State and County Quick Facts.* http://quickfacts.census.gov/qfd/states/36/3651000.html (last accessed 11 April 2013).

## RIO DE JANEIRO

***Robert Coates and Matt Richmond***

After struggling economically for many years Rio de Janeiro is booming. Sustained growth and the discovery of large offshore oil deposits have helped it attract rising levels of foreign investment and drive down unemployment (*The Economist*, 2011). Nonetheless, hosting the 2014 World Cup and 2016 Olympics challenge the city to face the global spotlight (Cano, 2011). Inequality and violence have both fallen in recent years, yet it remains one of the most unequal cities in the world, with insecurity perpetuated by drug-trafficking gangs, militias and the police (Zaluar, 2010). Consistent failures in land regulation and governance leave sections of Rio's population without essential public services, encouraging environmental degradation and increasing vulnerability. A series of fatal landslides across Rio de Janeiro state in 2010 and 2011 served to expose these problems further, leaving well over 1,000 dead and tens of thousands homeless.

### Urbanisation and climate change impacts

Lying alongside near-landlocked Guanabara Bay in Brazil's Southeast, Rio de Janeiro is the country's largest and most complex coastal urban centre. Its population of 6.3 million – 11 million in the wider metropolitan region – makes it the second largest in the country after São Paulo (IBGE Cidades@, 2012). Since peaking in the 1960s population growth has fallen considerably, registering 0.8 per cent per annum between 2000 and 2010 (ibid.). However, favelas are growing more quickly than the wider city – 2.8 per cent per year in the last decade – and now contain 22 per cent of the population (14.4 per cent of the metropolitan area) (IBGE, 2010). *Favelas* are officially labelled as 'subnormal agglomerations', defined as collections of over 50 housing units lacking essential public services, occupying land not (or not until recently) owned by the residents, and characterised by dense and disorganised construction (IBGE, 2010). Resident numbers in central areas have declined slightly, but urban sprawl and densification have extended the city to the west, along the Barra da Tijuca coastline and northwards into the municipalities of the Baixada Fluminense. These areas represent new economic growth poles and contain affluent areas, but are also seeing rapid favela expansion, with some of the lowest living standards and highest levels of poverty and insecurity in the city (Pero *et al.*, 2005).

Rio's pattern of urbanisation has had a major impact on the natural environment. A federal government-commissioned report on the 2011 disaster in Rio State points to a consistent lack of political will and capacity to enforce Brazil's Forest Code, which prohibits construction of dwellings in 'Areas of Permanent Preservation' (APPs) (Ministério do Meio Ambiente, 2011). These include vulnerable and sensitive locales such as riverbanks and forested slopes, including endangered Atlantic Rainforest. Another long-term trend is the destruction of mangroves, which originally fringed most of Guanabara Bay and now only cover 90 square kilometres (Lima and Legey, 2010). Lack of infrastructure and key services in newly settled and densifying areas, particularly sewerage and waste collection, leads to the dumping of waste on vacant public land and in lagoons and rivers (Silva, 2010).

Guanabara Bay in particular, which acts as a drainage basin for the entire metropolitan population, records dangerously high levels of pollution. Its inner shallow regions are characterised by hypertrophic conditions and witness frequent hypoxic events, while in some areas faecal coliform and chlorophyll counts are several times higher than acceptable levels (Lima and Legey, 2010). Despite this, the state environmental agency, Fundação Estadual de Engenharia do Meio Ambiente (FEEMA), believes that seasonal tidal mixing naturally regulates the content of the bay (ibid.). Consequently, a long-planned new metropolitan sewage system, including the construction of new treatment plants, should dramatically improve water quality, increasing the amount of treated water entering the bay from 30 per cent to 65 per cent by 2014 (Barbassa, 2011).

As Rio attempts to solve existing environmental problems, climate change represents an additional layer of risk. In 2011 the state environment department produced composite vulnerability indices by municipal government district right across the state. The study combined sophisticated vulnerability mapping of future climate models with historical disaster incidence and current social and environmental vulnerability (Macedo de Lima Barata, 2011). Perhaps unsurprisingly due to their coastal location, around half the municipalities in the metropolitan area come out as amongst the highest risk areas in the state (i.).

Flood and landslide events in 2010 and 2011 illuminated the startling reality of Rio de Janeiro's susceptibility to climate hazards and global environmental change. The combination of urbanised low-elevation coastal zones with residential and commercial building on inland flood plains, and unregulated self-construction of housing on steep slopes by the urban poor, together present fundamental challenges for future development and sustainability. Sea-level rise (SLR) is not only expected to increase flood incidence in low-lying zones but also intensify coastal erosion and storm surges, and cause saline intrusion of inland water systems (Nobre *et al.*, 2011; Neves and Muehe, 2008). Though disagreements persist, mainly due to the poor monitoring of the Brazilian coastal zone, SLR is likely to result in changes to the layout of beaches, destruction of coastal roads and serious drainage problems (Neves and Muehe, 2008).

## Policy response

Policy action on climate change mitigation was set in motion in 1998 with the publication of Rio's first GHG emissions inventory (Martins and Ferreira, 2011). With *Rio+20* offering an additional incentive for proving its sustainability credentials, 2011's *Municipal Climate Change Policy* targets emissions reductions of 16 per cent by 2016 and 20 per cent by 2020, using 2005 levels as the baseline (Prefeitura da Cidade do Rio de Janeiro, 2011a). Accounting for 39 per cent of the city's GHG emissions and long viewed as a key stumbling block to environmental quality, transport policy focuses on reducing car dependency through *metrô* extensions, doubling bicycle lane provision to 300 kilometres and gradual renewal of the bus fleet with biodiesel vehicles. A new solid waste treatment plant began operations in April 2011 and is expected to generate 1.9 million tonnes of biogas per year, to be negotiated through the CDM (Prefeitura da Cidade do Rio de Janeiro, 2010a). The 2016 Olympics is also driving political action on environmental sustainability, including a major reforestation programme: 130 hectares of critical areas of Rio's Tijuca forest have been replenished and another 1,200 hectares are planned for 2012.

Municipal administrations across Rio de Janeiro vary in their size and capacity, but balancing the difficult challenges presented by urbanisation and climate change with local party political and security constraints has proven to be a major challenge across the board (Allen, 1994; Hardoy and Pandiella, 2009). Partly as a result of this, municipal and state governments have been slow to initiate an integrated strategy on disaster risk. However, increased incidence and severity of natural hazards may now be adding impetus to public policy. Drainage capacity, particularly in the rainy season, is being upgraded in key districts (Prefeitura da Cidade do Rio de Janeiro, 2010b), and an early warning system via community sirens and SMS alerts has been introduced to counter landslide impact in a number of hillside favelas. Capacity building on emergencies and heavy rainfall in vulnerable areas is being rolled out through community council interaction with Brazil's Civil Defence, while better meteorological monitoring aims for faster emergency response from a cross-section of local government departments (Subsecretaria de Defesa Civil, 2011).

Vital though resilience-building activities such as these are, question marks still hang over their adequacy in tackling the root causes of vulnerability for the longer term – associated with informal settlement on floodplains and hillsides. Nevertheless, there are signs that the gap between public policy formulation and implementation in key areas, particularly on APPs and infrastructural management, is now beginning to be acknowledged (Nobre *et al.*, 2011). In meeting the demands of disaster risk reduction and climate change adaptation, closing this gap will be essential. It is promising that following a long history of flood and landslide events recent government-commissioned reports (such as those cited here) now go beyond physical geography to draw attention to the social and political aspects of vulnerability. The production of risk mapping since 2010 offers a powerful tool for targeted intervention in municipal areas and sub-districts most at risk from environmental change (Macedo de Lima Barata, 2011; Prefeitura da Cidade do Rio de Janeiro, 2011b). Acknowledgement of the depth of the problem does not in itself equate to political action on reducing vulnerability, but it does contribute to the mobilisation of political will and public opinion (Nobre *et al.*, 2011).

The management of social risk in Rio de Janeiro inevitably relies on both the solidity of informal communities' coping strategies *and* their place within the wider institutional framework of public policy (Perch, 2011). Risk reduction and climate adaptation in Brazil's largest coastal agglomeration may then be seen in the context of longer-term national processes of social reform, citizenship building and poverty alleviation. Though still insufficient, major investment is being made to improve access to services, which will help to improve community resilience. Neighbourhood infrastructure upgrades and regeneration schemes like Favela Bairro are now accompanied by reforms in health and education, conditional cash transfers through Bolsa Familia, and security policy in the form of the occupation of gang-controlled favelas by Unidades de Policias Pacificadores (UPPs). The combination of social policy and specific adaptation measures may collectively reduce community vulnerability to environmental hazards in the long term.

## References

Allen, E. (1994) Political Responses to Flood Disaster: The Example of Rio de Janeiro. In: Varley, A. (ed.) *Disasters, Development and Environment*, London: Wiley.

Barbassa, J. (2011) Pollution in Rio's bay poses Olympic challenge. *The Guardian*. 6 September. www.guardian.co.uk/world/feedarticle/9833006 (last accessed 11 April 2013).

Cano, I. (2011) Rio de Janeiro's Olympic challenge. *The Guardian*. 29 November. www.guardian.co.uk/commentisfree/2011/nov/29/rio-de-janeiro-world-cup (last accessed 11 April 2013).

Hardoy, J. and Pandiella, G. (2009) Urban Poverty and Vulnerability to Climate Change in Latin America. *Environment and Urbanization*, 21(1), 203–224.

IBGE Cidades@ 2012 *Instituto Brasileiro de Geografia e Estatistica* (cities database). www.ibge.gov.br/cidadesat (last accessed 10 March 2012).

Lima, E. and Legey, L.F.L. (2010) Water Quality Restoration in Rio de Janeiro: From a Piecemeal to a Systems Approach. *The Journal of Environment Development*, 19(3), 375–396.

Macedo de Lima Barata, M. (2011) *Mapa de Vulnerabilidade da População do Estado do Rio de Janeiro aos Impactos das Mudanças Climáticas nas Áreas Social, Saúde e Ambiente. FioCruz e Secretaria do Ambiente do Estado do Rio de Janeiro*. http://download.rj.gov.br/documentos/10112/364217/DLFE-40943.pdf/rel_vulbilidade.pdf (last accessed 12 March 2012).

Martins, R.D.A. and Ferreira, L.C. (2011) Climate change action at the city level: tales from two megacities in Brazil. *Management of Environmental Quality: An International Journal*, 22(3), 344–357.

Ministério do Meio Ambiente (2011) *Relatório de Inspeção, Área atingida pela tragédia das chuvas, Região Serrana do Rio de Janeiro: Áreas de Preservação Permanente & Áreas de Risco, Brasília*. www.defesacivil.sp.gov.br/v2010/portal_defesacivil/conteudo/documentos/relatorio-tragedia-rj.pdf (last accessed 12 March 2012).
Nobre, C., Young, A. and de Gusmão, P.P. (2011) *Megacidades, Vulnerabilidades e Mudanças Climáticas: Região Metropolitana do Rio de Janeiro*. Centro de Ciência do Sistema Terrestre (CCST), Instituto Nacional de Pesquisas Espaciais (INPE), Núcleo de Estudos de População, Universidade Estadual de Campinas (NEPO). www.poli.ufrj.br/noticias/arquivos/completo.pdf (last accessed 12 March 2012).
Perch, L. (2011) *Mitigation of what and by what? Adaptation by whom and for whom? Dilemmas in delivering for the poor and the vulnerable in international climate policy*. Working Paper No. 79, International Policy Centre for Inclusive Growth (IPC-IG), UNDP. www.ipc-undp.org/pub/IPCWorkingPaper79.pdf (last accessed 12 March 2012).
Pero, V., Cardoso, A. and Elias, P. (2005) Discriminac_a_o no mercado de Trabalho: o caso dos moradores de favelas cariocas. In: *Colec_a_o Estudos da Cidade, Instituto Pereira Passos*. Rio de Janeiro: Prefeitura da Cidade do Rio de Janeiro.
Prefeitura da Cidade do Rio de Janeiro (2010a) *Prefeitura dá início às obras de construção da Central de Tratamento de Resíduos de Seropédica*. www.rio.rj.gov.br/web/guest/exibeconteudo?article-id=1036688 (last accessed 10 March 2012).
Prefeitura da Cidade do Rio de Janeiro (2010b) Programa Integrado de Manutenção da Drenagem. www.rio.rj.gov.br/web/guest/exibeconteudo?article-id=981954 (last accessed 10 March 2012).
Prefeitura da Cidade do Rio de Janeiro (2011b) *Armazem de Dados, Mapa de Susceptabilidade ao Escorregamento*. http://portalgeo.rio.rj.gov.br/mapa_risco/ (last accessed 12 March 2012).
Silva, G.C. (2010) Environmental justice: A case of socio-environmental vulnerability in Rio de Janeiro, *Urban Environment*, 4, 27–48.
Souto, R.D., Polette, M., Kampel, M. (2009) Evaluation of Sustainability Indicators for the Coastal Zone of Rio de Janeiro, Brazil, *Journal of Coastal Research*, SI56, pp. 1311–15.
Subsecretaria de Defesa Civil (2011) *Defesa Civil Rio: A Estratégia de Redução de Desastres Humanos com a Aplicação do Sistema de Alerta e Alarme Comunitário para Chuvas Fortes na Cidade do Rio de Janeiro*. http://www0.rio.rj.gov.br/defesacivil (last accessed 12 March 2012).
*The Economist* (2011) Doing Business in Brazil: Rio or São Paulo? 24 August. www.economist.com/blogs/schumpeter/2011/08/doing-business-brazil (last accessed 11 April 2013).
Transparência Olímpica (2011) available at: http://www.transparenciaolimpica.com.br/meio_ambiente.html
Zaluar, A. (2010) Turf War in Rio de Janeiro: Youth, drug traffic and hyper-masculinity. *Vibrant*, 7(2), 7–27.

## SHANGHAI, CHINA

### *Ayden Cumming and Davina Layne*

It is Shanghai's political location as a Chinese city, subject to the country's one-party system, that distinguishes it from other megacities in the world. China's central government has an unrivalled ability to ensure compliance with its mandates; this suggests China could be a world leader in both adapting to and mitigating climate change. As appointments to top positions are made by one level above in the political hierarchy, lower levels of government know that non-compliance could result in their being removed from office. Thus, when the central government agrees on policy priority, compliance is extraordinary (Lieberthal, 1997). However, most policy issues are not granted such precedence. The prioritisation of economic growth and a decentralised structure of incentives and authority ensure that environmental protection remains an afterthought in Chinese policy, leaving its coastal cities to cope with increasingly dangerous hazards.

Bounded by the East China Sea and Hangzhou Bay, Shanghai is located on China's east coast at the delta of the Yangtze River (Yin *et al.*, 2011). It is one of the largest seaports in the world with a land area of 6,341 kilometres squared. A large temporary migrant labour force combined with permanent residents has pushed Shanghai's population in excess of 19 million (Shi and Cui, 2012), making it one of China's largest and most densely populated cities (Li *et al.*, 2004). Due to globalisation and foreign direct investment Shanghai has experienced rapid economic growth over the past 20 years, expanding at an annual rate of approximately 10 per cent (Wu, 2008; Yin *et al.*, 2011). In 2009 Shanghai's gross domestic product (GDP) was US$237.5 billion, or 4.5 per cent of China's total GDP. Shanghai is the largest economic city in China (Yin *et al.*, 2011) and its growth is expected to continue as the government-mandated export-orientated economic strategy favours coastal cities.

Shanghai, like other megacities that are rapidly urbanising and have high population densities, has had a negative impact on human and environmental health (Zhao *et al.*, 2006; Li *et al.*, 2004; Wu *et al.*, 1999; Cao *et al.*, 2012). Increasing demand for and use of fossil fuels has perpetuated air pollution and other environmental negatives such as haze and poor water and soil quality (Tan *et al.*, 2012; Wu *et al.*, 1999). Population size, density and a low-lying coastal location (De Sherbinin *et al.*, 2007) contribute to Shanghai's physical and socioeconomic vulnerability to climate change and city–coast risks (Yin *et al.*, 2011, McGranahan *et al.*, 2007). Fuelled by industrialisation, groundwater withdrawal has caused land subsidence (Chai *et al.*, 2004); the ground level in the city centre is now more than half a meter below sea level (Zong and Chen, 1999). Sitting in a bowl-shaped land area, the metropolitan districts are at great risk of flooding and waterlog hazards. For example, digital projections suggest that 97 per cent of Shanghai's total land area could be flooded by a tide level that is expected to occur within the next 100 years. Twenty-four million people would be directly affected by such an event, making Shanghai 'another New Orleans' (Yin *et al.*, 2011: 412).

Shanghai's challenges present it with a conundrum. Can such rapid urbanisation and economic growth be maintained despite the realities of climate-related hazards and the significant detrimental impacts on the environment? To date, efforts to mitigate the potentially devastating impacts of climate change on China's coastal cities have been limited. China has a 'no regrets' approach for climate policy; any measures chosen must be consistent with its economic strategy. Economic growth is prioritised over environmental policy because in the short-term, the costs of environmental policy are believed to outweigh the benefits. Consequently, the impact and scope of existing environmental policy is limited. For example, the reduction of greenhouse gases has mainly been a by-product of energy efficiency and cost-saving measures within energy and transportation departments, driven by a desire to overcome the detrimental impacts of energy scarcity on economic development. Thus, while climate policy is an important part of China's Five-Year Programs for National and Social Development, measures are taken only within the context of ensuring economic growth (Richerzhagen and Scholz, 2008).

When climate measures are adopted, their implementation is constrained by conflicting interests and poor incentives resulting from the political and instructional framework in China (Lieberthal, 1997). Local government has been granted authority over economic reforms, including certain power over industries. Rapid economic growth is often awarded with promotion, leading local officials to act as both administrators and entrepreneurs. As China's reforms are growth oriented, local governments prioritise economic goals over environmental responsibilities. In some cases, governments even try to limit the detrimental impacts of local environmental protection agencies on economic growth (EPAs). There are recorded episodes of EPAs issuing businesses with fines and passing the money to the local government, who then issue tax breaks to the offending businesses for amounts roughly equal to the environmental fines (Lieberthal, 1997). Due to the complexity of China's political structure, it is difficult for the central government to oversee actions such as these at the local level (Richerzhagen and Scholz, 2008; de Sherbinin *et al.*, 2007).

As a consequence, despite having the legal capacity to develop climate regulation, China lacks enforcement capability. Even when strong environmental policies are adopted, the lack of enforcement results in large discrepancies between policy and practice. In sum, China's efforts to mitigate climate change have been exceptionally restricted (Richerzhagen and Scholz, 2008). Cities on China's coast, however, are adopting coping strategies as they struggle with increasing climatic and environmental hazards. In Shanghai these take the form of engineering projects, control of undesirable human activities and disaster preparedness and emergency measures (Xu *et al.*, 2009).

The most prolific engineering projects in Shanghai are flood defence walls, such as those built along the Suzhou River. These walls have been raised three times since the 1960s due to the combined threat of land subsidence and sea-level rise. The flood defences of the Huangpu River have also been raised from 3.12 metres above sea level in the 1950s to 4.09 metres above sea level in the late 1990s. The continued need to raise the height of these defences demonstrates the inadequacies of structural solutions to coastal hazards (Xu *et al.*, 2009). It also indicated the vulnerability of Shanghai to subsidence. Arguably the most important environmental impact of urbanisation Shanghai needs to control is groundwater withdrawal to help prevent further subsidence. In response to public demand for regulation following an accident during the construction of a subway line, the government set up a groundwater level survey, prohibited construction on unstable hills and restricted sand excavation (Xu *et al.*, 2009). Finally, the Shanghai Municipal Defence Office formulates disaster relief planning for the city and the 284 city and civil defence organisations coordinate community volunteer teams (Shanghai Municipal Government, 2009; de Sherbinin *et al.*, 2007).

Shanghai's challenges and coping strategies are characteristic of coastal megacities. It is therefore the political location of the city within the People's Republic of China that makes it unique. China presents a political paradox; despite impressive enforcement capability, there remains a chasm between environmental policy and practice. The policies required to adequately address climate change mitigation are simply too tied up with competing economic interests to be granted the attention needed from the central government to ensure compliance by lower-level officials (Lieberthal, 1997). As long as reforms maintain a growth orientation, climate policy will continue to be neglected in favour of ensuring economic targets are met.

However, there are signs that the environmental policy gap could narrow in the future (Lieberthal, 1997). Moves toward a sharper distinction between government and enterprise may increase a willingness on the part of officials to enforce environmental standards. An improved understanding by central governmental officials about the potentially devastating economic impacts of environmental degradation and climate change could also lead to greater prioritisation of environmental policy. If these gradual changes were to mature and come into full effect, China would truly be the world leader in climate mitigation and adaption. Until then, Shanghai will have to cope with ever-increasing hazards.

## References


Cao, G.-Y., Chen, G., Pang L.-H., Zheng, X.-Y. and Nilsson, S. (2012) Urban Growth in China: Past, Prospect, and its Impacts. *Population Environment*, 33(2-3), 137–160.

Chai, J.-C., Shen, S.-L., Zhu, H.-H. and Zhang, X.-L. (2004) Land subsidence due to groundwater draw down in Shanghai. *Geotechnique*, 54(2), 143–147.

De Sherbinin, A., Schiller, A. and Pulsipher, A. (2007) The Vulnerability of Global Cities to Climate Hazards. *Environment and Urbanization*, 19(1), 39–74.

Li, J., Guttikunda, S.K., Carmichael, G.R., Streets, D.G., Chang, Y.S. and Fung, V. (2004) Quantifying the Human Health Benefits of Curbing Air Pollution in Shanghai. *Journal of Environmental Management*, 70(1), 49–62.

Lieberthal, K. (1997) *China's Governing System and its Impact on Environmental Policy Implementation*. China Environmental Series. Woodrow Wilson International Centre. www.wilsoncenter.org/sites/default/files/ACF4CF.PDF (last accessed 4 March 2012).

McGranahan, G., Balk, D. and Anderson, B. (2007) The Rising Tide: Assessing the Risks of Climate Change and Human Settlements in Low Level Coastal Zones. *Environment and Urbanization*, 19(1), 17–37.

Richerzhagen, C. and Scholz, I. (2007) China's Capacities for Mitigating Climate Change. *World Development*, 36(2), 308–324.

Shanghai Municipal Government (2009) Shanghai Municipal Defense Office. www.shanghai.gov.cn/shanghai/node17256/node17679/node17704/userobject22ai13001.html (last accessed 4 March 2012).

Shi, J. and Cui, L. (2012) Characteristics of High Impact Weather and Meteorological Disaster in Shanghai, China. *Natural Hazards*, 60(3), 951–969.

Tan, J., Zheng, Y., Tang, X., Guo, C., Li, L., Song, G., Zhen, X., Yuan, D., Kalkstein, A.J., Li, F. and Chen, H. (2010) The Urban Heat Island and its Impact on Heat Waves and Human Health in Shanghai. *International Journal of Biometeorol*, 54(1), 75–84.

Wu, J. (2008) The Peri-Urbanisation of Shanghai: Planning, Growth Pattern and Sustainable Development. *Asia Pacific Viewpoint*, 49(2), 244–253.

Wu, C., Maurer, C., Wang, Y., Xue, S. and Lee Davis, D.L. (1999) Water Pollution and Human Health in China. *Environmental Health Perspectives*, 107(4), 251–256.

Xu, Y., Zhang, D., Shen, S. and Chen, L. (2009) Geo-hazards with Characteristics and Prevention Measures along the Coast Regions of China. *Natural Hazards*, 49(3), 479–500.

Yin, J., Yin, Z., Hu, X., Xu, S., Wang, J., Li, Z., Zhong, H. and Gan, F. (2011) Multiple Scenario Analyses Forecasting the Confounding Impacts of Sea Level Rise and Tides from Storm Induced Coastal Flooding in the City of Shanghai, China. *Environmental Earth Sciences*, 63(2), 407–414.

Zhao, S., Da, L., Tang, Z., Fang, H., Song, K. and Fang, J. (2006) Ecological Consequences of Rapid Urban Expansion: Shanghai. *China Frontiers in Ecology and the Environment*, 4(7), 341–346.

Zong, Y. and Chen, X. (1999) Typhoon Hazards in the Shanghai Area. *Disasters*, 23(1), 66–80.

# Index

Note: page number in **bold** refer to tables.

Kolibri 360 GmbH
Ferdinand-Jühlke-Straße 7
99095 Erfurt, Deutschland
produktsicherheit@kolibri360.de